Preface

Welcome to "Robotics: From Fundamentals to Advanced Applications," your comprehensive guide to understanding and mastering the field of robotics. In an era where automation and intelligent systems are revolutionizing industries, robotics stands at the forefront, driving innovations across manufacturing, healthcare, exploration, and more. As we delve deeper into this transformative technology, it is essential for both beginners and seasoned professionals to grasp its fundamental concepts and applications thoroughly.

This book is meticulously crafted to serve as a complete learning resource, catering to the diverse needs of learners at all levels. Whether you are a student embarking on your first exploration into robotics or a professional seeking to enhance your expertise, this guide provides the essential tools and resources necessary to achieve your learning goals.

Table of Content

Chapter 1: Introduction to Robotics

Definition of Robotics: In a world characterized by technological marvels, the term "robotics" emerges as a dynamic synthesis of scientific principles and human ingenuity. Robotics represents a multidisciplinary field that integrates mechanical engineering, electronics, computer science, and artificial intelligence to create machines, referred to as robots, with the capacity to emulate, simulate, or extend human actions and thought processes. These machines can range from industrial arms meticulously assembling intricate components to autonomous drones navigating complex environments.

At its core, robotics encapsulates the art and science of designing, building, and operating robots that possess varying levels of autonomy. These machines may be programmed to perform repetitive tasks with precision on manufacturing lines, explore inhospitable environments where human presence is perilous, or even engage in creative endeavors like composing music or generating artwork. The versatility of robotics derives from its ability to transcend mere mechanization and embrace cognitive functions that challenge the boundaries of human achievement. The definition of robotics, within the realm of technological innovation and scientific exploration, encapsulates the essence of a multidisciplinary field that seamlessly merges mechanics, electronics, computer science, and artificial intelligence. It is a dynamic and evolving concept that transcends the boundaries of conventional disciplines, fostering the creation of machines that mimic, enhance, or exceed human capabilities while navigating a diverse spectrum of applications.

Robotics stands as the intersection where mechanical design, electronics, control theory, computer programming, and artificial intelligence converge. The essence of robotics is encapsulated in the development of machines, termed robots, which can execute tasks autonomously or under human guidance. These tasks span a wide array of complexity, ranging from mundane chores to intricate surgical procedures, from exploration of distant planets to enhancing manufacturing processes.

Three key components define the essence of robotics:

Mechanical Structure: The physical embodiment of a robot, which includes its body, limbs, sensors, and actuators. The mechanical structure defines the robot's range of motion, mobility, and interaction with the environment.

Electronics and Sensors: The nervous system of a robot, composed of sensors that gather data from the environment and feedback mechanisms that provide information about the robot's own state. These electronics facilitate the robot's perception of its surroundings.

Control and Intelligence: The brain of the robot, comprising algorithms, computer programs, and artificial intelligence techniques. This component processes sensor data, makes decisions, and generates commands to actuate the robot's movement and actions.

The definition of robotics comes to life through its applications in diverse fields:

Manufacturing: Industrial robots automate assembly lines, enhancing production efficiency, precision, and consistency.

Healthcare: Surgical robots enable minimally invasive procedures, reducing patient trauma and enhancing surgical precision.

Space Exploration: Robots venture into space, exploring celestial bodies and expanding our knowledge of the cosmos.

Agriculture: Agricultural robots automate tasks such as planting, harvesting, and pesticide application, optimizing crop yield and resource usage.

Autonomous Vehicles: Robotics plays a pivotal role in the development of self-driving cars and drones, transforming transportation and logistics.

The definition of robotics continues to evolve as technological advancements push the boundaries of what robots can achieve. Robots are no longer confined to rigid mechanical structures; they encompass soft and bio-inspired designs, enabling them to navigate complex environments with greater adaptability. In conclusion, the definition of robotics encapsulates a dynamic fusion of mechanics, electronics, computer science, and artificial intelligence, resulting in the creation of machines with the capacity to perform tasks ranging from mundane to extraordinary. Robotics stands as a testament to human creativity and innovation, opening doors to new possibilities, transforming industries, and shaping the future of human-robot collaboration.

Historical Overview: The tapestry of robotics is woven with threads of innovation, spanning across centuries and cultures. A historical exploration of robotics unveils a journey marked by visionary ideas, technological advancements, and a relentless human desire to transcend the limitations of the possible.

Ancient Foundations: Automata and Ingenious Devices

The seeds of robotics were sown in the fertile soils of ancient civilizations. Dating back to ancient Greece, inventors and thinkers like Hero of Alexandria conceived of intricate mechanical devices known as automata. These devices, powered by water or air pressure, exhibited rudimentary motions that fascinated and entertained audiences. Similarly, ancient Chinese inventors crafted mechanical puppets and clocks, showcasing a nascent understanding of mechanization.

Industrial Revolution: The Dawn of Machines - The Industrial Revolution of the 18th and 19th centuries ushered in a new era of mechanization, setting the stage for the modern concept of robotics. Textile mills and factories embraced machinery to automate labor-intensive tasks, heralding the mechanized age. Innovations such as the Jacquard loom, powered by punch cards, foreshadowed the integration of control mechanisms into machinery.

Early 20th Century: Robotic Dreams and Realities - The term "robot" itself was coined by Czech playwright Karel Capek in his 1920 play "R.U.R." (Rossum's Universal Robots), where robots were artificial beings manufactured for labor. This marked a conceptual shift, envisioning machines that resembled humans in their abilities. Concurrently, science fiction literature by authors like Isaac Asimov began exploring ethical dilemmas and the potential of intelligent machines.

Mid-20th Century: The Rise of Industrial Robots - The mid-20th century witnessed the transformation of robotics from fiction to reality. The development of programmable logic controllers (PLCs) in the 1960s paved the way for industrial automation. Animate, introduced in 1961, is widely regarded as the first industrial robot. It heralded a revolution in manufacturing by performing tasks such as die casting and spot welding with precision.

Late 20th Century: Robotics in Space and Beyond - Advancements in robotics extended beyond factories and into the cosmos. Space exploration missions, such as the Apollo lunar landings, introduced robots as extensions of human exploration. Robotic arms, like the Shuttle Remote Manipulator System (Canadarm), became integral to space shuttle missions. These robotic appendages underscored the potential of machines in environments hostile to human presence.

Contemporary Era: Integration and Autonomy - In the 21st century, robotics has become an integral part of our lives. From the surgical theater, where robotic-assisted surgery enhances precision, to household vacuum cleaners equipped with autonomous navigation, robots have woven themselves into our daily experiences. The amalgamation of advanced sensors, computational power, and artificial intelligence has enabled robots to navigate, perceive, and interact with their environments autonomously.

Conclusion: A Tapestry Unfinished - The historical journey of robotics reflects humanity's unyielding quest for progress and innovation. From ancient automata to cutting-edge humanoid robots, the evolution of robotics narrates a tale of human imagination turning into technological reality. As we stand on the cusp of an era characterized by AI-driven robotics and human-robot symbiosis, the historical foundations continue to shape the narrative of a future where the lines between creator and creation blur, and robots cease to be mere tools, but companions in our journey toward the unknown.

Importance and Applications of Robotics: The section on the importance and applications of robotics in the field of robotics literature serves as a captivating exploration into the transformative power that this interdisciplinary field wields across diverse industries. Robotics stands as a pinnacle of innovation, bridging the realms of mechanics, electronics, computer science, and artificial intelligence to shape the future of humanity's technological landscape.

Significance of Robotics: The importance of robotics transcends mere technological fascination; it embodies a paradigm shift in the way we interact with machines, automate tasks, and augment our capabilities. At its core, robotics is about creating intelligent agents that can replicate, simulate, or enhance human actions and cognition. The profound significance of robotics manifests in its ability to revolutionize industries, enhance efficiency, and address complex challenges.

Applications Across Industries: The applications of robotics span across a multitude of domains, each harnessing the potential of these machines to enhance processes, productivity, and innovation:

Manufacturing: Robotics has redefined manufacturing by introducing automation into assembly lines. Robots perform tasks with precision, consistency, and speed, leading to increased production rates and improved product quality.

Healthcare: Surgical robots enable minimally invasive procedures, reducing patient trauma, recovery times, and hospital stays. Robotic exoskeletons aid rehabilitation, helping individuals regain mobility and independence.

Exploration: In space exploration, robots venture into the unknown, gathering data from celestial bodies and expanding our understanding of the cosmos.

Agriculture: Robots automate tasks like planting, harvesting, and pesticide application, optimizing crop yields and reducing labor-intensive efforts.

Logistics: Warehouse robots streamline inventory management, sorting, and packaging, revolutionizing supply chains and e-commerce.

Entertainment: Robots are becoming performers in theaters and theme parks, captivating audiences with their unique abilities and interactions.

Research: In laboratories, robots assist scientists by performing repetitive experiments, data collection, and analysis, accelerating the pace of discovery.

Societal Impact: The societal impact of robotics is far-reaching:

Labor Transformation: While robots automate routine and physically demanding tasks, they also reshape the job landscape, necessitating the development of new skills and career paths.

Health and Safety: Robots enter hazardous environments, reducing human exposure to risks in fields like nuclear cleanup, disaster response, and deep-sea exploration.

Accessibility: Robots designed for individuals with disabilities enhance accessibility and quality of life, enabling greater independence.

Environmental Sustainability: Agricultural robots contribute to precision farming, reducing resource wastage and promoting sustainable agricultural practices.

Innovation and Creativity: The pursuit of robotics fosters innovation, inspiring researchers to push boundaries and develop breakthrough technologies.

Future Prospects: As robotics evolves, the applications continue to diversify and expand. From companion robots to assist the elderly to robots collaborating seamlessly with humans in shared workspaces, the future holds a dynamic landscape where robots become integral to various facets of human life.

Enhancing Efficiency in Industry: Automation Revolution - One of the paramount domains where robotics demonstrates its importance is industrial automation. Manufacturing processes are optimized through the integration of robotic systems. Robots excel in repetitive, precise tasks that demand high throughput and consistency. This translates into reduced production costs, enhanced product quality, and increased productivity. In industries spanning automotive, electronics, and pharmaceuticals, robotics plays a pivotal role in reshaping the landscape of production.

Healthcare and Medical Advancements: Precision and Assistance - Robotics has orchestrated transformative changes in the healthcare sector. In surgical theaters, robotic-assisted surgeries enable unprecedented precision and minimally invasive procedures. The surgeon's expertise, coupled with a robot's dexterity, translates into reduced patient trauma and quicker recovery times. Additionally, robotic prosthetics and exoskeletons empower individuals with disabilities, restoring mobility and enhancing their quality of life.

Exploration and Discovery: Expanding Horizons - The exploration of extreme environments, where human presence is challenging or perilous, has found an ally in robotics. Space agencies employ robots to navigate celestial bodies, gather data, and transmit insights back to Earth. Planetary rovers, like the Mars rovers Spirit, Opportunity, and Curiosity, have unveiled the secrets of distant worlds. Similarly, underwater robotic vehicles delve into ocean depths, unraveling mysteries of marine ecosystems and aiding in scientific research.

Logistics and Warehousing: Revolutionizing Supply Chains - Modern logistics and warehousing operations benefit immensely from robotic technologies. Automated guided vehicles (AGVs) navigate warehouse floors, autonomously transporting goods. Robots equipped with computer vision identify and sort packages, enhancing order accuracy and expediting delivery times. The synergy between robotics and e-commerce reshapes the way goods flow through global supply chains.

Education and Research: Nurturing Innovation - Robotics is a fertile ground for education and research, fostering innovation across disciplines. Educational robotics kits empower students to engage with STEM (science, technology, engineering, and mathematics) concepts in tangible ways. Academics explore advanced algorithms, AI, and control systems to push the boundaries of robot capabilities. These endeavors not only nurture future engineers but also drive technological breakthroughs.

Society and Social Impact: Ethical Considerations - As robotics becomes more integrated into society, ethical considerations emerge. Human-robot interaction raises questions about job displacement, privacy, and the boundaries between man and machine. Striking a balance between reaping the benefits of automation and addressing its potential downsides becomes a critical societal challenge.

Conclusion: Forging New Frontiers - The importance of robotics resonates through its multifaceted applications, shaping industries, accelerating scientific discovery, and transforming human experiences. From factories to operating rooms, from the depths of oceans to the vastness of space, robotics reshapes the contours of what is achievable. As the synergy between technology and human ingenuity continues to evolve, the trajectory of robotics surges toward uncharted territories, where the lines between science fiction and reality blur, and the potential of machines to augment and elevate human existence grows ever more profound.

Classification of Robots: In the vast landscape of robotics, a taxonomy emerges that categorizes robots based on their form, function, and capabilities. This classification system provides insights into the diverse roles robots play across industries and applications, illuminating the intricate nuances that define their characteristics.

Industrial Robots: Precision and Automation - Industrial robots form the backbone of modern manufacturing. These robots are engineered for precision, speed, and repeatability in tasks such as welding, painting, assembly, and material handling. They operate in controlled environments, often along production lines, enhancing efficiency and productivity. Industrial robots are designed with articulated arms, delta structures, and Cartesian configurations to execute tasks with high accuracy.

Service Robots: Enhancing Human Life - Service robots bridge the gap between machines and humans, serving as assistants in various environments. They encompass a diverse array of roles, from healthcare companions to customer service agents. Domestic service robots aid in tasks like vacuuming and lawn mowing, while healthcare robots provide assistance to elderly individuals and patients. These robots often possess mobility and perception capabilities, enabling them to interact safely and intelligently with their surroundings.

Mobile Robots: Navigating Dynamic Environments - Mobile robots possess the ability to move autonomously within their environments. This category encompasses ground-based robots like autonomous vehicles and drones, as well as underwater and aerial robots. Mobile robots often integrate sensor systems, including lidar, cameras, and GPS, to navigate and interact with their surroundings. They find applications in exploration, surveillance, logistics, and environmental monitoring.

Collaborative Robots (Cobots): Working alongside Humans - Collaborative robots, or cobots, are designed to work in close proximity to humans. They are equipped with advanced sensors to detect human presence and adjust their behavior accordingly. Cobots are employed in scenarios where human-robot collaboration enhances productivity and safety. From small-scale manufacturing tasks to healthcare applications, cobots epitomize the fusion of human expertise and robotic capabilities.

Entertainment and Research Robots: Nurturing Curiosity - Entertainment robots captivate audiences with their interactive and often anthropomorphic features. From robot pets that respond to touch and voice commands to humanoid robots capable of mimicry, these robots engage and delight. Additionally, research robots facilitate scientific exploration, serving as platforms for testing algorithms, studying animal behavior, or venturing into hazardous environments where human presence is challenging.

Military and Space Robots: Extending Frontiers - Military and space robots are deployed in environments where human intervention is fraught with risk. Military robots range from bomb disposal units to unmanned aerial vehicles used for reconnaissance. Space robots, like rovers and robotic arms, explore extraterrestrial landscapes and gather data in the cosmos. These robots push boundaries, expand human understanding, and contribute to technological advancements.

Educational and Hobbyist Robots: Inspiring Innovation - Educational robots introduce learners to the world of robotics, fostering STEM education. These kits enable students to build, program, and experiment with robots, cultivating essential skills in problem-solving and critical thinking. Hobbyist robots cater to enthusiasts who relish the art of crafting and programming machines, ranging from remote-controlled vehicles to intricate humanoid models.

Conclusion: A Tapestry of Diversity - The classification of robots illustrates the multifaceted nature of robotics and its integration into our lives. From shaping industries to venturing into unexplored territories, robots transcend boundaries and extend human capabilities. The taxonomy, fluid and evolving, reflects the harmonious coexistence of technology and human imagination, fostering a future where robots evolve as companions, collaborators, and catalysts of innovation.

Challenges and Trends in Robotics: The realm of robotics is a captivating intersection of innovation and complexity, encompassing a dynamic landscape of challenges and trends that drive the evolution of robotic technologies. As researchers, engineers, and enthusiasts push the boundaries of what is possible, they encounter hurdles and ride the waves of emerging trends that shape the future of robotics.

Technical Challenges: Bridging the Reality Gap: One of the enduring challenges in robotics is bridging the "reality gap" - the divide between the capabilities of robots in simulation versus the real world. Simulated environments provide controlled settings for testing and training, but translating these skills to unpredictable real-world scenarios remains a challenge. Simultaneously, issues like robust perception, accurate localization, and seamless interaction with dynamic environments continue to demand innovative solutions.

Human-Robot Interaction: Navigating Social Complexity: The interaction between humans and robots is a multifaceted challenge. Designing robots that can understand, interpret, and respond to human cues and emotions requires advancements in natural language processing, computer vision, and affective computing. Ensuring that robots can operate alongside humans safely, without causing discomfort or anxiety, is a critical consideration in collaborative settings.

Autonomy and AI: Striving for Intelligence - The pursuit of autonomous robots with cognitive abilities akin to human decision-making remains a focal point. Advancements in artificial intelligence, machine learning, and deep learning contribute to robots' ability to adapt to dynamic environments, learn from experiences, and make complex decisions. However, achieving true artificial general intelligence that can rival human intuition and reasoning is a monumental challenge that lies on the horizon.

Ethical and Social Implications: Balancing Innovation and Responsibility - As robots become more integrated into society, ethical considerations come to the forefront. Issues such as job displacement due to automation, privacy concerns related to surveillance robots, and the potential misuse of autonomous systems raise complex ethical dilemmas. Balancing the benefits of innovation with responsible development and deployment becomes paramount to ensure that robots contribute positively to society.

Soft and Bio-inspired Robotics: Emulating Nature - A burgeoning trend in robotics is the exploration of soft and bio-inspired robotics. Drawing inspiration from nature, researchers are developing robots with pliable, adaptable bodies that can navigate complex environments and interact safely with humans. These biomimetic robots mimic the mechanisms found in organisms, leading to breakthroughs in fields like medical robotics, search and rescue, and environmental monitoring.

Swarm Robotics: Collective Intelligence - Swarm robotics capitalizes on the power of collective behavior exhibited by groups of simple robots working together. Inspired by social insects like ants and bees, swarm robotics holds promise in scenarios where multiple robots collaborate to solve tasks that are beyond the capabilities of individual units. However, coordinating large numbers of robots without centralized control poses challenges in communication, synchronization, and resource allocation.

Human-Centered Design: Prioritizing User Experience - Trends in robotics are increasingly focusing on human-centered design, ensuring that robots are not just efficient machines, but also user-friendly and intuitive companions. Designing interfaces that facilitate seamless communication between humans and robots, considering ergonomic aspects, and accommodating diverse user needs are paramount to enhance the acceptance and usability of robotic technologies.

Conclusion: A Journey Unfolding - The challenges and trends in robotics paint a vivid tapestry of a field poised at the crossroads of possibility and complexity. The pursuit of overcoming technical hurdles, achieving harmonious human robot interaction, and navigating ethical considerations shapes the trajectory of robotics. Meanwhile, the emergence of novel paradigms like soft robotics, swarm intelligence, and human-centered design unveils horizons where robots transition from mere tools to transformative companions. The journey unfolds as roboticists and visionaries strive to surmount challenges, ride waves of innovation, and usher in an era where robotics enriches human lives, expands frontiers, and redefines our relationship with technology.

Exercise

Multiple-choice Questions

1. What is the primary focus of robotics?
a) Automating human tasks
b) Replicating human behavior
c) Enhancing artistic creativity
d) Integrating science and technology
2. Which ancient civilization is known for its early contributions to automata?
a) Ancient Egypt
b) Roman Empire
c) Mesopotamia
d) Ancient Greece
3. The term "robot" was first introduced by:
a) Nikola Tesla
b) Isaac Asimov
c) Karel Capek
d) Albert Einstein
4. What distinguishes robotics from traditional automation?
a) Robotics focuses on abstract tasks
b) Automation involves human-like decision-making
c) Robotics integrates creativity and science
d) Automation is exclusively used in factories
5. What role does artificial intelligence (AI) play in robotics?
a) AI limits robot capabilities
b) AI makes robots work independently of humans
c) AI enables robots to learn and adapt

d) AI only enhances robot aesthetics

Review Questions:

1. Define robotics and explain how it encompasses various scientific disciplines.
2. Discuss the historical origins of robotics and its evolution over time.
3. What distinguishes robotics from traditional automation, and why is this distinction important?
4. How has artificial intelligence influenced the capabilities of robots?
5. Provide examples of industries that have been transformed by the integration of robots.
6. Explain the concept of the "reality gap" in robotics and why it poses a challenge.
7. Describe the significance of human-robot interaction and collaboration in various applications.
8. Discuss the ethical considerations associated with the increasing integration of robots into society.
9. How does robotics contribute to the field of healthcare? Provide specific examples.
10. Define swarm robotics and explain its potential advantages in complex tasks.
11. Describe the concept of soft robotics and its applications in various fields.
12. How do robots play a role in disaster response and environmental monitoring?
13. Explain the concept of teleoperation and its relevance in hazardous environments.
14. Discuss the impact of robotics and automation on the future of work and employment.
15. How does the concept of human-centered design apply to robotics?
16. Provide examples of how robots are used in agriculture and logistics.
17. What are some potential benefits and challenges of using robots in the field of entertainment?
18. Describe the potential applications of robotics in space exploration.

Answers Multiple-choice Questions

Answers:

1. d)
2. d)
3. c),
4. c)
5. c)

Review Questions:

1. Robotics is the interdisciplinary field that combines mechanical engineering, electronics, computer science, and artificial intelligence to create machines known as robots. These robots are designed to replicate, simulate, or extend human actions and cognitive functions. Robotics blends scientific principles with creative innovation to develop machines that can perform tasks autonomously and intelligently.
2. The historical origins of robotics trace back to ancient civilizations, where automata were designed to mimic simple movements. In modern times, the concept evolved through the Industrial Revolution, where automation in manufacturing gained momentum. The term "robot" was introduced by Czech playwright Karel Čapek in his play "R.U.R." This concept expanded in science fiction and gradually led to the development of the diverse field of robotics we know today.
3. The primary distinction between robotics and traditional automation lies in adaptability and intelligence. While automation involves the mechanization of repetitive tasks, robotics goes beyond by integrating decision-making, learning, and interaction with the environment. Robots can adapt to changing conditions, making them suitable for tasks that demand flexibility and cognitive abilities.
4. Artificial intelligence (AI) has significantly expanded the capabilities of robots. It enables robots to learn from data, recognize patterns, and make informed decisions. Machine learning algorithms allow robots to improve their performance over time, adapting to different scenarios. AI empowers robots to handle complex tasks, interact with humans, and navigate dynamic environments.
5. Robots have transformed various industries, including manufacturing, healthcare, space exploration, agriculture, and entertainment. In manufacturing, robots streamline production processes and enhance precision. In healthcare, they assist surgeons, provide patient care, and aid in rehabilitation. Robotics also contributes to space exploration through autonomous rovers and spacecraft.
6. The "reality gap" refers to the challenge of translating robotic capabilities from simulated environments to real-world situations. While simulations offer controlled settings for testing, real-world scenarios introduce unpredictability. Bridging the reality gap involves developing algorithms and sensors that enable robots to adapt effectively to dynamic environments.

7. Human-robot interaction and collaboration are critical for robots to work alongside humans safely and efficiently. Collaborative robots, or cobots, are designed to interact with humans in shared workspaces. They incorporate sensors and algorithms to detect human presence and adjust their actions accordingly, promoting safe collaboration and productivity.

8. The integration of robots into society raises ethical concerns related to job displacement, privacy, and safety. Automation may lead to job loss in certain sectors, requiring workforce adaptation. Surveillance robots may infringe upon privacy rights. Ethical dilemmas also arise in situations where robots make autonomous decisions that impact human lives.

9. Robots play a significant role in healthcare by assisting in surgeries, providing rehabilitation exercises, and enhancing patient care. Surgical robots offer precision and minimally invasive procedures. Robotic exoskeletons aid individuals with mobility impairments in regaining movement. Robots also support patient monitoring and drug delivery.

10. Swarm robotics involves coordinating a group of simple robots to work together, similar to social insects in nature. This concept has applications in tasks such as environmental monitoring, disaster response, and exploration. Swarm robots can efficiently cover large areas, perform search and rescue operations, and adapt to changing conditions.

11. Soft robotics involves designing robots with flexible and deformable structures, often inspired by natural organisms. These robots can adapt to complex environments, interact safely with humans, and handle delicate objects. Applications of soft robotics span various fields, including medical surgery (minimally invasive procedures), search and rescue (navigating tight spaces), and wearable technology (exoskeletons for rehabilitation).

12. Robots play a crucial role in disaster scenarios where human safety is compromised. They can navigate through debris, hazardous materials, and unstable structures to locate survivors. Drones equipped with cameras provide real-time information to emergency responders, aiding in decision-making and resource allocation. In environmental monitoring, robots gather data on air quality, water pollution, and wildlife populations in hard-to-reach areas.

13. Teleoperation involves controlling a robot from a distance, often using haptic interfaces that provide tactile feedback. This technology is relevant in hazardous environments where direct human access is dangerous, such as deep-sea exploration, nuclear facilities, and space missions. Tele operated robots enable humans to perform tasks in such environments without risking their safety.

14. The integration of robotics and automation is transforming the nature of work. While it enhances efficiency and productivity in industries, it also raises concerns about job displacement. Routine and repetitive tasks are likely to be automated, leading to a shift in job requirements towards skills like creativity, problem-solving, and adaptability. The future of work involves reimagining roles and fostering human-machine collaboration.

15. Human-centered design focuses on creating robots that are user-friendly and intuitive for human interaction. It involves designing interfaces that facilitate seamless communication between humans and robots, considering ergonomic aspects for comfort, and accommodating diverse user needs. This approach ensures that robots enhance user experience, acceptance, and usability.

16. In agriculture, robots are used for tasks like planting, harvesting, and monitoring crops. Agricultural robots can identify ripe produce, apply pesticides precisely, and navigate fields autonomously. In logistics, robots automate tasks in warehouses, sorting and delivering packages efficiently. For instance, robots can move goods within fulfillment centers, optimizing inventory management.

17. Benefits of using robots in entertainment include captivating audiences with interactive performances, enhancing the immersive experience in theme parks, and enabling innovative forms of artistic expression. However, challenges include maintaining safety for human audiences, ensuring the reliability of complex robotic systems, and addressing potential ethical concerns when robots replace human performers.

18. Robotics is integral to space exploration. Robotic rovers like those on Mars gather data about the planet's surface. Robotic arms on spacecraft and space stations assist in tasks such as satellite deployment and maintenance. Autonomous drones can explore the atmospheres of other planets. Robots enable us to explore distant and hazardous environments without risking human lives.

Online Resources

Online Exam Paper

Link : https://forms.gle/MUwyNx9DfNq9zAbz9

QR Code :

YouTube Video Playlist Link

Link : https://www.youtube.com/playlist?list=PLzlh_31VWEFodq7iwXv5KWEWaFAf5S1EA

QR Code :

Chapter 2: Basics of Robotics

In an era where technology is reshaping the landscape of human endeavor, robotics emerges as a fascinating realm that bridges the gap between science fiction and reality. At its essence, robotics represents the culmination of human ingenuity, scientific curiosity, and engineering marvels, giving rise to machines that possess the ability to interact with the world, mimic human actions, and even make intelligent decisions. Delving into the basics of robotics unveils a captivating journey through the foundations of mechanics, electronics, computer science, and artificial intelligence, where machines cease to be mere tools and transform into partners in progress.

Intricately woven into the fabric of modern society, robotics is no longer confined to the realm of research labs and manufacturing floors. It permeates industries, enriches our lives, and shapes the trajectory of innovation. As we embark on this exploration, we peel back the layers of robotics, revealing the intricate components that constitute a robotic system – from mechanical structures that enable movement to sensors that imbue machines with perception, and from actuators that bring motion to life to control systems that orchestrate intricate maneuvers.

The evolution of robotics is an odyssey that traverses history, from the first automated looms of the Industrial Revolution to the cutting-edge humanoid robots that navigate dynamic environments with grace and precision. Beyond the mechanics, the incorporation of artificial intelligence adds an entirely new dimension, endowing robots with the capacity to learn, adapt, and interact with the world in ways that were once relegated to the realm of science fiction.

Applications of robotics are as diverse as the human imagination itself. From the medical domain, where robotic surgical assistants enhance precision and minimize invasiveness, to space exploration, where rovers traverse the surfaces of distant planets, and from manufacturing plants that see robots working alongside human counterparts to homes where robotic vacuum cleaners seamlessly maintain cleanliness, the reach of robotics knows no bounds.

However, the journey into the world of robotics is not without its complexities. Ethical considerations, societal implications, and the ever-evolving human-robot relationship take center stage in discussions. The role of robots in the workforce, their potential to alleviate or exacerbate social inequalities, and the questions surrounding robot autonomy raise critical ethical dilemmas that must be addressed as we navigate this frontier. In this exploration of the basics of robotics, we unravel not only the technical intricacies but also the profound impact that these machines wield in shaping the way we live, work, and interact. Robotics is no longer just a discipline; it is a narrative of human achievement, of collaboration between man and machine, and of a future where the boundaries of possibility continue to expand. So, as we journey through the fundamentals of robotics, we embark on a quest to understand not only the mechanics but also the essence of innovation that propels our world forward.

Robot Anatomy: Actuators, Sensors, and Controllers: In the intricate landscape of robotics, the concept of robot anatomy unfolds as a symphony of interconnected elements that breathe life into these technological marvels. Central to this anatomy are three essential components: actuators, sensors, and controllers. Actuators, often referred to as the "muscles" of a robot, drive its physical movements by converting electrical or pneumatic signals into mechanical action. Sensors, akin to sensory organs, enable robots to perceive and interact with their surroundings, gathering data about attributes like light, temperature, proximity, and touch. Controllers, serving as the robot's "brain," encompass algorithms that process sensor data, make informed decisions, and generate commands for the actuators. This seamless collaboration of actuators, sensors, and controllers creates a dynamic loop where robots perceive their environment, process information, and respond with precise actions. This integration forms the cornerstone of a robot's functionality, empowering it to navigate complex environments, manipulate objects, and execute tasks with dexterity and intelligence.

Actuators:Actuators serve as the "muscles" of a robot, responsible for generating movement and physical actions. They translate commands from the robot's controller into motion or manipulation. The diversity of robotic tasks demands various types of actuators. Electric motors are commonly used for their simplicity and controllability. Hydraulic and pneumatic actuators offer high power-to-weight ratios and are utilized in applications requiring heavy lifting. Piezoelectric actuators, with their rapid response and accuracy, find use in micro-robotics and precise positioning tasks. Actuators play acrucial role in endowing robots with the ability to perform tasks such as locomotion, gripping, and manipulation of objects.

Sensors:Sensors are the sensory organs of a robot, enabling it to gather information from its environment. These devices range from simple contact sensors to complex cameras and LiDAR systems. By detecting physical properties like light, sound, temperature, pressure, and distance, sensors provide the robot with an understanding of its surroundings. Vision sensors, including cameras and depth sensors, allow robots to recognize objects, navigate obstacles, and interact with humans. Tactile sensors enable robots to gauge force and pressure during interactions. Sensor data is processed by the robot's controller, allowing it to make informed decisions based on real-time feedback.

Controllers:Controllers function as the "brain" of the robot, processing sensory information and generating commands for the actuators. These are typically implemented using microcontrollers or embedded systems. The controller's algorithms analyze sensor data, interpret the robot's state, and determine appropriate actions based on pre-defined instructions or learning from experience. Proportional-Integral-Derivative (PID) controllers, neural networks, and reinforcement learning algorithms are common methods used to create effective controllers. As the heart of the robot's decision-making process, controllers enable robots to adapt to changing conditions and perform tasks autonomously.

The intricate interplay between actuators, sensors, and controllers grants robots the ability to interact with the world in dynamic and meaningful ways. Actuators execute tasks, sensors provide situational awareness, and controllers orchestrate the robot's responses. This synergy allows robots to fulfill a wide spectrum of roles, from precision manufacturing to autonomous exploration, showcasing the integration of mechanical, electronic, and computational principles in the field of robotics.

Degrees of Free: Degrees of Freedom, within the realm of robotics, encapsulate a fundamental concept that dictates the extent and versatility of a robot's motion and manipulation capabilities. At its core, Degrees of Freedom represent the number of independent parameters required to specify a robot's configuration, encompassing both position and orientation. It is akin to the robot's freedom to move and contort its body in various ways within its operational space.

Imagine a robot as a dynamic assemblage of interconnected components, each joint providing a potential axis of motion. The Degrees of Freedom denote the total count of these axes, portraying the robot's ability to perform distinct movements. For instance, a simple robot arm resembling a human elbow possesses one Degree of Freedom, allowing it to flex and extend like a hinge. In contrast, a more intricate robot arm might consist of several joints, including shoulder, elbow, and wrist joints, granting it multiple Degrees of Freedom for a richer range of motion.

The significance of Degrees of Freedom lies in its direct correlation with a robot's agility and adaptability. Robots endowed with a higher number of Degrees of Freedom can navigate complex paths, access varied positions, and orient themselves in intricate ways. This is crucial in applications demanding precision, such as manufacturing, where a robot's ability to approach objects from different angles streamlines production processes.

However, the augmentation of Degrees of Freedom also introduces complexities. More Degrees of Freedom require sophisticated control algorithms to orchestrate coordinated motion, and the risk of joint redundancy arises, demanding optimization strategies to ensure smooth and efficient movement.

The concept of Degrees of Freedom plays a pivotal role in designing robots tailored to specific tasks. Industrial robots, with their high Degrees of Freedom, excel in fine-tuned assembly tasks. Conversely, mobile robots prioritize maneuverability in constrained spaces, necessitating a balance between Degrees of Freedom and compact design.

In a broader context, humanoid robots aim to emulate human movement, necessitating a similar number of Degrees of Freedom. These robots imitate human arms, legs, and even facial expressions with a multitude of joints, enabling lifelike interactions and mobility.

In essence, Degrees of Freedom serve as the compass guiding a robot's motion possibilities. They dictate the intricacy of movement, the versatility of manipulation, and the extent of exploration a robot can undertake. The delicate interplay between Degrees of Freedom, control algorithms, and design considerations defines a robot's dexterity and autonomy, elucidating the elegance with which mechanics and mathematics unite to propel the frontiers of robotics.

Understanding Degrees of Freedom: Imagine a robot arm as an example. Each segment, or link, of the arm has a joint that allows movement. The robot arm's DoF is determined by the number of joints it possesses. A simple one-jointed arm, akin to a human elbow, has one DoF, allowing it to flex and extend. As complexity increases, a two-jointed arm (shoulder and elbow) boasts two DoF, enabling both flexion-extension and rotation. The addition of wrist joints further amplifies the DoF count.

Importance of Degrees of Freedom: DoF profoundly influences a robot's range of motion and maneuverability. Robots with higher DoF can access a broader range of positions and orientations, facilitating intricate and precise movements. This is particularly significant in applications such as manufacturing, where a robot's ability to approach and manipulate objects from various angles streamlines production processes.

Constraints and Challenges: While higher DoF empowers a robot with increased flexibility, it also introduces complexities in control, kinematics, and programming. Managing a robot's movement with numerous joints necessitates sophisticated control algorithms to ensure smooth and coordinated motion. Moreover, redundancy can arise in robots with excess DoF, demanding algorithms to optimize motion paths.

Applications in Robotics: Understanding DoF is pivotal in designing robots for specific tasks. Industrial robots, equipped with a high number of DoF, excel in precise assembly, welding, and painting. In contrast, mobile robots require fewer DoF but must account for dynamic environments. Legged robots, such as those used in exploration, mimic human and animal movement through multiple DoF in their limbs.

Degrees of Freedom in Humanoid Robots: Humanoid robots aim to replicate human movement, requiring a comparable number of DoF. These robots mimic human arms, legs, and even facial expressions with a multitude of joints. The goal is to achieve lifelike interactions with humans and the environment, enabling tasks ranging from basic locomotion to complex interactions in human-centered environments.

In essence, understanding Degrees of Freedom is paramount in designing robots tailored to specific tasks. It dictates a robot's versatility, motion capabilities, and complexity. The balance between DoF count, control algorithms, and practical applications underpins the intricate dance between robot design and function.

Robot Joints and Kinematics: In the realm of robotics, "2.3. Robot Joints and Kinematics" delves into the intricacies that govern a robot's motion and manipulation capabilities. At its core, this section elucidates the significance of robot joints and the mathematical principles of kinematics that orchestrate these movements. Robot joints, analogous to the articulations found in the human body, serve as pivotal connectors that enable motion and flexibility. This ranges from the revolute joints mimicking the hinge-like movement of an elbow to the spherical joints akin to a ball-and-socket joint. These joints collectively determine a robot's range of motion and interaction with its environment.

Complementing the study of joints, kinematics takes center stage as the mathematical framework that underpins a robot's motion. It encompasses both forward and inverse kinematics: the former traces the trajectory from joint angles to the robot's end effector position, while the latter calculates the joint angles needed to position the end effector at a desired location. Kinematics, thus, forms the essence of how robots navigate space, manipulate objects, and perform intricate tasks.

The section explores the profound importance of kinematics in tasks such as trajectory planning, where robots follow specific paths while avoiding obstacles. Understanding kinematics is imperative for programming and controlling robots effectively, enabling engineers to decipher how joint movements translate into the position and orientation of the robot's end effector. This foundational knowledge is vital not only for industrial robots streamlining manufacturing but also for sophisticated robotic systems performing delicate surgeries and exploring distant planets.

In essence, "Robot Joints and Kinematics" unravels the captivating interplay between robot joints and the mathematical principles of kinematics. This synergy forms the bedrock of robotic motion, driving the graceful movements and operational capabilities of robots across industries. From manufacturing to medicine, from space exploration to entertainment, the study of robot joints and kinematics elucidates the precision and ingenuity that fuel the dynamic field of robotics.

Robot Joints: Robot joints are pivotal components that enable motion and flexibility in robotic systems. They mimic the articulations found in the human body, granting robots the ability to navigate and manipulate their surroundings. The types of joints employed significantly impact a robot's motion capabilities. Common joint types include:

Revolute Joint: This joint facilitates rotational motion around a fixed axis. It is akin to a human joint like the elbow or knee, enabling movement in a circular arc.

Prismatic Joint: Prismatic joints enable linear motion along a single axis, mimicking movements akin to a sliding door or an elevator.

Spherical Joint: Spherical joints allow rotational motion in multiple axes, resembling the movement of a ball-and-socket joint like the human hip.

Cylindrical Joint: Cylindrical joints combine prismatic and revolute motions, enabling both linear and rotational movement.

Kinematics: Kinematics in robotics is the study of motion without considering the forces or torques that cause it. It deals with the mathematical representation of a robot's configuration, position, velocity, and acceleration. Forward kinematics traces the path from joint angles to the robot's end effector position, while inverse kinematics determines the joint angles needed to position the end effector at a desired location.

Importance of Kinematics: Understanding robot kinematics is essential for programming and controlling robots effectively. It aids in determining how joint movements affect the position and orientation of the robot's end effector. Kinematics is fundamental in tasks such as trajectory planning, where robots are programmed to follow specific paths while avoiding obstacles.

Robot Workspace and Singularity: The workspace of a robot refers to the range of positions it can reach. Understanding kinematics helps define this workspace, allowing engineers to optimize a robot's design and ensure it can fulfill its intended tasks. Singularity points are positions in which a robot's kinematic equations become unstable, causing irregularities in its motion. Knowledge of kinematics aids in identifying and avoiding these singularity points.

Applications and Complexity: Kinematics finds applications in various fields, from manufacturing to medical robotics. In manufacturing, it guides robots in assembling products with precision. In surgery, robots with accurate kinematic models aid in performing intricate procedures. Complex robots, like humanoid robots, employ intricate kinematic models to achieve lifelike movement and interaction.

In essence, the study of robot joints and kinematics forms the backbone of robotic motion. The synergy between different joint types, coupled with the mathematical principles of kinematics, orchestrates a robot's graceful movements and operational capabilities. This foundational knowledge underpins the ability to design, control, and program robots for a diverse array of tasks.

Coordinate Systems and Transformations: In the vast and dynamic realm of robotics, the mastery of coordinate systems and transformations is paramount for enabling robots to comprehend, navigate, and interact with their environment. These concepts serve as the linguistic bridge between the physical world and the mathematical models that guide a robot's actions. By understanding coordinate systems and transformations, robots can seamlessly transition between their internal representation of space and the external reality they inhabit.

Coordinate Systems: Coordinate systems are the foundation upon which robots perceive and navigate the world around them. In robotics, two primary types of coordinate systems are pivotal:

Global (World) Coordinate System: This serves as the overarching reference frame for all spatial calculations and interactions. The global coordinate system establishes a universal point of reference, often rooted at a fixed origin, against which the robot's movements, positions, and orientations are measured.

Local (Robot) Coordinate System: Each individual robot maintains its own local coordinate system, tailored to its specific structure and components. This local coordinate system allows the robot to interpret its surroundings based on its unique vantage point. This concept mirrors how humans use their own body as a reference frame.

Transformations: Transformations are the mathematical tools that facilitate seamless communication between different coordinate systems. They provide the means to convert positions and orientations from one reference frame to another. Transformations encompass both translation (movement) and rotation, and they are crucial for robotics applications that involve navigating, manipulating, or interacting with objects in various locations and orientations.

Homogeneous Transformations: A particularly powerful concept within transformations is homogeneous transformations. These transformations combine translation and rotation into a single matrix, encapsulating both spatial changes in one coherent mathematical entity. Homogeneous transformations enable the translation of points, vectors, and coordinates between different coordinate systems, simplifying complex mathematical operations and enhancing efficiency in robotic calculations.

Applications: Coordinate systems and transformations find wide-ranging applications in robotics across diverse industries:

Manufacturing: Robots in manufacturing rely on accurate transformations to precisely position and assemble components on assembly lines, enhancing production efficiency and product quality.

Autonomous Vehicles: In self-driving cars, transformations aid in mapping sensor data from multiple sources to a common reference frame, enabling safe navigation and collision avoidance.

Surgery: In robotic surgery, transformations ensure that surgical instruments are positioned accurately within a patient's body, facilitating precise and minimally invasive procedures.

Aerospace: Transformations play a key role in spacecraft navigation and orientation, allowing spacecraft to navigate through complex trajectories and reorient themselves in space.

Conclusion: Coordinate systems and transformations serve as the foundation upon which robotics operates. They empower robots to not only understand the world around them but also to interact with it in a meaningful and precise manner. By leveraging the principles of these concepts, robotic engineers and researchers unlock the potential for robots to navigate dynamic environments, manipulate objects, and perform tasks with a level of precision and adaptability that continues to shape the future of robotics.

Robot Programming: Offline and Online: Robot programming constitutes the orchestration of instructions and commands to guide robotic systems in performing a spectrum of tasks with accuracy and precision. This intricate process lies at the heart of robotics, transforming mechanical entities into intelligent agents capable of interacting with their surroundings. Robot programming encompasses a multifaceted journey that involves understanding the robot's capabilities, crafting algorithms, and facilitating autonomous actions.

At its core, robot programming is the art of translating human intentions into machine-executable actions. It serves as the bridge between human cognition and robotic motion, imbuing machines with the ability to navigate complex scenarios, manipulate objects, and interact with the environment.

Robot programming encompasses diverse paradigms to cater to the complexity and variability of tasks:

Manual Programming: This approach involves manual guidance, where operators control the robot in real time to demonstrate desired actions. This is particularly useful for tasks that require human intuition, adaptability, and intricate manipulation.

Offline Programming: Offline programming involves crafting robot programs away from the physical robot, often using computer simulations. This method offers the advantage of precise planning, optimization, and testing without risking the robot or its environment.

Automated Programming: Automated programming employs advanced algorithms and artificial intelligence to generate robot programs based on high-level task descriptions. This approach streamlines programming for repetitive or structured tasks.

Robot programming is not devoid of challenges:

Complexity: Programming robots for intricate tasks demands an understanding of kinematics, dynamics, sensor integration, and control algorithms. These complexities amplify as robots become more versatile and autonomous.

Human-Machine Interaction: Achieving seamless interaction between robots and humans requires programming robots to understand and respond to human input, whether through spoken language, gestures, or interfaces.

Dynamic Environments: Programming robots to function effectively in dynamic and unstructured environments presents challenges in adapting to changing conditions, avoiding obstacles, and interacting with unpredictable elements.

Robot programming influences an array of industries:

Manufacturing: Industrial robots execute tasks in assembly lines with precision and efficiency, boosting productivity and product quality.

Healthcare: Surgical robots assist in minimally invasive procedures, enhancing surgical precision and reducing patient recovery time.

Exploration: Robots programmed for space exploration gather data from distant planets and moons, enabling scientific discoveries.

Logistics: Robots navigate warehouses, sorting packages and optimizing inventory management, revolutionizing the supply chain.

The future of robot programming intertwines with emerging technologies:

Artificial Intelligence: Advanced AI algorithms contribute to autonomous robot programming, enabling machines to adapt and learn from their interactions.

Human-Robot Collaboration: Programming robots to collaborate safely and harmoniously with humans becomes crucial as robots integrate into various domains.

Programming Abstraction: Abstraction layers simplify robot programming, allowing users to focus on high-level tasks without delving into intricate technicalities.

In essence, robot programming is the poetic symphony orchestrating the movement and actions of machines. It combines engineering ingenuity, mathematical precision, and the potential of artificial intelligence to give rise to machines that seamlessly navigate our world, perform intricate tasks, and redefine the boundaries of what is possible. As technology advances, so too does the artistry and potential of robot programming, paving the way for a future where robots are integral collaborators in our daily lives.

Offline Programming: Offline programming is a method that entails crafting robot programs away from the physical presence of the robot, often within a computerized environment. In this approach, the robot's actions are meticulously designed, simulated, and verified before being executed on the actual machine. Offline programming offers a multitude of advantages:

Safety: By programming in a virtual environment, the risk of accidents or collisions is nullified, ensuring the robot's integrity and the safety of surrounding objects and individuals.

Efficiency: Engineers can fine-tune and optimize programs without halting or disrupting the robot's physical operations, maximizing productivity.

Complexity: Intricate tasks that demand precision, intricate movements, or intricate coordination can be devised and tested with higher accuracy in a controlled digital space.

Online Programming: Online programming, often referred to as teach programming, is a hands-on approach where a human operator directly guides the robot's actions. This interaction involves manually manipulating the robot through desired motions, which are then recorded and translated into executable commands. Online programming presents its own array of advantages:

Flexibility: In rapidly evolving environments or tasks with unpredictable variables, online programming allows operators to make real-time adjustments, adapting the robot's actions on the fly.

Intuitiveness: Online programming capitalizes on human intuition and adaptability, making it accessible even to non-experts who can guide the robot to perform desired actions.

Complex Environments: In scenarios where the complexities of the real world might challenge precise programming, online programming enables the robot to navigate and interact based on human judgment.
Hybrid Approaches: In practice, a hybrid approach often emerges, blending the strengths of both offline and online programming methodologies. Engineers may use offline programming for tasks that require meticulous planning and optimization, ensuring the highest degree of accuracy. Conversely, they might resort to online programming for tasks that demand real-time judgment, adaptability, and a more intuitive human touch.
Conclusion: The world of robot programming serves as the conduit through which human creativity and machine precision converge. The duality of offline and online programming approaches showcases the breadth of possibilities within this domain. Offline programming amplifies precision and optimization, while online programming infuses robots with adaptability and human-like interaction. As technology advances, the seamless integration of these methodologies promises a future where robots seamlessly collaborate with humans, performing tasks efficiently, intelligently, and with an innate understanding of their environment.

Exercise

Multiple-choice Questions

1. What is the purpose of actuators in a robot?
a) Provide sensory input
b) Execute physical movements
c) Process control algorithms
d) Generate power for sensors
2. Which component of a robot is responsible for detecting attributes like light, temperature, and touch?
a) Actuators
b) Controllers
c) Sensors
d) Kinematics
3. What role do controllers play in a robot?
a) Execute physical movements
b) Detect attributes from the environment
c) Provide power to sensors
d) Process sensor data and generate commands
4. What is the purpose of kinematics in robotics?
a) Providing power to actuators
b) Detecting sensor attributes
c) Generating commands for controllers
d) Describing robot motion and relationships between components
5. Robot joints are analogous to which human body part?
a) Heart
b) Brain
c) Muscles
d) Bones

Review Questions:

1. What is the role of actuators in robotics, and how do they contribute to a robot's functionality?
2. Explain the purpose of sensors in a robotic system. Provide examples of different types of sensors used in robotics.
3. How do controllers play a crucial role in enabling a robot to perform tasks autonomously? Describe their function in processing sensor data and generating commands.
4. Define kinematics in the context of robotics. Differentiate between forward kinematics and inverse kinematics.
5. Why are robot joints important, and how do they impact a robot's range of motion? Give examples of different types of robot joints.
6. How does an understanding of kinematics benefit robot programming and control? Provide a real-world example of kinematics application.
7. Differentiate between global and local coordinate systems in robotics. Explain the significance of using both types of coordinate systems.
8. What are transformations in robotics, and why are they essential? Describe how transformations are used to convert positions between coordinate systems.

9. Provide examples of how knowledge of coordinate systems and transformations is applied in real-world robotic applications.
10. How does the collaboration between actuators, sensors, and controllers create a cohesive robotic system capable of performing complex tasks?
11. Describe the relationship between sensors and perception in a robotic system. How do sensors enable a robot to gather information about its environment?
12. Explain how controllers process sensor data to generate meaningful commands for a robot's actuators. What role do algorithms play in this process?
13. How does kinematics help in calculating the relationships between joint angles and end effector positions in a robot?
14. What are the advantages of robots having multiple degrees of freedom in their joints? Give examples of tasks that benefit from robots with high degrees of freedom.
15. How do coordinate systems provide a spatial reference for robots? What purpose does a global coordinate system serve?
16. In what scenarios would a local coordinate system be more advantageous for a robot's operations?
17. Describe the concept of transformations in robotics. How do they enable seamless communication between different coordinate systems?
18. Provide examples of how transformations are used to convert positions and orientations from one coordinate system to another.
19. Explain how an understanding of coordinate systems and transformations enhances a robot's ability to interact with its environment.
20. How does the mastery of robot anatomy, kinematics, coordinate systems, and transformations contribute to the development of advanced robotic applications across industries?

Answers Multiple-choice Questions

Answers:
1. b)
2. c)
3. d)
4. d)
5. c)

Review Questions

1. Actuators serve as the mechanisms responsible for generating physical movement and action in a robotic system. They are the "muscles" of the robot, converting control signals into motion. Actuators contribute significantly to a robot's functionality by enabling it to perform a wide range of tasks, from locomotion to manipulation. Depending on the application, different types of actuators are used, such as motors, pneumatic cylinders, hydraulic systems, and even shape-memory alloys. For instance, in a robotic arm, actuators at each joint provide the necessary motion to position the arm accurately and carry out tasks with precision. Without actuators, a robot would remain stationary and unable to interact with its environment.
2. Sensors play a pivotal role in providing robots with the ability to perceive and gather information about their surroundings. They act as the robot's "senses," capturing data from the environment and converting it into usable signals that the robot's control system can process. Various types of sensors are employed in robotics, including proximity sensors, cameras, accelerometers, gyroscopes, force sensors, and temperature sensors. For example, a camera sensor enables a robot to recognize objects, navigate through environments, and even perform tasks such as image-guided manipulation. Force sensors, on the other hand, allow robots to interact delicately with objects by providing feedback on the force exerted during contact.
3. Controllers are the brain of the robotic system, responsible for interpreting sensor data and generating commands for the actuators. They play a crucial role in enabling a robot to perform tasks autonomously by processing sensory information and making decisions based on predefined algorithms. Controllers take into account the robot's current state, the desired task, and the sensory inputs to generate appropriate control signals. These signals drive the actuators to execute the desired actions. In a robotic vacuum cleaner, for instance, the controller processes input from sensors like cameras and obstacle detection sensors to determine the optimal path for cleaning while avoiding obstacles.
4. Kinematics in robotics refers to the study of the motion of robots and their constituent parts. It focuses on understanding the relationships between joint angles, positions, and orientations of the robot's components. Forward kinematics involves calculating the end effector's position and orientation based on the joint angles. In contrast, inverse kinematics involves finding the joint angles necessary to achieve a specific end effector position and

orientation. Forward kinematics is used to determine where the robot's end effector will be located given joint configurations, while inverse kinematics is vital for tasks such as motion planning and positioning.

5. Robot joints are pivotal components that facilitate movement and flexibility in robotic systems. They enable the articulation and rotation of various parts of the robot. Different types of robot joints include revolute joints (allowing rotation around an axis), prismatic joints (allowing linear motion along an axis), and spherical joints (enabling multi-axis rotation). The choice of joint type impacts a robot's range of motion and capabilities. For example, a robot arm with revolute joints can achieve complex manipulations and reach around obstacles due to its articulated nature, making it suitable for tasks like assembly or painting.

6. An understanding of kinematics is essential for robot programming and control as it forms the foundation for calculating and predicting the robot's movements. For instance, in a pick-and-place robot tasked with moving objects from one location to another, kinematics helps in determining the required joint angles to position the end effector accurately. This knowledge guides the creation of control algorithms that drive the actuators to achieve the desired motions. In industries like manufacturing, efficient programming of robots hinges on precise kinematic calculations to ensure accuracy and avoid collisions.

7. Global and local coordinate systems are fundamental concepts in robotics used to define positions and orientations of objects in space. A global coordinate system provides an absolute reference for positions, often tied to a fixed point in the environment. It enables robots to navigate and interact with the surroundings accurately. A local coordinate system, on the other hand, is specific to an object or a robot's frame of reference. It simplifies calculations within the robot's immediate vicinity, streamlining tasks like motion planning and control. Using both coordinate systems allows robots to comprehend their surroundings globally while performing localized tasks efficiently.

8. Transformations in robotics refer to the mathematical operations used to convert positions and orientations between different coordinate systems. They are essential for enabling communication and coordination between various components of a robotic system. Transformations involve translation (shifting positions), rotation (changing orientations), and scaling (resizing). By applying transformations, positions and orientations can be seamlessly converted from one reference frame to another, allowing a robot to navigate its environment while maintaining accurate spatial relationships.

9. Knowledge of coordinate systems and transformations finds practical application in numerous real-world robotic scenarios. In robotic surgery, precise transformations enable surgeons to control robotic instruments remotely while ensuring accurate positioning within the patient's body. In autonomous vehicles, transformations aid in translating sensor data from various sources (lidar, cameras, radar) into a unified coordinate system, enabling effective navigation and obstacle avoidance. In manufacturing, transformations are crucial for accurately placing components in assembly lines and guiding robotic arms to perform intricate tasks with precision.

10. The collaboration between actuators, sensors, and controllers forms a cohesive robotic system capable of performing complex tasks through a feedback loop. Sensors provide input about the environment, controllers process this data and generate control signals, and actuators execute the desired actions. For example, in a humanoid robot, sensors detect changes in terrain, controllers adjust joint angles for balance, and actuators execute leg movements to maintain stability, resulting in smooth and controlled locomotion.

11. Sensors and perception are intricately linked in a robotic system, with sensors providing the means for the robot to gather information about its environment. Sensors enable the robot to perceive aspects such as object presence, distance, temperature, force, and more. In an autonomous car, sensors like lidar and cameras detect pedestrians, vehicles, and road conditions. Perception algorithms process sensor data, allowing the robot to make informed decisions and adapt to dynamic surroundings.

12. Controllers process sensor data to generate meaningful commands for a robot's actuators by implementing algorithms that interpret sensory information. For instance, in a robotic arm assembling components, cameras provide data on the part's position, and the controller's algorithm calculates the required joint angles to pick up and place the part accurately. Algorithms play a critical role in optimizing control signals to achieve smooth and efficient motion, enhancing a robot's performance and precision.

13. Kinematics aids in calculating the relationships between joint angles and end effector positions by using geometric principles and mathematical transformations. This understanding allows engineers and programmers to design motion sequences that achieve specific tasks. In a delta robot used for high-speed pick-and-place operations, knowledge of kinematics enables precise control of the robot's movements, ensuring rapid and accurate handling of objects.

14. Robots with multiple degrees of freedom (DOF) in their joints possess enhanced flexibility and versatility in their motions. Additional DOF provide greater maneuverability, enabling robots to reach complex positions and

orientations. For instance, a humanoid robot with high DOF in its arms can mimic intricate human gestures, making it suitable for tasks like sign language interpretation or delicate assembly tasks that require dexterity and precision.

15. Coordinate systems provide a spatial reference for robots, allowing them to navigate and interact effectively with their environment. A global coordinate system establishes a consistent reference point for positioning objects or robots across an entire environment. In navigation, robots use a global coordinate system to understand their position relative to landmarks or map references, aiding in accurate path planning and localization.

16. A local coordinate system is advantageous in scenarios where a robot's operations are confined to a specific area or object. For example, a robotic arm tasked with assembling intricate components within a limited workspace may employ a local coordinate system centered around the assembly point. This simplifies calculations and control, optimizing the robot's efficiency within its designated scope.

17. Transformations in robotics are fundamental concepts that allow different coordinate systems to communicate seamlessly. They enable the conversion of positions, orientations, and measurements between various frames of reference. Transformations ensure that data from sensors, controllers, and actuators can be accurately interpreted and applied across different components of a robotic system, fostering efficient and coordinated operations.

18. Transformations are employed to convert positions and orientations from one coordinate system to another in various contexts. In robotic navigation, transformations enable accurate mapping by aligning sensor data with a global coordinate system. In 3D printing, transformations are used to position the printing head precisely in relation to the work piece, ensuring the creation of intricate and accurate designs layer by layer.

Online Resources

Online Exam Papers

Link: https://forms.gle/pLBBoXUvV81hQWLF7

QR Code:

YouTube Video Playlist Link

Link: https://www.youtube.com/playlist?list=PLzlh_31VWEFqxo25cjcb5wKne2YpQCZnp

QR Code:

Chapter 3: Robot Sensors and Perception

In the ever-evolving landscape of robotics, a crucial facet that has propelled the field to new heights is the domain of robot sensors and perception. In a world where machines are transcending their traditional roles and assuming roles as perceptive and adaptable entities, sensors have emerged as the eyes, ears, and touch of intelligent robots. Perception, fueled by advanced algorithms and sensory data interpretation, is rapidly reshaping the way robots interact with their surroundings, make decisions, and navigate complex environments. The realm of robot sensors and perception introduces us to a fascinating realm where machines cease to operate blindly and instead possess the ability to sense, comprehend, and respond to the intricacies of the world around them. This exploration delves into the intricacies of sensors, the algorithms that drive perception, and the profound impact this convergence of technology holds for robotics and the wider technological landscape.

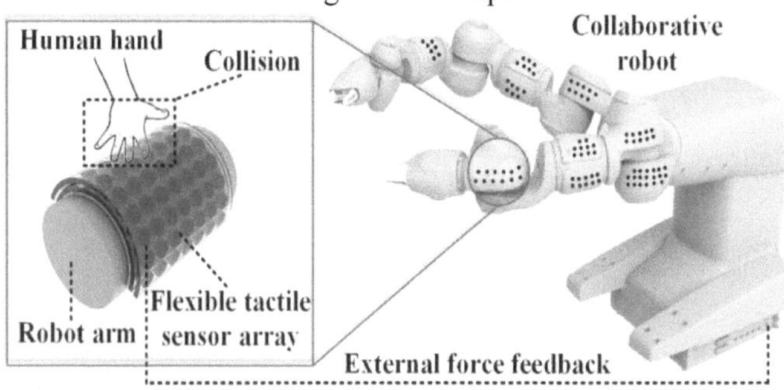

At the heart of robot sensors lies the ability to capture data from the environment. These sensors serve as the interface through which robots engage with the world, enabling them to perceive physical parameters such as distance, light, temperature, force, and more. From the depth-sensing capabilities of lidar to the image-capturing prowess of cameras, each sensor type contributes a unique perspective to the robot's perceptual repertoire. Through their data collection and interpretation, sensors offer robots the capacity to gather information that is crucial for decision-making, motion planning, and interaction with objects and humans.

Perception, on the other hand, transforms raw sensor data into meaningful insights. It encompasses the algorithms and computational methodologies that allow robots to process sensor inputs, recognize patterns, and generate an understanding of their environment. With the aid of machine learning techniques, robots can identify objects, classify scenes, and even predict outcomes based on historical data. Perception grants robots the ability to differentiate between obstacles and pathways, to understand the nuances of color and texture, and to interpret gestures and expressions, paving the way for natural and intuitive human-robot interaction.

The symbiotic relationship between sensors and perception is exemplified in the navigation of autonomous vehicles. Cameras and lidar sensors collect data about the surroundings, detecting pedestrians, traffic signals, and road conditions. Perception algorithms then process this data to identify objects, estimate distances, and predict the behavior of other vehicles and pedestrians. These insights are fed into control systems that enable the vehicle to make real-time decisions, ensuring safe and efficient navigation.

As robotics continues its expansion into diverse domains such as healthcare, manufacturing, and space exploration, the significance of sensors and perception becomes ever more apparent. Surgical robots equipped with precise sensors enhance the precision of minimally invasive procedures, while industrial robots rely on force and tactile sensors to manipulate delicate components. Exploration robots on distant planets employ sensors to gather data about the environment and make informed decisions in uncharted terrain.

However, the journey into the world of robot sensors and perception is not without its challenges. Sensor accuracy, data fusion, real-time processing, and the ability to adapt to dynamic environments are ongoing areas of research and development. Ethical considerations surrounding privacy and data collection also come to the forefront, as robots become more integrated into our daily lives.

Types of Sensors: Vision, Tactile, Range, and Force: In the intricate realm of robotics, sensors serve as the crucial sensory organs that endow machines with the ability to comprehend, interact with, and respond to their surroundings. These sensors encompass an array of types, each meticulously designed to capture specific dimensions of the environment. By deciphering the intricacies of these sensor categories, we gain insight into the mechanisms that enable robots to gather data, interpret the world, and execute tasks with precision.

Vision Sensors: Vision sensors, inspired by the human sense of sight, bestow robots with the ability to perceive the visual spectrum of the environment. Through the employment of cameras and advanced imaging technologies, robots equipped with vision sensors can capture the vibrancy of colors, the intricacy of patterns, and the geometry of shapes. However, these sensors transcend mere image acquisition, delving into the realm of computer vision. This convergence allows robots to not only capture visual data but also derive meaningful insights from it. From identifying objects to tracking their movements, from interpreting complex scenes to navigating intricate terrains, vision sensors form the bedrock of robots' visual perception. With the incorporation of stereoscopic vision, achieved through the deployment of multiple cameras simulating human binocular vision, robots gain the invaluable capability of depth perception, enabling them to discern distances and comprehend the spatial arrangement of their surroundings.

Tactile Sensors: Tactile sensors emulate the sense of touch, endowing robots with the capacity to interact with objects and environments in a manner akin to human touch. By quantifying forces, pressures, and even vibrations, tactile sensors grant robots haptic feedback, enabling them to engage with objects delicately and with precision. These sensors play an instrumental role in tasks necessitating intricate manipulation, where a gentle touch is paramount. Moreover, in the realm of human-robot interaction, tactile feedback becomes a safety net, allowing robots to recognize the presence of humans or unexpected obstacles, thereby triggering instant responses to ensure accident prevention. The incorporation of tactile sensors translates to enhanced dexterity, enabling robots to grasp objects with the appropriate amount of force, thereby preventing damage while ensuring secure manipulation.

Range Sensors: Range sensors, known as proximity sensors, confer spatial awareness upon robots by gauging the distances between the robot and its surroundings. Ultrasonic sensors, for instance, emit sound waves, measuring the interval between emission and reception to calculate distances. This data becomes instrumental in the obstacle detection realm, enabling robots to deftly navigate through cluttered environments by identifying and avoiding obstructions. Meanwhile, LIDAR sensors harness laser beams to meticulously measure distances, generating detailed 3D maps of the environment. These maps are pivotal in charting not only the terrain but also in planning efficient paths. Thus, range sensors serve as the eyes that facilitate robots' spatial understanding, ensuring that their movements are accurate, safe, and optimized for the context.

Force Sensors: Force sensors amplify robots' understanding of the physical forces encountered during interactions with objects. By quantifying these forces, robots are empowered to tailor their grip strength, pressure, and responses based on real-time requirements. This capability finds relevance in tasks necessitating finesse, such as the delicate handling of objects or the intricate assembly of components. Medical robotics, particularly surgical scenarios, benefits immensely from force sensors, which provide surgeons with haptic feedback, refining precision and control. Through force sensors, robots acquire the sensitivity to recognize nuances in pressure, leading to the modulation of their actions accordingly. This is invaluable in domains where tactile discernment is critical, allowing robots to navigate the intricate balance between effectiveness and fragility.

The amalgamation of vision, tactile, range, and force sensors transmutes robots into multi-sensory entities. Vision sensors provide sight, tactile sensors grant touch, range sensors offer spatial awareness, and force sensors replicate sensitivity to physical forces. These sensors collectively form the sensory ecosystem that empowers robots to comprehend their environment holistically. As robots progress, driven by these sensory insights, their capacity to engage, navigate, and interact with their surroundings evolves, underscoring the transformative role sensors play in advancing the capabilities of robotics.

Computer Vision: Object Detection, Image Processing: In the realm of robotics, computer vision emerges as a cornerstone technology, enabling machines to perceive and interpret the visual world much like human vision. Through a combination of image processing techniques and advanced algorithms, computer vision empowers robots to analyze images and extract meaningful information, revolutionizing their ability to navigate, interact, and comprehend their environment.

Image Processing: At the core of computer vision lies image processing, a discipline focused on manipulating digital images to enhance their quality and extract relevant features. Robots equipped with image sensors capture visual data, which then undergoes a sequence of operations such as noise reduction, image enhancement, and contrast adjustment. Filtering techniques, like convolution, enable the isolation of specific image characteristics, enhancing the clarity of vital elements. Morphological operations, such as dilation and erosion, aid in refining object boundaries and texture analysis. By fine-tuning these processes, robots can effectively enhance the visual data received from their surroundings, thus laying the foundation for accurate analysis.

Object Detection: One of the pivotal applications of computer vision is object detection – the ability to identify and locate specific objects within an image or a video stream. This capability transcends mere visual recognition; it involves identifying objects irrespective of their orientation, scale, or position within the frame. Convolutional Neural Networks (CNNs), a subset of machine learning algorithms, have revolutionized object detection by enabling robots to automatically learn and discern object features from extensive training data. By dividing the image into grids and analyzing each grid cell's potential to contain certain objects, CNNs can identify objects with remarkable accuracy. Furthermore, the integration of anchor boxes facilitates the identification of objects of varying sizes and shapes within the same image.

Applications in Robotics: The integration of computer vision in robotics unlocks a myriad of applications across domains. In industrial settings, robots endowed with computer vision can accurately identify and sort objects on assembly lines, improving efficiency and reducing errors. In autonomous vehicles, computer vision systems interpret road signs, traffic lights, and other vehicles, ensuring safe navigation. In healthcare, robots equipped with object detection capabilities can assist in surgical procedures by identifying critical anatomical structures. Additionally, service robots can use computer vision to recognize and interact with humans, enhancing their capability to perform tasks in dynamic environments.

Challenges and Advancements: While computer vision holds immense potential, it also presents challenges. Variations in lighting, occlusions, and varying perspectives can hinder accurate object detection. To counter these challenges, techniques such as feature extraction and image segmentation are employed. Deep learning models, like Faster R-CNN and YOLO (You Only Look Once), have paved the way for real-time object detection, revolutionizing the speed and accuracy of robotic perception.

In essence, computer vision empowers robots to see, interpret, and interact with the visual world. The amalgamation of image processing techniques and cutting-edge algorithms enables robots to identify objects, assess their context, and respond accordingly. This symbiotic relationship between robots and computer vision is driving automation to new heights, reshaping industries, and amplifying the scope of tasks robots can undertake with precision and intelligence.

Sensor Fusion: In the intricate landscape of robotics, the world is not perceived through the lens of a single sensor; instead, robots rely on an array of sensors, each capturing a distinct facet of the environment. Sensor fusion, a sophisticated technique, emerges as the conduit that seamlessly integrates data from multiple sensors, crafting a comprehensive and coherent understanding of the world. This integration not only enhances perception but also refines decision-making, enabling robots to navigate, interact, and respond with heightened accuracy and reliability.

The Essence of Sensor Fusion: Sensor fusion rests on the principle that a single sensor often falls short in delivering a holistic view of the environment. Different sensors possess strengths and weaknesses; for instance, vision sensors excel in capturing detailed visual information, while range sensors excel in detecting distances. Sensor fusion, therefore, harmonizes these diverse inputs, generating a composite representation that transcends the limitations of individual sensors. This synergistic approach transforms fragmented data into a comprehensive map, offering robots the ability to interpret the environment from multiple perspectives.

Types of Sensor Fusion: Two main paradigms govern sensor fusion: data-level fusion and feature-level fusion. Data-level fusion entails the direct integration of raw sensor data. This approach combines data streams from various sensors, consolidating them into a cohesive dataset that exhibits a broader scope of information. In contrast, feature-level fusion involves extracting relevant features from individual sensor data and combining these features to craft a richer representation. This approach leverages the strengths of each sensor, generating a comprehensive understanding while mitigating the impact of individual sensor weaknesses.

Enhancing Perception and Decision-Making: Sensor fusion serves as the backbone for an array of robotics applications, enriching robots' perception and decision-making capabilities. In autonomous vehicles, fusing data from cameras, LIDAR, and radar sensors offers a multi-dimensional view of the surroundings, ensuring accurate obstacle detection and safe navigation. In medical robotics, combining information from vision and tactile sensors can enhance the precision of surgical procedures, contributing to better patient outcomes. Furthermore, in industrial automation, sensor fusion refines the detection of defects in manufacturing processes, reducing errors and enhancing product quality.

Challenges and Advances: While sensor fusion bestows robots with remarkable capabilities, it also presents challenges. Calibration, alignment, and synchronization of sensors demand meticulous attention to ensure accurate data fusion. Moreover, managing the uncertainty associated with sensor measurements necessitates sophisticated algorithms that account for sensor noise, drift, and discrepancies. Advances in machine learning, particularly techniques like Kalman filtering and Bayesian inference, have elevated sensor fusion to new heights, enabling robots to not only fuse data but also make predictions based on fused data, enhancing their anticipation and responsiveness.

In essence, sensor fusion transcends the boundaries of individual sensors, forging a panoramic perception of the environment that empowers robots to navigate, interact, and perform tasks with heightened accuracy. By amalgamating the strengths of various sensors and mitigating their weaknesses, sensor fusion is instrumental in unlocking the full potential of robotics, propelling the field toward greater levels of sophistication and utility.

Perception Algorithms and Techniques: In the intricate realm of robotics, perception algorithms and techniques stand as the cognitive engines that enable machines to decipher their surroundings. These algorithms, inspired by human sensory processes, empower robots to process sensory data, derive meaningful insights, and make informed decisions. By delving into the intricacies of perception algorithms, robots elevate their ability to navigate, interact, and comprehend the world with greater accuracy and sophistication.

Perception Algorithm Principles: Perception algorithms draw inspiration from biological processes to imbue robots with sensory comprehension. These algorithms convert raw sensor data, often noisy and complex, into structured representations that encapsulate essential information. Filtering techniques, such as Gaussian filters, smooth out sensor noise, enhancing data quality. Edge detection algorithms, inspired by the human visual system, identify boundaries and contours within images, aiding in object recognition. Feature extraction algorithms spotlight unique attributes within data, enabling the identification of critical landmarks or objects. These algorithms collectively pave the way for robots to discern patterns, recognize objects, and interpret their environment.

Localization and Mapping (SLAM): One paramount application of perception algorithms is Simultaneous Localization and Mapping (SLAM). This technique allows robots to construct maps of their environment while simultaneously determining their own location within that environment. SLAM algorithms amalgamate data from sensors like LIDAR and cameras, generating a spatial representation of the surroundings. By comparing current sensor readings with historical data, robots estimate their position and orientation, navigating through unfamiliar terrain or updating maps in dynamic environments. SLAM plays a pivotal role in autonomous navigation, enabling robots to traverse complex spaces with precision.

Object Recognition and Tracking: Perception algorithms also excel in object recognition and tracking. Leveraging deep learning techniques, convolutional neural networks (CNNs) are adept at identifying and categorizing objects within images or video streams. These algorithms can not only identify objects but also track their movements across frames, crucial in scenarios such as autonomous vehicles navigating through traffic or robots interacting with moving objects in industrial settings. This capability enhances robots' understanding of their environment and their ability to respond effectively to dynamic situations.

Semantic Segmentation: Another transformative perception technique is semantic segmentation, which involves classifying each pixel within an image to a specific object category. This technique goes beyond object recognition, enabling robots to discern object boundaries and segment scenes into meaningful regions. Semantic segmentation is instrumental in applications like environmental monitoring, where robots can differentiate between different types of terrain, vegetation, and structures, aiding in tasks such as precision agriculture and disaster response.

Challenges and Advances: Despite the remarkable potential of perception algorithms, challenges persist. Variations in lighting, occlusions, and cluttered environments can impede object recognition. The development of algorithms that can operate in real-time while handling complex data is an ongoing endeavor. Advances in machine learning, particularly deep learning architectures like CNNs, have propelled the field, enabling robots to learn from vast datasets and improving their recognition and understanding capabilities.

In essence, perception algorithms are the cognitive interpreters that enable robots to bridge the gap between raw sensor data and meaningful understanding. These algorithms replicate the intricacies of human perception, allowing robots to navigate intricate terrains, identify objects, and respond effectively to dynamic environments. By continually advancing perception algorithms, the field of robotics evolves, ushering in an era of machines that can perceive and interpret the world around them with unprecedented clarity and intelligence.

Exercise

Multiple-choice Questions

1. What is the primary purpose of sensor fusion in robotics?
a) To eliminate the need for sensors
b) To replace outdated sensors with new ones
c) To integrate data from multiple sensors for enhanced perception
d) To reduce the complexity of robotic systems

2. Which type of sensor emulates the sense of touch in robots?
a) Vision sensor
b) Range sensor
c) Tactile sensor
d) Force sensor

3. Which technique combines data streams from various sensors into a cohesive dataset?
a) Sensor processing
b) Sensor integration
c) Sensor fusion
d) Sensor calibration

4. What is the purpose of Simultaneous Localization and Mapping (SLAM) in robotics?
a) Object detection
b) Data visualization
c) Creating maps while determining robot's location

d) Controlling robot motion
5. What is the main application of semantic segmentation in robotics?
a) Object recognition
b) Object tracking
c) Determining sensor accuracy
d) Classifying pixels in an image to object categories

Review Questions:

1. What is the purpose of sensor fusion in robotics, and why is it important?
2. Explain the difference between data-level fusion and feature-level fusion in sensor fusion.
3. How does computer vision enable robots to process visual data and derive meaningful information?
4. Describe the principles behind image processing and how it enhances the quality of sensor data.
5. What is Simultaneous Localization and Mapping (SLAM), and how does it benefit robotic navigation?
6. How do convolutional neural networks (CNNs) contribute to object recognition and tracking in robotics?
7. Define semantic segmentation and provide an example of its application in robotics.
8. What are some challenges that perception algorithms face in recognizing objects in various environments?
9. How do perception algorithms replicate human sensory processes in robots?
10. Explain the significance of perception algorithms in enhancing robotic decision-making and interaction with the environment.
11. What is the purpose of sensor fusion in robotics, and why is it important?
12. Explain the difference between data-level fusion and feature-level fusion in sensor fusion.
13. How does computer vision enable robots to process visual data and derive meaningful information?
14. Describe the principles behind image processing and how it enhances the quality of sensor data.
15. What is Simultaneous Localization and Mapping (SLAM), and how does it benefit robotic navigation?
16. How do convolutional neural networks (CNNs) contribute to object recognition and tracking in robotics?
17. Define semantic segmentation and provide an example of its application in robotics.
18. What are some challenges that perception algorithms face in recognizing objects in various environments?
19. How do perception algorithms replicate human sensory processes in robots?
20. Explain the significance of perception algorithms in enhancing robotic decision-making and interaction with the environment.

Answers Multiple-choice Questions

Answers:

1. c)
2. c)
3. c)
4. c)
5. d)

Review Questions:

1. Sensor fusion integrates data from multiple sensors to provide a more comprehensive and accurate understanding of the environment. It is important because individual sensors may have limitations, and fusing their data helps overcome these limitations and provides a holistic perception for robots. Sensor fusion in robotics refers to the process of combining data from multiple sensors to improve the accuracy, reliability, and comprehensiveness of information gathered about the environment. It is essential because individual sensors may have limitations, inaccuracies, or blind spots in their measurements. By fusing data from various sensors, a more comprehensive and accurate representation of the environment can be obtained. For example, combining data from cameras, lidar, and radar sensors in an autonomous vehicle allows the vehicle to have a more robust perception of its surroundings, enhancing its ability to make informed decisions and navigate safely.
2. Data-level fusion and feature-level fusion are two approaches to sensor fusion. Data-level fusion involves directly combining raw data from multiple sensors. This can involve averaging, weighted averaging, or more sophisticated techniques such as Kalman filters. Feature-level fusion, on the other hand, involves extracting relevant features or information from each sensor's data before combining them. This approach can lead to a reduction in data dimensionality and computational complexity. For instance, in a robot navigating an environment, data-level fusion might involve merging raw distance measurements from different sensors, while feature-level fusion could involve extracting specific object characteristics from each sensor's data before combining them.
3. Computer vision empowers robots to process visual data captured by cameras and other imaging sensors. Through computer vision algorithms, robots can analyze images and videos to detect objects, recognize patterns, and extract meaningful information. This enables robots to understand their surroundings, identify objects of interest, and make

decisions based on visual input. For instance, in a manufacturing setting, robots equipped with computer vision can identify defective products on an assembly line and take appropriate actions.

4. Image processing involves manipulating and enhancing images to improve their quality, extract useful information, or prepare them for further analysis. It encompasses operations like noise reduction, image enhancement, edge detection, and object segmentation. Image processing enhances sensor data quality by removing noise, enhancing contrasts, and isolating relevant features. This is particularly crucial in environments where lighting conditions or sensor limitations might affect the clarity of captured images.

5. Simultaneous Localization and Mapping (SLAM) is a technique used by robots to create maps of unknown environments while simultaneously determining their own location within those environments. It is vital for robotic navigation, especially in scenarios where the robot lacks prior knowledge of the environment. SLAM integrates sensor data from sources like cameras and lidar to build a map of the surroundings and estimate the robot's position relative to that map. This information is pivotal for safe and accurate navigation, especially in autonomous vehicles and exploration robots.

6. Convolutional Neural Networks (CNNs) are a class of deep learning models widely used in robotics for object recognition and tracking. They are adept at learning hierarchical features from visual data, making them powerful tools for tasks like detecting objects, recognizing patterns, and tracking moving targets. In robotics, CNNs analyze sensor data, such as images or video frames, to identify objects of interest and track their movements. For example, a robot equipped with CNNs can identify pedestrians and vehicles in real-time, ensuring safe navigation in urban environments.

7. Semantic segmentation is a computer vision technique that involves assigning a semantic label to each pixel in an image, effectively categorizing different regions based on their content. This allows robots to understand the composition of an image at a detailed level, distinguishing between different objects and their boundaries. In robotics, semantic segmentation has applications in autonomous vehicles, where it helps the vehicle differentiate between road, pedestrians, other vehicles, and obstacles, contributing to safe navigation and decision-making.

8. Perception algorithms in robotics face challenges such as occlusions (partial obstruction of objects), varying lighting conditions, cluttered environments, and distinguishing objects with similar appearances. Robust perception requires addressing these challenges to ensure accurate object detection and recognition. For instance, in a warehouse environment, perception algorithms must cope with varying object arrangements, shadows, and reflective surfaces while identifying items for picking or sorting.

9. Perception algorithms replicate human sensory processes in robots by emulating how humans process sensory information to understand the environment. Just as humans rely on visual cues, auditory signals, and tactile sensations to comprehend their surroundings, robots use sensors and algorithms to gather data from the environment and interpret it into meaningful insights. This replication involves extracting relevant features, recognizing patterns, and making decisions based on sensory inputs.

10. Perception algorithms play a pivotal role in enhancing robotic decision-making and interaction with the environment. By enabling robots to gather information from their surroundings, identify objects, and recognize patterns, these algorithms equip robots with the ability to make informed decisions autonomously. This is crucial for tasks like navigating through complex environments, interacting with humans, and performing intricate manipulations. Perception algorithms ensure that robots can operate effectively and safely in dynamic and unstructured surroundings.

11. Sensor fusion in robotics is the process of integrating data from multiple sensors to enhance the accuracy, reliability, and comprehensiveness of information about the environment. It is vital because individual sensors may have limitations or inaccuracies in their measurements. By combining data from different sensors, robots can create a more accurate and robust perception of their surroundings, enabling them to make informed decisions and navigate complex environments effectively.

12. Data-level fusion and feature-level fusion are two approaches to sensor fusion. Data-level fusion involves directly combining raw data from multiple sensors. This approach can be effective when sensor measurements are complementary or have similar data formats. Feature-level fusion, on the other hand, involves extracting relevant features or information from each sensor's data before combining them. This approach reduces the dimensionality of the data and can enhance computational efficiency.

13. Computer vision empowers robots to process visual data captured by cameras and other imaging sensors. Through computer vision algorithms, robots can analyze images and videos to detect objects, recognize patterns, and extract meaningful information. This enables robots to understand their surroundings, identify objects of interest, and make decisions based on visual input. For example, in agriculture, robots equipped with computer vision can identify ripe fruits for harvesting or detect crop diseases for targeted treatment.

14. Image processing involves manipulating and enhancing images to improve their quality, extract useful information, or prepare them for further analysis. Image processing techniques include noise reduction, contrast enhancement, edge detection, and image restoration. These techniques enhance the quality of sensor data by reducing noise, improving image clarity, and emphasizing relevant features. Image processing is especially beneficial in scenarios where sensor limitations or environmental factors can degrade the quality of captured images.
15. Simultaneous Localization and Mapping (SLAM) is a technique used by robots to create maps of unknown environments while simultaneously estimating their own position within those environments. SLAM is essential for robotic navigation, particularly in scenarios where the robot lacks prior knowledge of the environment. By integrating data from sensors like lidar, cameras, and IMUs, SLAM enables robots to build accurate maps of their surroundings and localize themselves within those maps. This information is crucial for safe and precise navigation, particularly in autonomous vehicles and exploration robots.
16. Convolutional Neural Networks (CNNs) are a class of deep learning models that excel at object recognition and tracking in visual data. They utilize convolutional layers to automatically learn hierarchical features from images, making them powerful tools for identifying objects, detecting patterns, and tracking moving targets. In robotics, CNNs are used to analyze sensor data, such as camera images or lidar point clouds, to identify and track objects. For example, in surveillance robots, CNNs can detect intruders or monitor changes in an environment.
17. Semantic segmentation is a computer vision technique that involves assigning a semantic label to each pixel in an image, effectively categorizing different regions based on their content. This technique enables robots to understand the composition of an image at a detailed level, distinguishing between different objects and their boundaries. In robotics, semantic segmentation finds applications in autonomous vehicles, where it helps the vehicle perceive the road, pedestrians, other vehicles, and obstacles separately. This information aids in navigation and decision-making.
18. Perception algorithms in robotics face challenges such as occlusions, varying lighting conditions, object variability, and limited training data. Occlusions occur when objects are partially hidden from view, requiring algorithms to infer their presence based on available information. Lighting conditions can affect sensor accuracy, and objects with similar appearances may be challenging to differentiate. Overcoming these challenges demands the development of robust algorithms capable of handling diverse and complex scenarios.
19. Perception algorithms replicate human sensory processes in robots by utilizing sensors and computational methodologies to gather data from the environment and derive meaningful insights. Like humans rely on their senses to navigate and understand the world, robots use perception algorithms to process data from sensors, recognize patterns, and make informed decisions. This replication enables robots to perceive the environment, identify objects, and interact effectively with their surroundings.
20. Perception algorithms hold significant importance in enhancing robotic decision-making and interaction with the environment. By enabling robots to gather information from their surroundings, identify objects, and understand their context, these algorithms enable robots to autonomously make informed decisions. This capability is invaluable for tasks ranging from autonomous driving, where robots must navigate complex traffic scenarios, to warehouse logistics, where robots must identify and manipulate objects efficiently. Perception algorithms bridge the gap between sensors and intelligent actions, enabling robots to operate seamlessly and effectively in dynamic and unstructured environments.

Online Resources

Online Exam Papers
Link: https://forms.gle/b7TSjDUXriT3q8917

QR Code:

YouTube Video Playlist Link
Link: https://www.youtube.com/playlist?list=PLzlh_31VWEFpi8HZ83Iu2vs6G3hGvznSs

QR Code:

Chapter 4: Robot Control Systems

Robot control systems form the backbone of the intricate dance between a machine's physical actions and its intelligent decision-making. At the core of robotics, these systems are the architects of precision, seamlessly translating high-level commands into intricate movements, ensuring accurate manipulations, and facilitating dynamic interactions with the environment. As robots evolve from performing repetitive tasks to engaging in complex and adaptive behaviors, control systems have become increasingly sophisticated, incorporating advanced algorithms, sensor data integration, and real-time adjustments. The essence of robot control lies in its ability to strike a harmonious balance between autonomy and human guidance, enabling machines to execute tasks with a finesse that parallels human capabilities.

Control systems are responsible for driving the actuators that propel robots into motion. They receive input from various sources, including sensors that capture the state of the robot and its environment. Through mathematical models and algorithms, control systems process this data, generating commands that govern the behavior of actuators. In industrial automation, for instance, control systems oversee the precise movement of robotic arms in assembly lines, ensuring components are positioned with accuracy and consistency.

As robots venture into unstructured environments and dynamic scenarios, adaptive control becomes essential. Adaptive control systems employ machine learning techniques to adjust the control parameters in real-time, enabling robots to respond to changing conditions and uncertainties. These systems learn from experience, optimizing performance and enhancing the robot's ability to navigate complex terrains or collaborate with human operators.

The integration of control systems with artificial intelligence and machine learning further elevates robot capabilities. Reinforcement learning, for example, allows robots to learn optimal actions through trial and error, enabling them to refine their behavior over time. Control systems infused with AI empower robots to make decisions, adapt to novel situations, and even predict outcomes based on historical data.

Human-robot interaction relies heavily on intuitive and safe control systems. Teleoperation interfaces enable remote control of robots, suitable for tasks in hazardous environments or space exploration. Haptic feedback systems provide operators with a sense of touch, allowing them to manipulate objects remotely while feeling resistance and texture. These control paradigms merge the precision of robotics with human expertise, underscoring the role of control systems in fostering collaboration between humans and machines.

The evolution of control systems extends beyond individual robots, encompassing multi-robot coordination and swarm robotics. Centralized and decentralized control architectures orchestrate the collective behavior of multiple robots, enabling them to perform collaborative tasks such as search and rescue missions or environmental monitoring. These systems encapsulate the intricacies of communication, cooperation, and decision-sharing among robots.

However, the journey towards perfecting control systems is not without challenges. Ensuring stability, responsiveness, and safety in real-time operations, mitigating delays in communication, and addressing unforeseen scenarios are ongoing areas of research. The advent of autonomous systems raises ethical considerations, demanding control mechanisms that align with human values and safety norms.

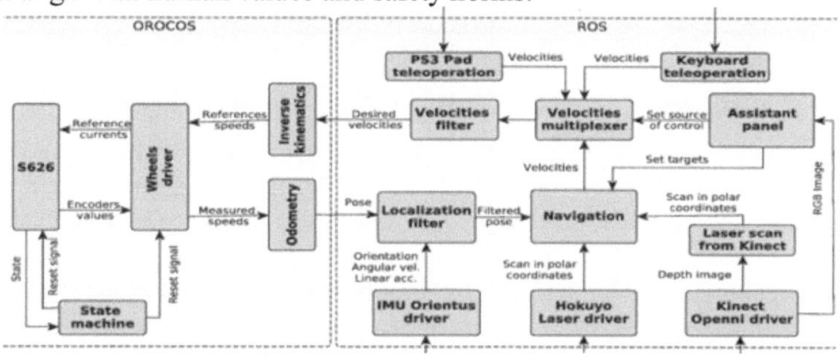

Open-Loop vs. Closed-Loop Control: In the domain of robotics, the distinction between open-loop and closed-loop control mechanisms plays a fundamental role in shaping a robot's behavior and its interaction with the environment. Open-loop control, often referred to as feedforward control, involves the execution of predefined commands without considering feedback from the surroundings. In this approach, the robot's actions are determined solely by the input commands, regardless of any unforeseen changes or disturbances that might occur during the process. While open-loop control is efficient for tasks in stable and predictable environments, it lacks the ability to adapt to real-time changes, potentially leading to inaccuracies and errors when confronted with uncertainties.

Conversely, closed-loop control integrates real-time feedback from the robot's sensors to continually adjust its actions. This control mechanism establishes a feedback loop where the robot's performance is monitored, compared to the desired outcome, and any deviations are identified and corrected. By incorporating feedback, closed-loop control empowers robots to adapt to dynamic conditions, compensate for disturbances, and maintain accuracy even in the face of uncertainties. This adaptability is particularly advantageous in tasks that require precision, adaptability, and interaction with changing environments.

The selection between open-loop and closed-loop control hinges on the specific requirements of the task at hand and the environment in which the robot operates. Open-loop control is well-suited for tasks that follow defined processes and environments, where repeatability and speed are paramount. Examples include repetitive manufacturing processes or predefined motion patterns. In contrast, closed-loop control shines in tasks that demand real-time adaptation, accuracy, and robust performance. Applications such as autonomous navigation, object manipulation, and interactions with dynamic surroundings benefit from the responsiveness and adaptability of closed-loop control.

However, closed-loop control introduces challenges related to sensor accuracy, system latency, and computational complexity. Accurate sensor data is essential for reliable feedback, and any delays in sensor response can affect the control system's effectiveness. Additionally, the design and implementation of intricate closed-loop control algorithms require computational resources, underscoring the importance of efficient hardware and software solutions.

Open-Loop Control: Open-loop control, also known as feedforward control, involves a direct sequence of commands or actions sent to the robot without incorporating feedback from the environment. In this mechanism, the robot's actions are determined solely by the predetermined commands, regardless of any deviations or disturbances that might occur during execution. While open-loop control can be efficient for tasks with predictable and static environments, it lacks adaptability. Robots operating under open-loop control cannot rectify errors or respond to unexpected changes in their surroundings. As a result, their accuracy and reliability can be compromised when confronted with uncertainties.

Closed-Loop Control: Closed-loop control, on the other hand, integrates feedback from the robot's sensors to adjust and refine its actions in real-time. This mechanism involves a continuous loop where the robot's actions are compared with the desired outcome, and any deviations are sensed and corrected. The feedback loop enables robots to adapt to changing conditions and maintain accurate performance even in the presence of disturbances or uncertainties. Closed-loop control systems enhance a robot's reliability, as they can detect errors and adjust their actions accordingly, ensuring that the robot's behavior aligns with the desired outcomes.

Advantages and Applications: The choice between open-loop and closed-loop control hinges on the task's requirements and the environment in which the robot operates. Open-loop control is suitable for tasks with well-defined processes and predictable conditions, where repeatability and speed are paramount. Examples include assembly line processes or predefined motion patterns. In contrast, closed-loop control shines in tasks that demand adaptability and accuracy, such as autonomous navigation, object manipulation, and interactions with dynamic environments. In these scenarios, the robot's ability to respond to changes in real-time enhances its efficiency and effectiveness.

Challenges and Considerations: While closed-loop control offers greater adaptability, it introduces challenges related to sensor accuracy, latency, and computational requirements. Accurate sensor data is crucial for precise feedback, and delays in sensor response can impact the control system's effectiveness. Moreover, intricate closed-loop control algorithms demand computational resources, necessitating efficient hardware and software implementations.

In essence, the distinction between open-loop and closed-loop control mechanisms lies at the heart of a robot's interaction with its environment. While open-loop control offers simplicity and efficiency, closed-loop control endows robots with adaptability, accuracy, and the ability to respond to changes in real-time. The selection of control strategy hinges on the application's demands and the robot's role, underscoring the importance of understanding and leveraging these control mechanisms to create intelligent and capable robotic systems.

PID Controllers: In the realm of control systems for robotics, Proportional-Integral-Derivative (PID) controllers emerge as a fundamental and widely used mechanism for achieving precise and stable control over a robot's behavior. A PID controller is designed to regulate the error between a desired set point and the actual state of a system by adjusting control inputs. The three components of a PID controller—proportional, integral, and derivative terms—work harmoniously to respond to deviations, eliminate errors, and optimize performance.

The proportional term enables the controller to respond proportionally to the current error, generating a control output that is directly proportional to the error magnitude. This term facilitates a rapid initial response and helps reduce steady-state errors, but it may lead to oscillations around the desired set point. The integral term, on the other hand, accumulates past errors over time and corrects any residual steady-state errors, enhancing the controller's ability to eliminate long-term deviations from the set point. The derivative term introduces anticipatory control by considering the rate of error change. This term counteracts the system's tendency to overshoot the set point by

applying a damping effect. It aids in stabilizing the system and improving response times, particularly in scenarios with sudden changes or disturbances.
The integration of these three terms in a PID controller offers a balance between rapid response, steady-state accuracy, and stability. By fine-tuning the coefficients of each term, control engineers can customize the PID controller's behavior to match the characteristics of the robot and the specific task at hand. This adaptability is a crucial feature, allowing PID controllers to cater to a wide range of applications in robotics, from precise motion control to stabilization in dynamic environments. Nevertheless, PID controllers are not without limitations. They may struggle to handle nonlinearities and complex dynamics in certain systems. Additionally, tuning PID parameters can be a challenging task, requiring a deep understanding of the system's behavior and careful adjustment to achieve desired performance.
In the domain of robotics, PID controllers, which stand for Proportional-Integral-Derivative controllers, play a pivotal role in orchestrating precise and adaptable control over a robot's actions. The essence of PID control lies in its ability to adjust control inputs based on three key factors: proportionality, integration, and differentiation. These components work collaboratively to enable robots to respond dynamically to inputs, disturbances, and desired outcomes, ensuring accuracy and efficiency in various applications.
The proportional term, a fundamental element of PID control, calculates the discrepancy between the target setpoint and the present system state. By multiplying this error with a proportional constant, the controller generates an output proportional to the error magnitude. This immediate response allows the robot to correct deviations swiftly, resulting in a rapid approach towards the desired state.

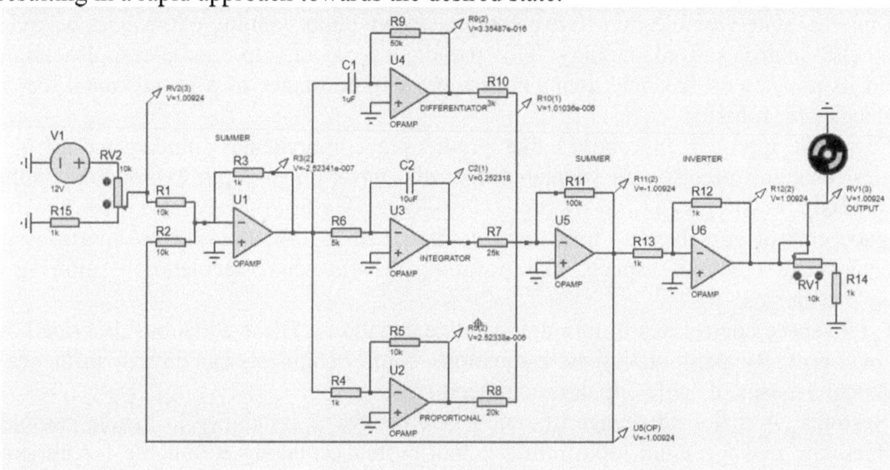

Building on the proportional component, the integral term addresses persistent errors that might accumulate over time. It sums up historical errors, thus addressing steady-state inaccuracies that the proportional term might not entirely resolve. The integral term's contribution lies in its ability to mitigate long-term errors and enhance stability by considering the broader history of deviations.
The derivative term introduces an anticipatory aspect to the control mechanism. By examining the rate of error change, the derivative term predicts how the system's state will evolve. It then computes control input adjustments to counteract potential overshooting of the setpoint. This predictive behavior improves responsiveness and aids in damping oscillations or sudden changes in the system.
The amalgamation of these three terms within a PID controller equips it with the capacity to adapt control inputs based on real-time feedback. The nuanced adjustment of the coefficients associated with each term allows engineers to tailor the controller's behavior to suit the robot's dynamics and specific tasks, balancing responsiveness, stability, and precision.
The versatility of PID controllers renders them indispensable in robotics. They find utility in diverse applications, ranging from motor control and robotic arm manipulation to mobile robot navigation. However, it's crucial to acknowledge that while PID controllers excel in numerous scenarios, they might face limitations when dealing with complex nonlinear systems or abrupt environmental changes. Nonetheless, PID controllers remain a valuable tool for achieving accuracy and adaptability in robotic control, underpinning the efficacy and versatility of robotic systems.
State-Space Control: In the realm of control systems for robotics, the state-space control paradigm emerges as a comprehensive approach that offers a deeper insight into a robot's behavior and dynamics. Unlike traditional control methods, which operate in the realm of input-output relationships, state-space control delves into the internal state variables of a system, providing a holistic understanding of its dynamics.

State-space control encapsulates the system's behavior through a set of state equations that describe how the system's state variables evolve over time. These state variables are a compact representation of the system's internal characteristics and include factors such as position, velocity, and acceleration. By manipulating these state equations, control engineers can directly influence the system's behavior, providing a more intuitive and versatile means of achieving desired outcomes.

A central advantage of state-space control lies in its ability to handle multi-input, multi-output systems seamlessly. This flexibility allows control engineers to manage complex robotic systems with multiple actuators and sensors, accommodating the interplay between various elements of the system. By analyzing the system's internal state variables, state-space control enables the design of control strategies that optimize performance and stability across a range of operating conditions.

One of the key components of state-space control is the concept of feedback. By incorporating feedback loops, control engineers can continually monitor the system's state and adjust control inputs in real-time. This capability facilitates error correction and adaptation to changing conditions, enhancing the precision and reliability of robotic systems.

While state-space control offers numerous advantages, its implementation requires a solid understanding of linear algebra and control theory. Engineers must model the system's dynamics, define state equations, and design appropriate feedback control laws. However, the investment in understanding and implementing state-space control can yield significant benefits in terms of control performance, adaptability, and robustness.

In conclusion, state-space control represents a sophisticated and insightful approach to controlling robotic systems. By operating in the domain of internal state variables, it offers a comprehensive understanding of a robot's behavior and dynamics, allowing for precise control and adaptability. This paradigm's capacity to handle complex multi-input, multi-output systems and its reliance on feedback loops underscore its importance as a foundational tool in achieving advanced control strategies for robotics.

State-space control is a sophisticated approach in robotics that provides a comprehensive understanding of a system's behavior and dynamics, enabling precise and adaptable control. Here's an in-depth exploration of this control paradigm presented in points:

Holistic Perspective: State-space control goes beyond traditional control methods by focusing on internal state variables. These variables encapsulate essential aspects like position, velocity, and acceleration, offering a comprehensive view of a system's dynamics.

State Equations: The core of state-space control lies in formulating state equations. These equations describe how internal state variables evolve over time. By manipulating these equations, control engineers can directly influence a system's behavior, offering a versatile means of achieving desired outcomes.

Multi-Input, Multi-Output Systems: A major advantage of state-space control is its ability to handle complex systems seamlessly. It can effectively manage multi-input, multi-output systems, making it suitable for intricate robotic setups with multiple actuators and sensors.

Optimization across Conditions: State-space control facilitates the design of control strategies that optimize performance and stability across diverse operating conditions. The paradigm captures the intricate interplay between system elements, enabling refined control mechanisms.

Feedback Integration: Feedback loops play a crucial role in state-space control. These loops enable real-time monitoring of a system's state, allowing for continuous adjustments to control inputs. This dynamic mechanism empowers robots to correct errors and adapt to changing conditions on the fly.

Enhanced Precision and Adaptability: By working with internal state variables and incorporating feedback loops, state-space control enhances precision, reliability, and adaptability. It enables robots to respond effectively to deviations and uncertainties, leading to more robust control outcomes.

Linear Algebra and Control Theory: Implementing state-space control demands a solid grasp of linear algebra and control theory. Engineers need to accurately model system dynamics, define precise state equations, and design appropriate feedback control laws.

Upfront Investment for Long-Term Benefits: While mastering state-space control might be challenging, the investment in understanding and implementing it can yield significant benefits. These benefits include improved control performance, adaptability to various scenarios, and enhanced control robustness.

Foundational Tool for Advanced Control: State-space control is a foundational tool for achieving advanced control strategies in robotics. Its capability to handle complex systems and its reliance on feedback loops underscore its significance in enhancing control precision and adaptability.

In summary, state-space control offers a comprehensive and insightful approach to controlling robotic systems. By exploring internal state variables and integrating feedback, it empowers robots to achieve precise control and adaptability, making it a pivotal tool in shaping robotic behavior across various domains.

Trajectory Planning and Motion Control: In the intricate landscape of robotics, the concepts of trajectory planning and motion control emerge as pivotal components, orchestrating a robot's ability to navigate its environment and execute tasks with precision. Delving deeply into this topic, trajectory planning involves the meticulous design of paths that robots follow to reach destinations or perform tasks. These paths are crafted with consideration for robot dynamics, environmental constraints, and task-specific requirements. The ultimate goal is to generate trajectories that optimize efficiency, smoothness, and obstacle avoidance, facilitating a seamless interaction between the robot and its surroundings. Several key considerations underscore the importance of effective trajectory planning. Smoothness is a critical factor, as a well-constructed trajectory ensures that the robot's movements remain devoid of abrupt changes or sudden jerks. This not only enhances stability but also contributes to longevity by reducing wear and tear on the robot's components. Additionally, collision avoidance plays a crucial role. Modern trajectory planning algorithms employ intricate analyses of the environment to predict and circumvent potential collisions, ensuring safe and obstacle-free trajectories. Considering the robot's dynamics, including factors such as acceleration, deceleration, and maximum speed, is imperative to ensure that the robot's movements adhere to physical limits, guaranteeing feasible execution.

Complementing trajectory planning, motion control translates these meticulously planned paths into precise robot movements. This involves managing the robot's actuators and motors to execute the intended trajectory with accuracy and fluidity. Here, the Proportional-Integral-Derivative (PID) controllers, previously discussed, find their application. These controllers facilitate real-time adjustments based on sensor feedback, allowing the robot to correct deviations from the planned trajectory, ultimately achieving accurate execution.

Interpolation techniques further refine the execution process by ensuring smooth transitions between waypoints on a trajectory. By calculating intermediate points, interpolation minimizes abrupt changes, thereby optimizing the continuity of movement. However, trajectory planning and motion control present their challenges, particularly in environments with dynamic obstacles or complex movement requirements. Real-time adjustments, precise sensor data, and an intricate understanding of robot dynamics are crucial to achieving seamless execution.

The applications of trajectory planning and motion control are far-reaching and impactful. In manufacturing, robots meticulously follow trajectories for tasks such as welding or assembly, contributing to enhanced productivity and quality. Autonomous vehicles rely on trajectory planning to navigate roadways safely and efficiently. Medical robotics leverage motion control to perform intricate and delicate procedures, where precision is paramount.

In summation, trajectory planning and motion control are the keystones of effective robotic movement. By synergizing the creation of well-defined trajectories with the precision execution of those paths, these concepts enable robots to maneuver through their environments, interact with objects, and accomplish tasks with unparalleled finesse and accuracy, revolutionizing the landscape of robotics across diverse applications.

The realm of robotics is intrinsically tied to the ability to plan and control intricate movements seamlessly. In this context, the concept of trajectory planning and motion control emerges as a critical component, enabling robots to navigate their environment, execute tasks, and interact with objects effectively. Let's delve deeply into this topic:

Trajectory Planning: Trajectory planning involves the art of designing a path or trajectory that a robot follows to reach a specific destination or perform a task. It encompasses a multitude of factors such as robot dynamics, environment constraints, and task requirements. The objective is to generate a trajectory that optimizes aspects like efficiency, smoothness, and avoidance of obstacles.

Key Considerations:

Smoothness: A well-designed trajectory exhibits smoothness to ensure that the robot's movements are free from sudden jerks or abrupt changes. This is crucial for maintaining stability, reducing wear and tear, and delivering a polished performance.

Collision Avoidance: Trajectory planning must account for obstacles in the environment to avoid collisions. This involves sophisticated algorithms that analyze the robot's surroundings, predict potential collisions, and generate paths that circumvent obstacles.

Dynamics: Taking into account the robot's physical dynamics is essential. Trajectories should consider factors like acceleration, deceleration, and maximum speed to ensure the robot's movements remain within feasible limits.

Motion Control: Motion control complements trajectory planning by executing the calculated trajectory with precision. It involves managing the robot's actuators and motors to ensure that the intended movements are executed accurately and smoothly.

PID Control in Motion Control: Proportional-Integral-Derivative (PID) controllers, discussed earlier, find significant application in motion control. They enable robots to track trajectories with precision by adjusting control inputs based on feedback from sensors. This allows the robot to correct deviations from the planned trajectory in real-time, ensuring accurate execution.

Interpolation and Path Following: Interpolation techniques are employed to ensure smooth transitions between waypoints on a trajectory. By calculating intermediate points, interpolation reduces abrupt changes and optimizes movement continuity.

Challenges and Complexity: Trajectory planning and motion control pose challenges, particularly in environments with dynamic obstacles or complex movements. Real-time adjustments, sensor accuracy, and intricate modeling of robot dynamics are crucial for achieving seamless execution.

Applications: Trajectory planning and motion control are vital in various robotic applications, such as manufacturing, autonomous vehicles, and medical robotics. In manufacturing, robots follow precise trajectories for tasks like welding or assembly. Autonomous vehicles use trajectory planning to navigate roads safely. Medical robots rely on accurate motion control to perform delicate procedures.

In conclusion, trajectory planning and motion control are instrumental in enabling robots to perform tasks effectively and safely. By combining well-designed trajectories with precise motion execution, robots can navigate environments, interact with objects, and achieve desired outcomes with finesse and accuracy.

Exercise

Multiple-choice Questions

1. What is the primary focus of trajectory planning in robotics?
a) Efficient data storage
b) Real-time communication
c) Designing paths for robot movement
d) Optimizing sensor accuracy

2. Which term refers to the smoothness of a trajectory in robotics?
a) Velocity control
b) Collision avoidance
c) Jerk reduction
d) Obstacle detection

3. How does motion control complement trajectory planning?
a) By generating random paths
b) By managing robot actuators for precise movement
c) By optimizing sensor accuracy
d) By planning environmental interactions

4. What does PID stand for in the context of motion control?
a) Proportional-Inertial-Derivative
b) Precision-Inertia-Damping
c) Proportional-Integral-Derivative
d) Planning-Interaction-Dynamics

5. Which component of PID control anticipates the rate of error change?
a) Proportional
b) Integral
c) Derivative
d) Feedback

Review Questions:

1. What is the primary purpose of trajectory planning in robotics, and how does it contribute to efficient robotic movements?
2. Explain the concept of smoothness in trajectory planning. How does ensuring smooth trajectories benefit robotic operations?
3. How does collision avoidance factor into trajectory planning for robots? Provide an example scenario illustrating its importance.
4. Describe the role of motion control in executing trajectories. How does motion control ensure accurate and precise robot movements?
5. What are PID controllers, and how do they enhance motion control in robots? Provide a real-world application where PID controllers are used.
6. What is interpolation in the context of trajectory planning? How does interpolation improve the execution of complex robotic movements?
7. Define state-space control and highlight its advantages over traditional control methods. How does it contribute to effective robot control?

8. How does feedback contribute to motion control accuracy? Describe how feedback loops are utilized in trajectory planning and execution.
9. Discuss the challenges that can arise in trajectory planning and motion control when dealing with dynamic and unpredictable environments.
10. Provide examples of industries where trajectory planning and motion control are crucial for achieving desired outcomes. Explain their impact.
11. Compare and contrast open-loop and closed-loop control systems. How does each approach influence trajectory planning and execution?
12. How does the integration of advanced sensors, such as LiDAR or computer vision, enhance trajectory planning and motion control in robotics?
13. What is the primary purpose of trajectory planning in robotics, and how does it contribute to efficient robotic movements?
14. Explain the concept of smoothness in trajectory planning. How does ensuring smooth trajectories benefit robotic operations?
15. How does collision avoidance factor into trajectory planning for robots? Provide an example scenario illustrating its importance.
16. Describe the role of motion control in executing trajectories. How does motion control ensure accurate and precise robot movements?
17. What are PID controllers, and how do they enhance motion control in robots? Provide a real-world application where PID controllers are used.
18. What is interpolation in the context of trajectory planning? How does interpolation improve the execution of complex robotic movements?
19. Define state-space control and highlight its advantages over traditional control methods. How does it contribute to effective robot control?
20. How does feedback contribute to motion control accuracy? Describe how feedback loops are utilized in trajectory planning and execution.
21. Discuss the challenges that can arise in trajectory planning and motion control when dealing with dynamic and unpredictable environments.
22. Provide examples of industries where trajectory planning and motion control are crucial for achieving desired outcomes. Explain their impact.
23. Compare and contrast open-loop and closed-loop control systems. How does each approach influence trajectory planning and execution?
24. How does the integration of advanced sensors, such as LiDAR or computer vision, enhance trajectory planning and motion control in robotics?

Answers Multiple-choice Questions

Answers:

1. c)
2. c)
3. b)
4. c)
5. c)

Review Questions:

1. Trajectory planning involves designing paths for robots to follow during tasks or movements. It contributes to efficient robotic movements by ensuring optimized paths that consider factors like smoothness, collision avoidance, and dynamics. Efficient trajectory planning results in precise, safe, and resource-effective robot motions.Trajectory planning is a fundamental aspect of robotics that involves determining the path a robot should follow to reach its desired destination or perform a specific task. The primary purpose of trajectory planning is to optimize robot movements, ensuring efficient and safe traversal through its environment. By generating well-defined trajectories, robots can avoid obstacles, minimize unnecessary movements, and achieve tasks with higher accuracy and speed. Trajectory planning contributes to efficient robotic movements by reducing energy consumption, minimizing wear and tear on mechanical components, and enabling robots to perform complex tasks in a precise and controlled manner.
2. Smoothness in trajectory planning refers to the continuity and absence of sudden changes in robot movement parameters, such as velocity and acceleration. Ensuring smooth trajectories is essential because abrupt changes in motion can lead to mechanical stress, reduced efficiency, and instability in robot operations. Smooth trajectories facilitate gradual transitions between different states, resulting in reduced jerk and mechanical vibrations. This

benefits robotic operations by improving the quality of movement, enhancing stability, and making the robot's behavior more human-like, which is crucial for applications that involve human-robot interaction or tasks requiring delicate manipulations.

3. Collision avoidance is a critical consideration in trajectory planning for robots, aiming to prevent collisions with obstacles or other entities in the environment. For instance, in an industrial setting, a robot tasked with picking and placing objects on a conveyor belt needs to avoid collisions with other robots or objects that might unexpectedly enter its path. Through collision avoidance mechanisms, trajectory planning ensures that robots choose paths that steer clear of obstacles, thereby enhancing safety, preventing damage to both the robot and its surroundings, and maintaining the integrity of the task execution.

4. Motion control plays a pivotal role in executing trajectories by translating the desired trajectories into precise movements of the robot's actuators. It involves generating control signals that regulate the velocity, acceleration, and position of the robot along the planned trajectory. Motion control ensures accurate and precise robot movements by accounting for factors like actuator dynamics, sensor feedback, and environmental variations. By continuously adjusting the control signals, motion control compensates for errors and disturbances, resulting in the robot faithfully following the planned trajectory and achieving the desired outcome with high accuracy.

5. PID controllers, or Proportional-Integral-Derivative controllers, are widely used in motion control for robots. They enhance motion control by providing a mechanism to regulate the robot's behavior based on error signals between the desired and actual states. PID controllers adjust the control signals proportionally to the error, integrate the accumulated error over time, and take into account the rate of change of the error. This allows robots to achieve smooth and precise movements while minimizing overshooting and oscillations. In applications like line-following robots, PID controllers adjust wheel speeds to keep the robot on a predefined path, ensuring accurate tracking.

6. Interpolation in trajectory planning involves generating intermediate points along a planned path to create a smoother transition between waypoints. By interpolating additional points, the robot's motion becomes more continuous, reducing sudden changes in direction or speed. This improves the execution of complex robotic movements by eliminating jerky or abrupt motions. For instance, in CNC (Computer Numerical Control) machining, interpolation ensures that the tool smoothly follows a complex toolpath, resulting in precise and intricate cuts on the work piece.

7. State-space control is an advanced control methodology that represents the dynamics of a system in a multidimensional state space, encompassing both its states and inputs. It offers advantages over traditional control methods by providing a comprehensive framework for modeling and control design. State-space control contributes to effective robot control by allowing for the integration of complex dynamics, the incorporation of feedback loops, and the design of controllers that can handle a range of operating conditions. This flexibility and adaptability are crucial in scenarios where robots need to operate in varied environments or execute tasks with changing requirements.

8. Feedback is integral to motion control accuracy as it allows the robot to continuously adjust its actions based on the comparison between desired and actual states. In trajectory planning and execution, feedback loops involve sensors that provide real-time information about the robot's position, velocity, and other relevant parameters. Controllers then use this feedback to make dynamic adjustments to the control signals, compensating for errors and disturbances. Feedback loops ensure that the robot's movements align with the planned trajectory, enhancing accuracy and robustness in the face of uncertainties.

9. Trajectory planning and motion control encounter challenges when operating in dynamic and unpredictable environments. Unexpected obstacles, changing conditions, or the presence of moving entities can disrupt planned trajectories and require rapid adjustments. Ensuring collision avoidance, adapting to varying terrains, and handling sudden changes in the environment demand sophisticated algorithms that can react in real-time. Additionally, unpredictable disturbances may affect the accuracy of motion control, necessitating adaptive strategies to maintain the desired trajectory.

10. Trajectory planning and motion control find applications in various industries where precision, efficiency, and safety are paramount. In manufacturing, robots equipped with trajectory planning execute complex assembly tasks with high precision. In healthcare, robotic surgical systems rely on accurate motion control for delicate procedures. The automotive industry benefits from trajectory planning in autonomous vehicles, ensuring safe and efficient navigation. Industries such as logistics and agriculture also leverage these technologies for tasks like material handling and autonomous field operations, respectively.

11. Open-loop and closed-loop control systems represent distinct approaches in trajectory planning and execution. Open-loop control involves pre-determining control signals based on the desired trajectory, without incorporating real-time feedback. This approach lacks adaptability and doesn't account for disturbances or errors. Closed-loop control, on the other hand, uses feedback from sensors to adjust control signals dynamically, ensuring that the robot

follows the trajectory accurately even in the presence of uncertainties. Closed-loop control systems enhance trajectory execution accuracy and robustness, making them suitable for complex and unpredictable environments.

12. The integration of advanced sensors like LiDAR (Light Detection and Ranging) or computer vision enhances trajectory planning and motion control by providing richer and more accurate information about the environment. LiDAR sensors generate detailed 3D maps of the surroundings, aiding in obstacle detection and precise localization. Computer vision systems enable robots to recognize objects, interpret visual cues, and adapt to changes in the environment. These sensors complement traditional feedback mechanisms, enabling robots to navigate, avoid collisions, and execute trajectories with a higher level of situational awareness.

13. Trajectory planning's primary purpose in robotics is to determine the optimal path for a robot to follow, contributing to efficient and safe movements. By generating well-defined trajectories, robots can navigate through their environments while avoiding obstacles, minimizing unnecessary movements, and achieving tasks with higher accuracy and speed. Trajectory planning enhances efficiency by reducing energy consumption, optimizing mechanical wear, and enabling precise manipulations and interactions.

14. Smoothness in trajectory planning refers to the continuity and absence of sudden changes in motion parameters, such as velocity and acceleration. Ensuring smooth trajectories benefits robotic operations by mitigating mechanical stress, reducing inefficiencies, and maintaining stability. Sudden changes in motion can lead to wear and tear on mechanical components, energy inefficiency, and instability in robot movements. Smooth trajectories promote stability, accuracy, and human-like behavior in robots, particularly in tasks that involve human-robot interaction or delicate actions.

15. Collision avoidance is a crucial aspect of trajectory planning, aiming to prevent robots from colliding with obstacles or other entities in the environment. For instance, consider an autonomous delivery robot navigating a busy urban street. The trajectory planning system must account for pedestrians, vehicles, and unexpected obstacles to ensure the robot can reach its destination safely without causing accidents or disruptions. Collision avoidance ensures the robot's movements are both efficient and safe, enabling it to navigate complex environments.

16. Motion control plays a pivotal role in executing trajectories by translating planned paths into precise robot movements. It involves generating control signals that regulate the robot's velocity, acceleration, and position along the trajectory. Motion control ensures accuracy and precision by accounting for factors like actuator dynamics, sensor feedback, and environmental variations. By continuously adjusting control signals, motion control compensates for errors and disturbances, allowing the robot to faithfully follow the planned trajectory and achieve tasks with high accuracy.

17. PID controllers, or Proportional-Integral-Derivative controllers, enhance motion control by providing a mechanism to regulate the robot's behavior based on error signals between desired and actual states. PID controllers adjust control signals proportionally to the error, integrate accumulated error over time, and consider the rate of change of the error. A real-world application of PID controllers is in self-balancing robots. The controllers adjust the robot's motors based on its inclination, enabling it to maintain an upright position.

18. Interpolation in trajectory planning involves generating intermediate points between waypoints to create smoother transitions along the path. By interpolating additional points, robotic motion becomes more continuous, reducing abrupt changes in direction or speed. This improves the execution of complex movements by eliminating jerky or sudden motions. In the context of a robotic arm, interpolation ensures that the arm follows a fluid path while reaching for an object, resulting in precise and controlled movements.

19. State-space control is an advanced approach that represents a system's dynamics in a multidimensional state space, encompassing states and inputs. Its advantages over traditional control methods include the ability to model complex dynamics, incorporate feedback, and adapt to changing conditions. State-space control contributes to effective robot control by providing a flexible framework for handling diverse operating conditions, accommodating uncertainties, and designing controllers that ensure stability and desired performance.

20. Feedback is crucial for motion control accuracy as it enables the robot to continuously adjust its actions based on the comparison between desired and actual states. Feedback loops involve sensors that provide real-time information about the robot's position, velocity, and other parameters. Controllers utilize this feedback to dynamically adjust control signals, compensating for errors and disturbances. Feedback loops enhance accuracy and robustness, allowing the robot to closely follow the planned trajectory, even in the presence of uncertainties.

21. Trajectory planning and motion control encounter challenges when dealing with dynamic and unpredictable environments. Unexpected obstacles, changing conditions, or moving entities can disrupt planned trajectories and require rapid adjustments. Ensuring collision avoidance, adapting to varying terrains, and handling sudden environmental changes demand sophisticated algorithms that can react in real-time. Additionally, unpredictable disturbances may affect the accuracy of motion control, necessitating adaptive strategies to maintain the desired trajectory.

22. Trajectory planning and motion control are crucial in industries where precision, efficiency, and safety are paramount. In manufacturing, trajectory planning enables robots to execute intricate assembly tasks with high accuracy. In healthcare, robotic surgical systems rely on accurate motion control for delicate procedures. The automotive industry benefits from trajectory planning in autonomous vehicles, ensuring safe and efficient navigation. Industries such as logistics and agriculture leverage these technologies for material handling and autonomous field operations, respectively, leading to increased efficiency and productivity.

Online Resources

Online Exam Papers

Link: https://forms.gle/atAD6uPSuP1eVgHs7

QR Code:

YouTube Video Playlist Link

Link: https://www.youtube.com/playlist?list=PLzlh_31VWEFoYSvfssB7G8aX1b4qe36xc

QR Code:

Chapter 5: Robot Kinematics and Dynamics

At the heart of the mesmerizing world of robotics lies a profound understanding of kinematics and dynamics – the intricate mechanics that orchestrate the graceful motions of robotic marvels. The fusion of art and science, kinematics and dynamics form the backbone of a robot's movement, imbuing it with the ability to navigate, manipulate, and interact within its environment. This realm of study unravels the secrets of how robots transition from one pose to another, how they elegantly coordinate multi-joint movements, and how they respond to external forces with finesse. A deep dive into robot kinematics unveils the intricate web of linkages and joints, akin to the bones and joints of living creatures, while dynamics unfurls the laws governing the interplay between forces, accelerations, and motions, as if uncovering the heartbeat of robotic creatures. The symphony of kinematics and dynamics lies at the core of designing, controlling, and optimizing robotic systems, bridging the chasm between theoretical elegance and real-world prowess. Through this exploration, we venture into the realm where mathematical rigor and engineering ingenuity converge to breathe life into the machines that redefine human interaction with the world around us. In the realm of robotics, where imagination meets engineering, the concepts of kinematics and dynamics stand as pillars that breathe life into mechanical creations. Just as a conductor orchestrates a symphony, kinematics orchestrates the graceful movements of robots, while dynamics governs the forces that propel these marvels into action. This intricate dance between form and function lies at the heart of a robot's ability to manipulate objects, traverse uneven terrains, and navigate complex environments.

Kinematics delves into the study of motion without delving into the forces that cause it, providing the essential blueprint for a robot's movements. It unravels the geometry of robot parts, elucidating how various components interact to generate motion. This branch of knowledge deciphers the intricate language of joint angles, link lengths, and end-effector positions, enabling us to predict how robots traverse space and time. Kinematics lays the foundation for precise control and graceful motions, whether it's a robotic arm on an assembly line or a humanoid robot executing intricate dance moves.

On the other hand, dynamics steps onto the stage with a focus on the forces and torques that drive robotic motion. It brings the laws of physics into the spotlight, addressing questions of how robots respond to external influences and how they generate the power to accomplish tasks. By examining the relationship between forces, accelerations, and velocities, dynamics unravels the mysteries behind a robot's ability to lift objects, maintain balance, and perform complex actions. From bipedal locomotion to the agility of robotic aerial vehicles, dynamics is the science that propels robots to navigate the physical world.

Together, kinematics and dynamics create a harmonious duet that choreographs robotic movements with precision and grace. Engineers and researchers leverage these principles to design mechanisms that mimic human limbs, animal locomotion, or even entirely novel forms of motion. By mastering kinematics, they devise robotic arms that mimic human dexterity, while delving into dynamics enables the creation of robotic platforms that navigate terrains as diverse as factory floors, planetary surfaces, or disaster-stricken environments.

As we journey through the complexities of kinematics and dynamics, we uncover the mathematics that underpins robotic motion. Equations describing rotations, translations, and coordinate transformations become our tools for understanding and controlling the motion of robots. Simulations that combine these principles with computational power allow us to predict how a robot will behave in different scenarios, enabling us to optimize designs and anticipate challenges.

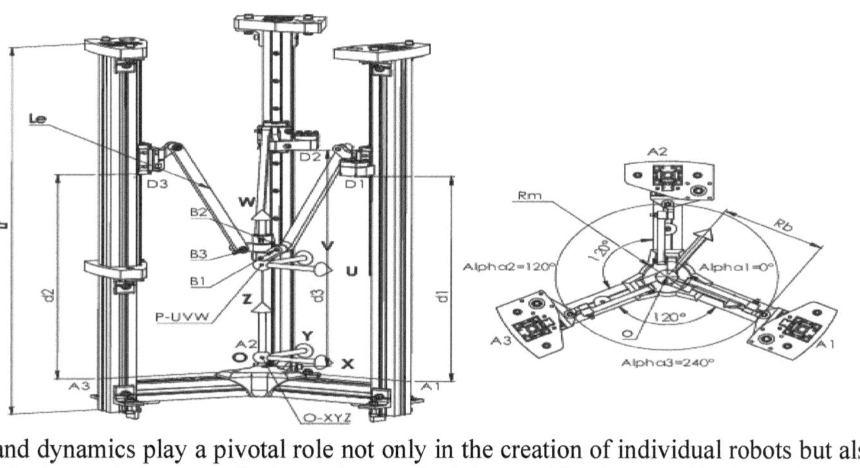

Robot kinematics and dynamics play a pivotal role not only in the creation of individual robots but also in the realm of collaborative robotics and swarm robotics. These principles enable the coordination of multiple robots to accomplish tasks that a single entity cannot achieve alone. From autonomous vehicles navigating through traffic to teams of drones mapping disaster-stricken areas, the understanding of kinematics and dynamics becomes a cornerstone for efficient and intelligent collective behavior.

Forward and Inverse Kinematics: Forward kinematics is a foundational concept in robotics that deals with determining the position and orientation of a robot's end effector in relation to its base frame. It involves understanding how the robot's various joints, which connect its links, contribute to the overall movement and placement of the end effector. The process of forward kinematics starts by defining the geometric parameters of the robot's kinematic structure, such as the lengths of its links and the types of joints they possess. Through a series of transformations involving rotations and translations, the positions and orientations of the end effector can be calculated based on the given joint angles. Forward kinematics is essential for visualizing and predicting the robot's movements, enabling path planning, and validating its behavior in a virtual environment. It serves as a crucial step in the control and simulation of robotic systems, enabling engineers to design, optimize, and program robots for various tasks and applications. And Inverse kinematics is a fundamental aspect of robotics that focuses on determining the joint angles necessary to achieve a desired position and orientation of a robot's end effector. Unlike forward kinematics, which calculates the end effector's pose based on given joint angles, inverse kinematics tackles the challenge of working backward from the desired end effector position to find the joint configurations that would result in that specific pose. This problem can be more intricate due to the non-linear and trigonometric nature of robot mechanisms. It often involves solving complex equations, utilizing optimization algorithms, numerical methods, or closed-form solutions. Inverse kinematics plays a critical role in applications where precision and accuracy are vital, such as robotic arm control, pick-and-place tasks, and humanoid robots mimicking human movements. It enables robots to effectively interact with their environment by accurately positioning their end effectors, facilitating tasks that require specific spatial arrangements.

Forward Kinematics: Forward kinematics is a fundamental concept in robotics that involves determining the position and orientation of a robot's end effector (such as its gripper or tool) based on the values of its joint angles. In other words, forward kinematics answers the question: "Given the joint angles, where is the end effector located?" The forward kinematics process typically starts with defining the robot's kinematic structure, which includes the lengths of its links and the types of joints connecting them. Using geometric transformations and matrices, the transformation from the robot's base to its end effector is calculated. This involves combining rotations and translations associated with each joint to determine the end effector's pose in relation to the robot's base frame. Forward kinematics is crucial for tasks like path planning, where the robot needs to reach specific points in its workspace. It's also essential for visualizing and verifying the robot's movements and positions during simulation and control.

Inverse Kinematics: Inverse kinematics is the reverse process of forward kinematics. It involves determining the joint angles required to achieve a desired position and orientation of the robot's end effector. In other words, inverse kinematics answers the question: "Given the desired end effector position, what should the joint angles be?" Inverse kinematics can be more complex than forward kinematics due to the non-linearity of trigonometric functions and the existence of multiple solutions or singularities. Solving the inverse kinematics problem often requires advanced mathematical techniques, including optimization algorithms, numerical methods, and closed-form solutions. Inverse kinematics is critical for tasks where precise positioning is required, such as manipulating objects with the robot's end effector. It's commonly used in applications like pick-and-place tasks, robotic arm control, and articulated robotic systems.

In summary, forward kinematics calculates the end effector's position and orientation based on joint angles, while inverse kinematics determines the joint angles needed to achieve a desired end effector pose. These concepts are foundational for controlling robotic movements, enabling robots to interact effectively with their environment and perform tasks accurately.

Denavit-Hartenberg Parameters: In the intricate world of robotics, understanding and representing a robot's kinematic structure accurately is pivotal for tasks ranging from planning its movements to controlling its actions. Among the many methodologies available, the Denavit-Hartenberg (DH) parameters stand as a robust and widely embraced framework for describing the relationships between the joints and links in a robotic manipulator's kinematic chain. This approach has proven its mettle by offering a systematic and standardized way to compute both forward and inverse kinematics, making it an invaluable tool in the field of robotics.

At the heart of the Denavit-Hartenberg parameters are four essential components that define the geometric properties of each joint in the robot's chain. These parameters are the link length (a), link offset (d), joint angle (θ), and twist angle (α). Each of these parameters is meticulously crafted to capture distinct aspects of the robot's configuration and movement. The link length (a) is the distance along the X-axis between the axes of two consecutive joints, providing insights into the extension or contraction of the robot's links. The link offset (d) represents the displacement along the Z-axis between the intersection of the Z-axes of two adjacent joints and the intersection of their X-axes. This parameter contributes to understanding the vertical shifts in the robot's structure. The joint angle (θ) signifies the rotation about the common normal between the Z-axes of adjacent joints, measured along the X-axis. This component is pivotal in comprehending the rotational motion of the robot's joints. Finally, the twist angle (α) encapsulates the rotation around the Z-axis that occurs when transitioning from the old X-axis to the new X-axis while moving from one joint to the next. This parameter sheds light on the rotational differences between consecutive joints.

The beauty of the DH parameters lies in their ability to create transformation matrices that connect consecutive links in the robot's kinematic chain. These matrices encapsulate both the translation and rotation between the corresponding joints, making them a powerful representation of the robot's movement. By repeatedly applying these transformation matrices, the forward kinematics of the robot can be computed. Forward kinematics involves calculating the position and orientation of the robot's end effector – a critical aspect for tasks like path planning, simulation, and control. Visualizing the robot's movement, verifying its motion path, and ensuring that it meets its intended goals become achievable through the understanding of forward kinematics facilitated by the DH parameters.

The significance of the Denavit-Hartenberg parameters extends further into solving the inverse kinematics problem. Inverse kinematics entails determining the joint angles that yield a desired position and orientation for the robot's end effector. This problem can be notably challenging due to its non-linear nature and the potential existence of multiple solutions or singularities. The DH parameters offer a structured and systematic approach to tackle this challenge. By leveraging the interplay of the parameters and their relationships, engineers and researchers can systematically compute the joint angles necessary to achieve a specific pose of the end effector. This capability is instrumental in scenarios where precise positioning is required, such as robotic arm control, pick-and-place tasks, and even humanoid robots emulating human-like movements. The Denavit-Hartenberg parameters have evolved to become a fundamental pillar in the field of robotics, enabling practitioners to model and analyze the kinematic behavior of manipulators with clarity and precision. By adhering to standardized conventions and formulas, the framework streamlines the mathematical representation of intricate robotic structures. This, in turn, empowers engineers and researchers to focus on higher-level objectives such as control algorithms, system design, and applications, rather than getting bogged down in complex mathematical derivations.

In conclusion, the Denavit-Hartenberg parameters offer a systematic and standardized approach to model the kinematic relationships within robotic manipulators. By encapsulating the geometric properties of joints and links, these parameters provide a structured way to compute forward and inverse kinematics, facilitating tasks like motion planning, control, and simulation. As robotics continues to advance, the Denavit-Hartenberg parameters remain an indispensable tool that empowers engineers to unlock the full potential of robotic manipulators.

Jacobian Matrix: The Jacobian matrix is a fundamental mathematical tool in robotics that plays a pivotal role in understanding and controlling the movement of robotic systems. Named after the mathematician Carl Gustav Jacob Jacobi, this matrix provides insights into the relationship between joint velocities and the resulting end effector velocities in a robotic manipulator. As robotic systems become increasingly sophisticated and versatile, the Jacobian matrix stands as a cornerstone for tasks ranging from trajectory planning to real-time control.

The Jacobian matrix encapsulates the partial derivatives of the end effector's position and orientation with respect to each joint's motion. It essentially quantifies how small changes in joint velocities translate to changes in the velocity of the end effector. This matrix is instrumental in mapping the intricate interplay between the robot's joint space and its operational space, providing a bridge between kinematics and dynamics.

One of the primary applications of the Jacobian matrix is in solving the inverse kinematics problem. When determining the joint velocities required to achieve a specific end effector velocity, the Jacobian matrix is utilized to solve for the joint velocity vector. In essence, it enables the translation of desired operational space velocities into the corresponding joint space velocities, enabling the robot to move accurately and effectively.

The Jacobian matrix also plays a critical role in singularity analysis. Singularities are points in a robot's configuration where its mobility is reduced, leading to challenges in controlling and planning its movements. By analyzing the Jacobian matrix, engineers can identify singularities and implement strategies to avoid or navigate them, ensuring the robot's stability and robustness.

In real-time applications, such as teleoperation or collaborative robotics, the Jacobian matrix enables precise control of the robot's movements. By mapping the operator's movements to the robot's joint velocities, the matrix ensures that the robot mimics the operator's actions with accuracy. This capability is vital in scenarios where human-robot interaction demands precise and intuitive control.

The Jacobian matrix is not limited to serial manipulators; it extends to parallel and hybrid manipulators as well. For complex manipulator architectures, the Jacobian matrix provides a comprehensive insight into the relationships between various components of the robot's structure, enhancing the understanding of its movement capabilities.

Furthermore, the Jacobian matrix has applications in dynamic simulations. By incorporating it into dynamic equations, engineers can analyze how external forces or disturbances affect the robot's behavior. This is particularly valuable for tasks like trajectory optimization and collision detection.

Introduction to the Jacobian Matrix: The Jacobian matrix is a fundamental concept in robotics and mathematics that serves as a bridge between the joint space and operational space of a robotic system. Named after Carl Gustav Jacob Jacobi, this matrix offers crucial insights into how small changes in joint velocities translate into corresponding changes in the velocity of the robot's end effector. In essence, it provides a mathematical tool to link the robot's movements at the joint level with its movement in the operational space.

Mathematical Formulation and Interpretation: The Jacobian matrix is a square matrix with rows representing the dimensions of the operational space (typically position and orientation) and columns representing the joint velocities. Each element of the matrix represents the rate of change of a particular operational space parameter with respect to a specific joint velocity. This matrix encapsulates the robot's kinematics, quantifying how joint motions contribute to end effector motions. When multiplied by the vector of joint velocities, the Jacobian matrix produces the vector of end effector velocities.

Applications in Inverse Kinematics: One of the primary applications of the Jacobian matrix lies in solving the inverse kinematics problem. Inverse kinematics involves finding the joint velocities required to achieve a desired end effector velocity. By using the Jacobian matrix, engineers can map the desired end effector velocities to corresponding joint velocities. This allows the robot to execute movements accurately and achieve the desired operational space motion. In essence, the Jacobian matrix provides a way to translate operational space goals into joint space actions.

Singularity Analysis and Avoidance: The Jacobian matrix is also instrumental in analyzing and avoiding singularities, which are points in the robot's configuration space where its mobility is reduced. A singularity can lead to challenges in control and planning, as well as unexpected robot behavior. By examining the Jacobian matrix, engineers can identify singular configurations and develop strategies to navigate or avoid them. This is particularly important for ensuring the stability and reliability of the robotic system, especially during critical tasks.

Real-Time Control and Human-Robot Interaction: In real-time control scenarios and human-robot interaction, the Jacobian matrix is a powerful tool. It enables precise control of the robot's movements based on the operator's input. By mapping the operator's desired operational space velocities to joint space velocities using the Jacobian matrix, the robot can mimic the operator's actions accurately. This is crucial for scenarios where intuitive and precise control is essential, such as teleoperation or collaborative robotics.

Dynamic Simulations and Force Analysis: Beyond kinematics, the Jacobian matrix also has applications in dynamic simulations and force analysis. By incorporating the Jacobian matrix into dynamic equations, engineers can understand how external forces, torques, or disturbances affect the robot's behavior. This is valuable for tasks like trajectory optimization, collision detection, and evaluating the robot's response to external influences.

Extensions to Complex Manipulators: The concept of the Jacobian matrix extends beyond traditional serial manipulators. It also applies to more complex manipulator architectures, including parallel and hybrid manipulators. In these cases, the Jacobian matrix offers insights into the intricate relationships between various components of the robot's structure, enhancing the understanding of its movement capabilities and aiding in control strategies.

In conclusion, the Jacobian matrix is a versatile and essential mathematical tool in robotics. It bridges the gap between joint space and operational space, enabling precise control, inverse kinematics solutions, and singularity analysis. With the rapid evolution of robotics and the increasing complexity of robotic systems, the Jacobian matrix continues to be a cornerstone for engineers and researchers seeking to optimize robot movement, control, and interaction in a wide range of applications.

Lagrangian Dynamics: Lagrangian dynamics is a powerful mathematical framework used in robotics to describe the motion and behavior of mechanical systems, including robotic manipulators. Named after the mathematician Joseph-Louis Lagrange, this approach provides a concise and elegant way to formulate the equations of motion for complex robotic systems. By expressing the system's dynamics in terms of generalized coordinates and minimizing a specific function called the Lagrangian, engineers gain insight into how forces and torques affect the robot's behavior.

Generalized Coordinates and Degrees of Freedom: In Lagrangian dynamics, the concept of generalized coordinates takes center stage. Generalized coordinates are a set of variables that uniquely define the configuration of a mechanical system. In robotics, these often correspond to joint angles, joint displacements, or other parameters that characterize the robot's state. Generalized coordinates significantly simplify the process of deriving equations of motion by reducing the number of variables that need to be considered.

Principle of Least Action: A cornerstone of Lagrangian dynamics is the principle of least action, also known as Hamilton's principle. This principle states that the true path of a dynamic system is the one that minimizes the action integral, a quantity that combines the difference between kinetic and potential energy over time. By minimizing the action integral, engineers can derive the equations of motion that govern the system's behavior.

Lagrangian Formulation and Equations of Motion: The Lagrangian is a mathematical function that encapsulates both the systems kinetic and potential energies. It serves as a compact representation of the system's dynamics and behavior. The equations of motion, known as Euler-Lagrange equations, are derived by applying the principle of least action to the Lagrangian. These equations express how forces, torques, and constraints influence the system's generalized coordinates over time, providing a systematic and elegant way to understand its dynamics.

Advantages in Robotic Systems: Lagrangian dynamics offers several advantages when applied to robotic systems. It provides a unified framework for modeling robots with multiple degrees of freedom, complex linkages, and various types of joints. This approach enables engineers to capture intricate dynamics without resorting to complex derivations. Lagrangian dynamics also offers insight into energy conservation, enabling the analysis of system stability and behavior over time.

Simulation, Control, and Optimization: Lagrangian dynamics plays a significant role in simulating and controlling robotic systems. By formulating accurate dynamic models using Lagrangian dynamics, engineers can simulate how robots respond to different inputs and disturbances. This aids in optimizing control strategies and fine-tuning parameters for optimal performance. Additionally, Lagrangian-based control strategies, such as computed torque control, utilize the equations of motion derived from this framework to achieve precise and stable control of robotic manipulators.

Complex Robotic Systems and Redundancy: Lagrangian dynamics is particularly beneficial for analyzing complex robotic systems and addressing redundancy. Complex mechanisms with multiple links and joints can be challenging to model accurately. Lagrangian dynamics provides a systematic approach to handling these complexities and understanding how forces and torques propagate through the system. Furthermore, it helps in addressing redundancy issues by providing a clear understanding of how additional degrees of freedom influence the overall behavior.

Conclusion: In conclusion, Lagrangian dynamics is a powerful mathematical framework that plays a vital role in understanding the motion and behavior of robotic systems. By utilizing generalized coordinates and the principle of least action, engineers can derive elegant equations of motion that govern the robot's dynamics. This approach is particularly advantageous for modeling complex robotic systems, simulating behavior, designing control strategies, and optimizing performance. Lagrangian dynamics stands as a foundational tool in robotics, enabling engineers to explore the intricate interplay between forces, torques, and robot movements in a systematic and insightful manner.

Exercise

Multiple-choice Questions

1. What does robot kinematics primarily focus on?
a) Robot dynamics
b) Robot control
c) Robot motion without considering forces
d) Robot programming
2. Which type of kinematics studies the relationship between joint angles and end effector position?
a) Forward kinematics
b) Inverse kinematics
c) Differential kinematics
d) Statics kinematics
3. What is the main objective of inverse kinematics?
a) Determining joint angles from end effector position
b) Calculating forward kinematics equations
c) Analyzing forces and torques in robotic systems
d) Programming robot controllers
4. Differential kinematics deals with:
a) Calculating joint angles from end effector position
b) Relating forces and torques to robot motion
c) Analyzing the dynamics of robotic systems
d) Studying the rates of change of joint angles and their effects on end effector velocity

5. Robot dynamics primarily involve the study of:
a) Joint angles and positions
b) Kinematic equations
c) Forces, torques, and their effects on robot motion
d) Robot programming languages

Review Questions:

1. What is the main focus of robot kinematics, and how does it differ from robot dynamics?
2. Explain the concept of forward kinematics in robotics. How does it relate to end effector position?
3. Describe the significance of inverse kinematics in robotic systems. What information does it provide?
4. What does differential kinematics study, and how does it contribute to robot motion analysis?
5. Define the term "end effector" in the context of robotics. Why is it important in kinematics?
6. How are joint angles and end effector position related in the context of robot motion analysis?
7. Briefly explain the concept of Lagrangian dynamics and its significance in analyzing robotic systems.
8. What is the role of generalized coordinates in Lagrangian dynamics? How do they simplify analysis?
9. Describe Euler-Lagrange equations and their importance in deriving equations of motion for robots.
10. How does Lagrangian dynamics handle complex systems with multiple degrees of freedom?
11. Discuss the advantages of using Lagrangian dynamics in analyzing robotic systems.
12. Explain the difference between forward kinematics and inverse kinematics.
13. How does differential kinematics contribute to understanding the velocity and motion of robotic systems?
14. What are the primary components studied in robot dynamics, and why are they essential?
15. How do forces and torques affect robot motion, and how does dynamics analysis help address them?
16. Describe the significance of the Newton-Euler equations of motion in robotics.
17. What is the Lagrangian of a mechanical system, and how is it related to kinetic and potential energy?
18. Explain the concept of joint space and its role in robot kinematics and dynamics.
19. How do the equations of motion derived from Euler-Lagrange equations help in understanding robotic behavior?
20. In what scenarios can Lagrangian dynamics become complex, and how can numerical techniques be employed to solve them?

Answers

Multiple-choice Questions

Answers:

1. c)
2. a)
3. a)
4. d)
5. c)

Review Questions:

1. Robot kinematics and dynamics constitute two fundamental branches of robotic mechanics, each focusing on distinct aspects of robotic behavior. Robot kinematics is primarily concerned with the study of how a robot's joints and links move in relation to each other, without delving into the forces causing the motion. It deals with geometric aspects, such as determining the position, orientation, and path that a robot takes as it moves. On the other hand, robot dynamics delve into the forces, torques, and accelerations that affect a robot's motion. Dynamics accounts for the interactions between a robot's components and external forces, enabling a deeper understanding of how forces drive robotic motion and influence the robot's overall behavior.
2. Forward kinematics in robotics refers to the process of determining the position and orientation of a robot's end effector based on the known joint angles. It establishes a relationship between the robot's joint variables and the position of its end effector in the workspace. Forward kinematics involves the application of transformation matrices to calculate the end effector's pose. This process enables engineers and researchers to visualize and predict where the robot's end effector will be located in the physical world for a given set of joint angles, aiding in tasks like path planning and trajectory optimization.
3. Inverse kinematics is of paramount significance in robotic systems, especially when it comes to real-world applications. Unlike forward kinematics, which calculates end effector position from joint angles, inverse kinematics calculates the required joint angles to achieve a desired end effector position. This information is crucial for tasks such as reaching a specific point in space, orienting the end effector correctly, or manipulating objects with precision. Inverse kinematics enables robotic systems to accurately plan and execute movements that accomplish specific objectives in the workspace, allowing robots to interact with their environment in a controlled manner.

4. Differential kinematics is a branch of robotics that focuses on studying the relationship between joint velocities and end effector velocities. It plays a critical role in analyzing the motion of robotic systems by providing insights into how the velocities of individual joints translate into the overall motion of the robot's end effector. This analysis helps researchers and engineers understand how changes in joint velocities affect the robot's ability to perform tasks, move along different paths, or maintain stability, making it an indispensable tool for motion planning and control.

5. The term "end effector" in the context of robotics refers to the device, tool, or mechanism attached to the last link of a robotic arm or manipulator. It is the part of the robot that interacts with the external world to perform specific tasks. The end effector can be anything from a gripper to a specialized tool, depending on the robot's intended application. The importance of the end effector in kinematics lies in its role as the interface through which the robot interacts with its environment. The kinematics of the end effector determine how its position and orientation change as the robot's joints move, directly influencing the robot's ability to accomplish tasks accurately and efficiently.

6. In robot motion analysis, joint angles and end effector position are intricately related through the robot's kinematic chain. The joint angles determine the configuration of the robot's links and joints, ultimately affecting the position and orientation of the end effector. Through the principles of forward kinematics, engineers can calculate how changes in joint angles translate into corresponding changes in end effector position. Conversely, inverse kinematics provides the necessary insight to determine the joint angles required to achieve a desired end effector position. This relationship is foundational in understanding and controlling a robot's motion.

7. Lagrangian dynamics is a powerful framework used to analyze the motion of mechanical systems, including robotic systems. It is based on the principle of virtual work and the concept of energy conservation. In Lagrangian dynamics, the behavior of a system is described in terms of generalized coordinates and their corresponding generalized forces. This framework simplifies the process of deriving equations of motion for complex systems with multiple degrees of freedom, allowing researchers to systematically analyze the forces, torques, and accelerations that influence the system's motion. Lagrangian dynamics offers a unified approach that elegantly captures the complexities of robotic systems, making it a cornerstone of advanced motion analysis.

8. Generalized coordinates are a set of variables that uniquely describe the configuration of a mechanical system. In Lagrangian dynamics, they replace the traditional Cartesian coordinates, simplifying the analysis by reducing the number of variables needed to describe the system's motion. Generalized coordinates offer a more compact and efficient representation of a robot's configuration, making it easier to formulate and solve equations of motion. By utilizing these coordinates, engineers can focus on the essential aspects of motion analysis, such as the system's energy and the forces acting on it, without getting bogged down in the details of individual joint positions.

9. Euler-Lagrange equations play a pivotal role in robotics as they are used to derive the equations of motion for a mechanical system. These equations relate the system's Lagrangian (the difference between its kinetic and potential energy) to the generalized coordinates and their time derivatives. By solving these equations, engineers can determine how forces and torques influence the motion of robotic systems. Euler-Lagrange equations are a cornerstone of dynamics analysis, providing a systematic and rigorous method for deriving the equations that govern a robot's behavior. They form the foundation for understanding how external forces and joint motions interact to shape a robot's motion.

10. Lagrangian dynamics excels in handling complex systems with multiple degrees of freedom, making it well-suited for analyzing the motion of robotic systems. As robots become increasingly sophisticated with a higher number of joints and links, traditional methods can quickly become unwieldy. Lagrangian dynamics provides a streamlined approach that simplifies the analysis by abstracting the system's behavior into generalized coordinates and forces. Numerical techniques, such as numerical integration or simulation, can be employed to solve the resulting equations of motion, enabling researchers to understand how a robot will behave in various scenarios and configurations.

11. The advantages of using Lagrangian dynamics in analyzing robotic systems are numerous. Firstly, it provides a unified framework that accommodates both kinematics and dynamics, enabling a comprehensive understanding of robot behavior. Secondly, Lagrangian dynamics offers a systematic approach to deriving equations of motion, which is crucial for predicting how a robot will respond to external forces and torques. Additionally, the use of generalized coordinates simplifies complex systems, making the analysis more manageable. Finally, by relying on principles of energy conservation, Lagrangian dynamics can capture the intricate interplay between kinetic and potential energy, shedding light on the underlying physics of robotic motion.

12. Forward kinematics and inverse kinematics are two fundamental concepts in robotics that address distinct aspects of robot motion. Forward kinematics involves determining the position and orientation of a robot's end effector based on the known joint angles. It enables engineers to predict where the end effector will be located in space for a given configuration of joint angles. In contrast, inverse kinematics deals with the reverse problem – given a desired end effector position, it calculates the required joint angles to achieve that position. This

understanding is vital for tasks such as positioning the robot's end effector accurately to grasp objects or navigate through specific points in the environment.

13. Differential kinematics, also known as velocity kinematics, is a branch of robotics that explores the relationship between joint velocities and end effector velocities. It focuses on understanding how changes in joint angles over time translate into changes in the robot's end effector velocity. Differential kinematics provides insights into how a robot's joints influence its overall motion, allowing engineers to analyze the robot's speed, direction, and responsiveness. This analysis is crucial for tasks such as real-time motion control, path planning, and optimizing robot trajectories to ensure smooth and accurate movements.

14. Robot dynamics encompasses the study of forces, torques, accelerations, and their interactions that drive a robot's motion. It delves into the underlying physics that dictate how a robot responds to external influences, both in terms of movement and stability. The primary components studied in robot dynamics include inertia, which represents a body's resistance to changes in motion; forces and torques applied to the robot's joints and links; and the resulting accelerations that propel the robot's movements. Understanding these components is essential for designing control strategies that ensure stable and accurate robot behavior.

15. Forces and torques play a pivotal role in influencing robot motion, and dynamics analysis provides a framework to comprehend their impact. Forces exerted on a robot's joints and links can either generate or hinder motion, while torques can cause rotations around joints. Dynamics analysis helps address these influences by revealing how forces and torques interact with the robot's components, leading to changes in its motion. Engineers use this information to design control algorithms that compensate for external disturbances, maintain stability, and achieve desired trajectories, ensuring that the robot performs tasks accurately and efficiently.

16. The Newton-Euler equations of motion are fundamental equations in robotics that enable the analysis of a robot's dynamic behavior. These equations extend Newton's second law of motion to robotic systems, connecting forces, torques, and accelerations. The Newton-Euler equations provide a systematic approach to calculating the forces and torques at each joint and link in a robot, taking into account both external forces and inertial effects. By solving these equations, engineers can gain insights into how a robot will respond to different types of forces, enabling them to design control strategies that optimize performance and stability.

17. The Lagrangian of a mechanical system represents the difference between its kinetic energy and potential energy. In the context of robotics, the Lagrangian captures the system's energy distribution, reflecting how kinetic and potential energies change as the robot moves. By analyzing the Lagrangian, engineers can gain a deeper understanding of the system's behavior, including its dynamics and interactions with external forces. The Lagrangian is a powerful tool that unifies the concepts of energy and motion, allowing researchers to derive equations of motion, predict robot behavior, and design control strategies.

18. Joint space refers to the configuration space defined by the joint angles of a robotic system. In this space, each point represents a specific combination of joint angles that uniquely define the robot's position. Joint space is pivotal in both robot kinematics and dynamics. In kinematics, joint space provides a concise representation of a robot's configuration, enabling researchers to calculate the end effector's position based on the joint angles. In dynamics, joint space serves as the foundation for analyzing how forces and torques influence the motion of the robot, allowing engineers to derive equations of motion and predict the robot's behavior.

19. The equations of motion derived from Euler-Lagrange equations provide a comprehensive understanding of a robot's behavior by relating forces, torques, accelerations, and the system's configuration. These equations capture the underlying physics of robotic motion, allowing engineers to predict how the robot will respond to various external influences. By solving these equations, researchers can uncover insights into the robot's stability, responsiveness, and ability to perform tasks accurately. The equations of motion derived from Euler-Lagrange equations are the cornerstone of dynamics analysis, enabling engineers to design control strategies that optimize robot performance in different scenarios.

20. Lagrangian dynamics can become complex when dealing with systems with numerous degrees of freedom, intricate interactions, or non-linearities. In such scenarios, analytical solutions to the equations of motion may be challenging or even infeasible to derive. Numerical techniques step in to solve these complex equations through computational methods. Numerical integration, such as the Euler method or Runge-Kutta methods, can approximate solutions by breaking down the equations of motion into smaller time steps. These techniques offer a practical way to simulate and analyze robot behavior in scenarios where analytical solutions are elusive, providing valuable insights into how robots respond in complex and dynamic environments.

Online Resources

Online Exam Papers

Link:https://forms.gle/wvyzyqY2Ed8PBX3TA

QR Code

YouTube Video Playlist Link

Link: https://www.youtube.com/playlist?list=PLzlh_31VWEFo9X7VDS4WD76-qa-htboBp

QR Code

Chapter 6: Mobile Robotics

In the ever-evolving landscape of robotics, the emergence of mobile robotics stands as a defining technological leap, reshaping our perception of autonomy, navigation, and interaction with the world around us. Mobile robotics transcends the conventional confines of static machinery, breathing life into robots that traverse landscapes, negotiate obstacles, and adapt to dynamic environments with unparalleled agility. At the heart of this technological marvel lies the fusion of cutting-edge hardware and sophisticated algorithms, enabling robots to explore terrains ranging from the vastness of outer space to the intricate corners of our homes. Mobile robotics isn't just about locomotion; it's a symphony of sensors, computation, and decision-making that empowers robots to embark on journeys of exploration, inspection, delivery, and assistance. With applications spanning industries from logistics to healthcare, environmental monitoring to disaster response, mobile robotics embodies the spirit of innovation that propels human ingenuity to new horizons. As we venture into the realm of mobile robotics, we uncover a world where machines not only learn to move but also to perceive, understand, and contribute to the advancement of society and technology.

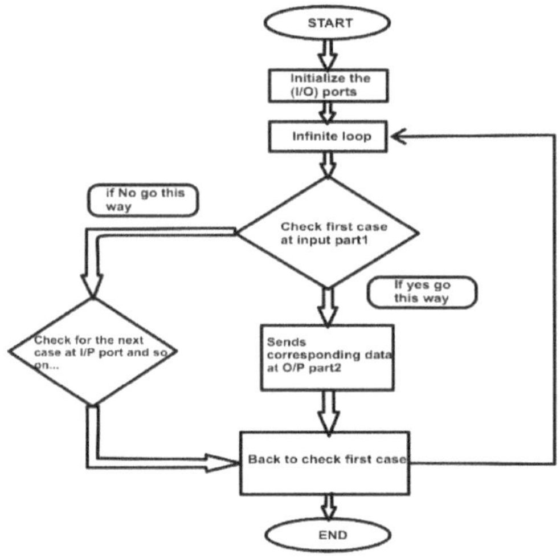

In an era defined by technological marvels, mobile robotics emerges as a transformative force, ushering in a new era of automation, adaptability, and intelligence. Unlike their stationary counterparts, mobile robots transcend the boundaries of fixed positions, embarking on journeys of discovery and utility across diverse landscapes. At the core of mobile robotics lies a captivating synergy between mechanical innovation, artificial intelligence, and sensor integration, converging to create autonomous agents capable of navigating and interacting with the complex environments that surround us. Beyond mere locomotion, these robots embody a realm of possibilities where they sense, process, and comprehend their surroundings, all while making informed decisions that shape their trajectory.

Mobile robotics unfolds as a remarkable tapestry of cutting-edge advancements. Robotic platforms, equipped with wheels, legs, or even aerial propellers, acquire the ability to traverse challenging terrains – from rugged terrains to indoor spaces – with a dexterity that mirrors the adaptability of living organisms. This mobility isn't just a technical achievement; it's an evolution that reshapes industries and reshuffles the boundaries of what we thought robots could accomplish. From autonomous delivery vehicles weaving through city streets to drones surveying remote landscapes, mobile robotics redefines the possibilities of exploration, efficiency, and intervention.

Central to the success of mobile robotics are the intricate sensors that serve as their eyes and ears. LiDAR, cameras, ultrasonic sensors, and a plethora of other technologies work in concert to gather data from the environment, painting a multidimensional canvas of the world around the robot. These sensors offer the raw material for algorithms to decipher, analyze, and derive meaning, enabling robots to perceive obstacles, identify landmarks, and understand spatial relationships. With the fusion of sensor data and machine learning techniques, mobile robots transcend mere motion; they become intelligent entities capable of real-time decision-making, route optimization, and even self-correction.

The allure of mobile robotics lies not only in their capacity to explore remote or hazardous locations, but also in their potential to revolutionize industries and services. In logistics, autonomous mobile robots navigate vast warehouses, orchestrating the efficient movement of goods with precision and reliability. Healthcare benefits from robotic platforms that transport medical supplies within hospitals or even perform delicate surgeries with enhanced precision. Environmental monitoring witnesses a paradigm shift as drones gather critical data on deforestation, climate change, and disaster-stricken areas. The interplay between robotics and society accelerates as mobile robots find their roles in agriculture, construction, entertainment, and countless other domains.

As we delve into the realm of mobile robotics, we uncover a multidisciplinary arena where mechanical engineering, computer science, artificial intelligence, and human-robot interaction converge. The challenges are formidable: creating algorithms that enable robots to map unknown environments, devise optimal paths, and adapt to unforeseen changes. Ensuring safety during navigation, collision avoidance, and robustness in dynamic settings poses intricate engineering puzzles. Ethical and regulatory considerations surrounding autonomy and privacy further enrich the discourse.

In conclusion, mobile robotics signifies a pivotal moment in the journey of human-machine collaboration. It's an intersection of innovation and aspiration, where robots evolve from stationary tools to autonomous explorers. Through their mobility, these machines venture where humans may not tread, aiding us in understanding our world better and alleviating us from repetitive tasks. As mobile robots continue to evolve, driven by the convergence of technology and imagination, they inspire us to reimagine what it means to interact with the world, and how, together, we can propel ourselves into a future of unprecedented possibilities.

Wheeled Robots: Wheeled robots represent a fundamental and versatile class of mobile robotic systems that have played a pivotal role in various applications across industries. These robots are characterized by their use of wheels as their primary mode of locomotion, enabling them to move efficiently across a wide range of terrains and environments. Wheeled robots have been employed in tasks that range from simple navigation to complex manipulation and exploration missions. This section delves deeply into the key aspects of wheeled robots, including their design, locomotion mechanisms, and applications.

Design and Components: The design of wheeled robots is influenced by their intended application and operational environment. A wheeled robot typically consists of several key components, including the chassis, wheels, actuators, sensors, and control systems. The chassis serves as the structural framework that houses the robot's internal components and provides the necessary support for the wheels and other attachments. The wheels are responsible for facilitating movement and traction. Depending on the application, the wheels can be of varying sizes, shapes, and configurations. For instance, differential drive robots employ two large drive wheels, while omni-wheeled robots feature multiple smaller wheels arranged in a circular pattern to enable versatile motion.

Locomotion Mechanisms: Wheeled robots utilize different locomotion mechanisms to navigate through their environment. One of the most common mechanisms is differential drive, where two drive wheels are placed on either side of the robot's chassis. By controlling the speed and direction of each wheel independently, the robot can achieve straight-line motion, as well as turns by varying the speeds of the wheels. This mechanism is simple yet effective, making it widely used in various applications such as indoor robots and small exploration rovers.Another locomotion mechanism is holonomic motion, achievable through Omni-wheels or mecanum wheels. Omni-wheels enable the robot to move in any direction without the need to change its orientation. Mecanum wheels, on the other hand, provide a combination of lateral and diagonal movements, granting robots enhanced maneuverability in constrained spaces.

Applications: Wheeled robots find applications across a spectrum of domains due to their adaptability and efficiency. In industrial settings, they are employed in logistics, material handling, and warehouse automation. These robots can transport goods autonomously, optimizing workflows and reducing human intervention. In healthcare, wheeled robots serve as mobile platforms for delivering supplies, conducting patient monitoring, and even assisting in surgical procedures.Exploration is another domain where wheeled robots excel. Planetary rovers, such as those used in Mars exploration missions, utilize robust wheel systems to traverse challenging terrains while collecting data and samples. These rovers navigate diverse landscapes, overcoming obstacles and adapting to changing surface conditions.Wheeled robots also play a crucial role in education and research. Many universities and research institutions employ wheeled platforms to develop and test algorithms for navigation, mapping, and path planning. These platforms provide students and researchers with hands-on experience in robotics and autonomous systems.

Challenges and Future Directions: While wheeled robots offer numerous advantages, they also face challenges that researchers and engineers continually work to address. Terrain adaptability is one such challenge, as some environments may be too rugged or uneven for traditional wheeled locomotion. This has led to the exploration of hybrid systems that combine wheels with other forms of locomotion, such as legged mechanisms or tracks.Another challenge is the integration of advanced sensing and perception technologies to enhance navigation and obstacle avoidance. Wheeled robots require accurate and real-time information about their surroundings to make informed decisions and ensure safe operation.In the future, advancements in materials, sensors, and AI-driven algorithms are expected to drive innovations in wheeled robot design. Swarms of miniature wheeled robots could collaborate to achieve complex tasks, while AI-based path planning algorithms could enable more efficient and adaptive navigation.

Conclusion: Wheeled robots represent a cornerstone of mobile robotics with a wide array of applications in industries ranging from logistics to exploration. Their design, locomotion mechanisms, and adaptability make them invaluable tools for various tasks. As technology advances, wheeled robots will continue to evolve, shaping the way we interact with and explore the world around us.

Legged Robots: Legged robots are a fascinating and dynamic category of robotic systems designed to replicate the locomotion of creatures with legs, such as humans, animals, and insects. These robots emulate the remarkable ability of legged organisms to traverse diverse terrains, adapt to uneven surfaces, and navigate challenging environments. From biomimicry to advanced control algorithms, legged robots have captured the attention of researchers and engineers for their potential to revolutionize fields ranging from search and rescue to exploration. This section delves deeply into the intricacies of legged robots, exploring their design principles, locomotion strategies, applications, challenges, and future prospects.

Design Principles: The design of legged robots is intricately tied to the morphology and biomechanics of living organisms. Engineers draw inspiration from nature's evolutionary solutions, aiming to replicate the versatility, stability, and energy efficiency exhibited by animals. The key components of a legged robot typically include the body or chassis, legs with joints and actuators, sensors, and advanced control systems. The number, arrangement, and type of legs vary widely based on the intended application and terrain.

Locomotion Strategies: Legged robots employ various locomotion strategies to achieve movement similar to their biological counterparts. These strategies range from bipedal and quadrupedal gaits to more complex hexapod and octopod configurations. Bipedal robots, with two legs, emulate human-like walking and running motions. Quadrupedal robots, possessing four legs, excel in stability and adaptability, resembling mammals like dogs and horses. Hexapod and octopod robots, with six and eight legs respectively, offer enhanced maneuverability and load-bearing capabilities. The choice of locomotion strategy is influenced by factors such as terrain type, payload capacity, energy efficiency, and application-specific requirements. Advanced control algorithms play a critical role in coordinating leg movements and ensuring stable locomotion.

Applications: Legged robots hold immense promise in a multitude of applications due to their potential to navigate challenging and complex environments. In search and rescue scenarios, these robots can access disaster-stricken areas where wheeled or tracked robots might struggle. Their ability to traverse rubble, climb stairs, and cross uneven terrain makes them valuable tools for locating survivors and assessing damage. Exploration is another domain where legged robots shine. Planetary exploration missions could benefit from legged rovers capable of navigating rough and rocky surfaces on distant celestial bodies. These robots could collect samples, analyze geological features, and transmit valuable data back to Earth. In healthcare, legged robots have the potential to assist individuals with mobility impairments by providing ambulatory support or rehabilitation exercises. Additionally, legged robotic exoskeletons offer the possibility of enhancing human capabilities by augmenting strength and endurance.

Challenges and Future Prospects: Legged robots face a unique set of challenges due to their complex mechanical design and intricate control systems. Achieving stable and adaptive locomotion across various terrains demands sophisticated control algorithms that can dynamically adjust leg movements based on sensory feedback. Maintaining energy efficiency while emulating natural gaits is another challenge, as the energy requirements for legged locomotion can be substantial. The integration of perception and sensor technologies is crucial for enabling legged robots to autonomously navigate and respond to their environment. Accurate sensor data, such as terrain topology and obstacle detection, informs control algorithms and ensures safe and effective movement. The future of legged robots holds exciting possibilities. Advancements in materials, actuation, and AI-driven algorithms could lead to more agile and versatile legged platforms. The field of soft robotics, which focuses on flexible and compliant structures, could offer innovative solutions for achieving natural and adaptive legged locomotion.

Conclusion: Legged robots represent a remarkable convergence of biology, engineering, and robotics. Drawing inspiration from the animal kingdom, these robots aim to emulate the incredible locomotion abilities of legged organisms. From disaster response to planetary exploration, the potential applications of legged robots are vast and diverse. While challenges in control, energy efficiency, and perception persist, ongoing research and technological advancements hold the promise of unlocking new capabilities for legged robots, shaping their role in various industries and pushing the boundaries of robotic innovation.

Autonomous Navigation: Autonomous navigation is a cornerstone of modern robotics, enabling robots to move, perceive, and interact with their environment without human intervention. This capability is crucial for a wide range of applications, from self-driving cars to planetary exploration rovers. Autonomous navigation involves a complex interplay of sensors, algorithms, and decision-making processes that empower robots to safely and effectively traverse diverse terrains, avoid obstacles, and reach predefined goals. This section delves deeply into the intricacies of autonomous navigation, exploring its key components, challenges, strategies, and real-world applications.

Key Components of Autonomous Navigation: Autonomous navigation relies on an integrated system of components working harmoniously to enable robotic motion. These components include perception sensors, localization systems, mapping algorithms, path planning techniques, and control systems.

Perception Sensors: Perception sensors, such as cameras, LiDAR (Light Detection and Ranging), and radar, provide robots with real-time information about their surroundings. These sensors help identify obstacles, road features, pedestrians, and other vehicles, allowing the robot to make informed decisions.

Localization Systems: Localization is the process of determining a robot's position in a known environment. GPS (Global Positioning System), inertial measurement units (IMUs), and odometer are commonly used techniques to estimate a robot's location.

Mapping Algorithms: Mapping involves creating a representation of the robot's environment. Simultaneous Localization and Mapping (SLAM) algorithms allow the robot to build a map while simultaneously localizing itself within that map.

Path Planning Techniques: Path planning involves finding a safe and efficient path from the robot's current location to its desired destination. Techniques such as A* search, D* Lite, and rapidly-exploring random trees (RRT) help robots navigate around obstacles and optimize their routes.

Control Systems: Control algorithms ensure that the robot executes its planned path accurately. These algorithms adjust motor commands and steering angles to account for uncertainties and disturbances in the environment.

Challenges in Autonomous Navigation: Achieving robust and reliable autonomous navigation is fraught with challenges that researchers and engineers continuously strive to address:

Sensor Limitations: Perception sensors may have limitations in certain conditions, such as poor lighting, adverse weather, or reflective surfaces, which can affect their accuracy.

Uncertainty and Noise: Sensors and algorithms introduce uncertainty and noise into the navigation process. Robust algorithms must account for these uncertainties to avoid erroneous decisions.

Dynamic Environments: Navigating in environments with moving obstacles, pedestrians, and other vehicles requires real-time adaptation and collision avoidance strategies.

Map Accuracy: Mapping errors can accumulate over time, leading to inaccuracies in robot localization and navigation.

Complex Terrain: Autonomous navigation should handle various terrains, including rough, slippery, or uneven surfaces, which demand adaptable motion strategies.

Real-Time Decision Making: Robots must make split-second decisions to navigate safely and efficiently, requiring robust decision-making algorithms.

Strategies for Autonomous Navigation: Several strategies and algorithms are employed to enable autonomous navigation:

Sensor Fusion: Integrating data from multiple sensors improves perception accuracy and redundancy, enhancing the robot's understanding of the environment.

SLAM: Simultaneous Localization and Mapping algorithms enable robots to build maps of their environment while tracking their own position within the map.

Path Planning: Algorithms like A* search and RRT enable robots to find optimal paths through complex environments while avoiding obstacles.

Reactive Navigation: Reactive approaches involve making real-time decisions based on immediate sensor data to avoid obstacles and navigate safely.

Machine Learning: Machine learning techniques, including deep neural networks, enable robots to learn navigation behaviors from data and adapt to various scenarios.

Real-World Applications: Autonomous navigation has transformed various industries and domains:

Self-Driving Cars: Autonomous vehicles use a combination of sensors, mapping, and AI algorithms to navigate roads, avoid collisions, and follow traffic rules.

Agriculture: Autonomous agricultural robots navigate fields, identify crops, and perform tasks such as planting, spraying, and harvesting.

Search and Rescue: Robots navigate disaster-stricken areas to locate survivors and assess damage, providing vital information to first responders.

Warehouse Automation: Robots autonomously navigate warehouses, picking and transporting goods, optimizing logistics, and reducing human labor.

Planetary Exploration: Rovers like those used in Mars missions navigate extraterrestrial terrains, conducting scientific experiments and collecting data.

Future Prospects and Conclusion: Autonomous navigation continues to evolve with advancements in sensor technology, AI algorithms, and robotics research. The integration of artificial intelligence and machine learning will lead to smarter robots capable of learning from their experiences and adapting to dynamic environments.The future of autonomous navigation holds promises of enhanced safety, improved efficiency, and greater accessibility. As robots become more capable of navigating complex and challenging scenarios, they will reshape industries, improve quality of life, and contribute to our understanding of the world and beyond. While challenges persist, the relentless pursuit of innovation ensures that the field of autonomous navigation will continue to make remarkable strides in the years to come.

Simultaneous Localization and Mapping (SLAM): Simultaneous Localization and Mapping (SLAM) is a foundational concept in robotics that addresses the challenge of enabling a robot to build a map of an unknown environment while simultaneously determining its own location within that map. This capability is essential for autonomous navigation, enabling robots to navigate and interact with their surroundings accurately and safely without prior knowledge. SLAM has revolutionized various applications, from mobile robotics to augmented reality, and its significance continues to grow as technology advances. In this section, we will delve deeply into the intricacies of SLAM, exploring its key components, working principles, challenges, strategies, and real-world applications.

Key Components of SLAM: SLAM systems involve a complex interplay of sensors, algorithms, and estimation techniques to achieve accurate mapping and localization. Several key components contribute to the functioning of SLAM:

Sensors: SLAM relies on a combination of sensors such as cameras, LiDAR, odometry, GPS, and IMUs. These sensors provide information about the robot's movement, its environment, and the features it encounters.

Feature Extraction: Features are distinct landmarks in the environment, such as corners or edges. SLAM algorithms identify and track these features to create a map.

Data Association: Matching features observed in different sensor measurements to corresponding features in the map is a critical step in SLAM. Proper data association ensures accurate mapping and localization.

Estimation: SLAM involves estimating the robot's pose (position and orientation) as well as the locations of the mapped features. This estimation process uses probabilistic techniques to account for uncertainties.

Optimization: SLAM algorithms optimize the estimated poses and feature locations to minimize errors and inconsistencies in the map and localization estimates.

Working Principles of SLAM: The SLAM process involves the following steps:

Initialization: The robot starts in an unknown environment and begins collecting sensor data.

Feature Extraction: The robot extracts features from the sensor data, identifying distinct landmarks in the environment.

Data Association: The robot associates the observed features in different sensor measurements with the corresponding features in the map.

Estimation: The robot estimates its current pose and the locations of the features using probabilistic techniques such as the Extended Kalman Filter (EKF) or Particle Filter.

Map Update: As the robot moves, new sensor data is collected, and the map is updated with the locations of new features.

Optimization: The map and robot poses are optimized to minimize errors and inconsistencies using optimization techniques like Graph SLAM or Bundle Adjustment.

Loop Closure: The robot detects when it revisits a previously visited location (loop closure) and uses this information to refine the map and improve localization accuracy.

Challenges in SLAM: SLAM poses several challenges due to the complexities of real-world environments and sensor limitations:

Sensor Noise and Uncertainty: Sensors introduce noise and uncertainties in measurements, affecting the accuracy of mapping and localization.

Data Association: Matching observed features to map features accurately, especially in cluttered or dynamic environments, can be challenging.

Scale Ambiguity: Determining the scale of the map is challenging, especially when using monocular cameras or ambiguous sensor data.

Degeneracy: SLAM algorithms can face situations where uncertainty becomes large, leading to convergence issues or incorrect solutions.

Computational Complexity: SLAM involves complex computations, especially in large-scale environments, demanding efficient algorithms and hardware.

Strategies for SLAM: Various strategies and algorithms are employed to address the challenges of SLAM:

EKF-SLAM: The Extended Kalman Filter (EKF) is used to estimate the robot's pose and map features using linearization and Gaussian assumptions.

Graph SLAM: This approach models the SLAM problem as a graph, where robot poses and map features are nodes connected by constraints representing sensor measurements. Optimization techniques are used to refine the graph.
Particle Filter SLAM: Particle filters use a set of particles to represent possible robot poses and map configurations. This approach is effective in handling non-Gaussian uncertainties.
Feature-Based vs. Direct Methods: SLAM algorithms can be categorized into feature-based methods, which use extracted features, and direct methods, which directly optimize for pixel intensities in images.
Real-World Applications:SLAM has transformed various domains and industries:
Autonomous Vehicles: Self-driving cars utilize SLAM to build maps of their surroundings and localize themselves, enabling safe and efficient navigation.
Robotics: Mobile robots, drones, and underwater vehicles use SLAM for exploration, mapping, and navigation in unknown environments.
Augmented Reality: SLAM is used in AR applications to overlay virtual objects onto the real world accurately, enhancing user experiences.
Mapping and Inspection: SLAM is employed in mapping indoor environments, infrastructure inspection, and creating 3D models.
Future Prospects and Conclusion:SLAM continues to advance with the integration of improved sensors, powerful computing, and machine learning techniques. Future developments may focus on addressing challenges like real-time performance, robustness in dynamic environments, and efficient large-scale mapping. SLAM's significance will only grow as robots become more integrated into our lives, shaping industries, enhancing exploration, and contributing to a future where autonomous systems thrive.

Exercise

Multiple-choice Questions

1. What is the primary goal of Simultaneous Localization and Mapping (SLAM) in mobile robotics?
a) Path planning
b) Obstacle avoidance
c) Mapping the environment while localizing the robot
d) Communication with other robots
2. Which sensor emits laser beams to measure distances and create a 3D representation of the environment?
a) Camera
b) Sonar
c) LiDAR
d) GPS
3. What component of a mobile robot helps determine its position and orientation?
a) Actuators
b) GPS
c) Sensors
d) Control algorithms
4. What is the primary challenge in achieving autonomous navigation for mobile robots?
a) Developing complex actuators
b) Building high-definition maps
c) Navigating without human intervention
d) efficiently storing sensor data
5. Which algorithm helps mobile robots find optimal paths through environments while avoiding obstacles?
a) SLAM
b) A* search
c) Kalman Filter
d) PID control

Review Questions:

1. What is the primary purpose of Simultaneous Localization and Mapping (SLAM) in mobile robotics?
2. Which sensor technology emits laser beams to measure distances and create 3D environmental representations?
3. What component of a mobile robot helps determine its position and orientation within an environment?
4. What is the central challenge in achieving autonomous navigation for mobile robots?
5. Which algorithm is commonly used by mobile robots to find optimal paths through environments while avoiding obstacles?
6. What does the acronym "SLAM" stand for in the context of mobile robotics?

7. Which strategy involves representing possible robot poses and map configurations using particles?
8. In which domain is SLAM technology applied to overlay virtual objects onto the real world?
9. What advantage do legged robots offer in search and rescue scenarios compared to wheeled robots?
10. What type of motion allows legged robots to move in any direction without altering their orientation?
11. What term refers to the sensors and algorithms that enable a robot to perceive its environment?
12. Which navigation challenge involves determining a robot's location within a known map?
13. What role do localization systems play in mobile robotics?
14. Which sensor type is effective for measuring distances and detecting obstacles in close proximity to a robot?
15. How does SLAM address the simultaneous tasks of mapping and localization in mobile robotics?
16. What is the primary purpose of path planning algorithms in autonomous navigation?
17. What does the acronym "LIDAR" stand for, and how does it contribute to mobile robot perception?
18. Explain the concept of reactive navigation in the context of mobile robots.
19. How does sensor fusion enhance the accuracy and reliability of mobile robot perception?
20. Provide an example of an industry or application where mobile robotics and SLAM technology are extensively used.

Answers

Multiple-choice Questions

Answers:

1. c)
2. c)
3. b)
4. c)
5. b)

Review Questions:

1. Simultaneous Localization and Mapping (SLAM) is a critical technology in mobile robotics that serves the dual purpose of mapping an unknown environment while simultaneously localizing the robot within that environment. The primary goal of SLAM is to enable a robot to navigate autonomously in unfamiliar surroundings, even without prior knowledge of the environment's layout or the robot's initial position. SLAM algorithms utilize sensor data, such as laser range measurements or camera images, to build a map of the surroundings. At the same time, these algorithms estimate the robot's position and orientation relative to the map. By iteratively refining these estimates, SLAM ensures that the robot's map aligns with its observations, allowing the robot to accurately localize itself as it explores and interacts with the environment. SLAM is essential in various applications, including autonomous vehicles, warehouse automation, and search and rescue missions.

2. The sensor technology that emits laser beams to measure distances and create 3D environmental representations is known as LiDAR, which stands for Light Detection and Ranging. LiDAR sensors work by emitting laser pulses and measuring the time it takes for the light to bounce off objects and return to the sensor. By analyzing the time-of-flight and the angles at which the laser beams return, LiDAR sensors create detailed point cloud data, which can be used to generate accurate 3D maps of the environment. LiDAR technology plays a crucial role in mobile robotics by providing high-resolution spatial data that aids in mapping, localization, and obstacle detection, making it a foundational tool in enabling robots to perceive and navigate their surroundings effectively.

3. The component of a mobile robot that helps determine its position and orientation within an environment is the localization system. The localization system combines data from various sensors, such as GPS, LiDAR, cameras, and inertial measurement units (IMUs), to estimate the robot's pose (position and orientation) relative to a known coordinate frame. This estimation process involves sensor fusion algorithms that integrate the different sensor measurements while accounting for their uncertainties and inaccuracies. By continuously updating and refining the robot's pose estimation, the localization system provides the robot with an accurate understanding of its location and orientation within its environment. This information is essential for enabling autonomous navigation, path planning, and interaction with the surroundings.

4. The central challenge in achieving autonomous navigation for mobile robots lies in the complex task of enabling robots to navigate safely and effectively in diverse and dynamic environments without human intervention. Autonomous navigation involves making real-time decisions to plan paths, avoid obstacles, and reach designated destinations while adhering to safety constraints. The challenge arises from factors such as uncertainty in sensor measurements, changing environments, unpredictable obstacles, and the need to account for dynamic elements like moving objects or pedestrians. Overcoming this challenge requires developing robust algorithms for perception, localization, path planning, and control, which must collectively enable the robot to make informed decisions and adapt to the environment's nuances while avoiding collisions and achieving its objectives.

5. The algorithm commonly used by mobile robots to find optimal paths through environments while avoiding obstacles is the A* (A-star) search algorithm. A* search is a popular path planning algorithm that efficiently explores the search space to identify the shortest path from a starting point to a goal while considering obstacles and terrain constraints. A* assigns a cost to each potential path based on the distance traveled and a heuristic function that estimates the remaining distance to the goal. By selecting paths with lower total cost, A* search algorithm identifies optimal routes while efficiently avoiding obstacles and navigating complex environments. This algorithm's effectiveness and versatility make it a valuable tool for enabling mobile robots to plan safe and efficient paths in real time.

6. The acronym "SLAM" stands for "Simultaneous Localization and Mapping" in the context of mobile robotics. SLAM refers to the integrated process of creating a map of an unknown environment while concurrently determining the robot's position and orientation within that environment. The goal of SLAM is to enable robots to navigate autonomously in unfamiliar surroundings, even without prior knowledge of the environment's layout or the robot's initial position. SLAM algorithms utilize sensor data, such as laser range measurements or camera images, to build a map of the surroundings. Simultaneously, these algorithms estimate the robot's pose by refining the position and orientation estimates based on observed data. SLAM plays a pivotal role in enabling robots to understand and interact with their surroundings, making it a foundational technology in mobile robotics.

7. The strategy that involves representing possible robot poses and map configurations using particles is known as "Particle Filter SLAM" or "Monte Carlo Localization." This strategy is a probabilistic approach used in Simultaneous Localization and Mapping (SLAM) algorithms. In Particle Filter SLAM, the robot's pose and map are represented by a set of particles, each particle representing a possible hypothesis about the robot's position and the environment's features. These particles are propagated and updated based on sensor measurements and control inputs. Over time, as the robot's knowledge improves, particles that align with the sensor data become more likely, leading to an accurate estimate of the robot's pose and the map. Particle Filter SLAM is particularly effective in handling uncertainties and non-linearity's associated with robot localization and mapping.

8. SLAM technology is applied in the domain of Augmented Reality (AR) to overlay virtual objects onto the real world. Augmented Reality involves enhancing the perception of the physical world by superimposing computer-generated images, videos, or other digital content onto the user's view of the environment. SLAM algorithms are used to precisely align virtual objects with the real-world scene, ensuring that they appear to interact realistically with the surroundings. By accurately registering virtual objects within the physical space, SLAM technology enhances the immersive and interactive experience of Augmented Reality applications, ranging from gaming and entertainment to industrial and educational contexts.

9. Legged robots offer a significant advantage in search and rescue scenarios compared to wheeled robots due to their enhanced mobility and adaptability in challenging environments. Search and rescue operations often involve navigating through complex, unstructured terrains, such as rubble, debris, or disaster-stricken areas. Legged robots, inspired by animal locomotion, can traverse irregular surfaces, climb over obstacles, and cross terrains that would be challenging for traditional wheeled robots. Their ability to navigate uneven terrain, ascend stairs, and maneuver through confined spaces makes them well-suited for accessing and exploring disaster-stricken areas where human intervention might be risky or challenging. Legged robots can provide valuable assistance in locating survivors and assessing the extent of damage in scenarios where traditional methods might be limited.

10. Legged robots can achieve holonomic motion, which enables them to move in any direction without altering their orientation. Holonomic motion is characterized by the robot's ability to translate and rotate simultaneously, allowing it to achieve complex movements and precise positioning. This capability is particularly advantageous in environments with tight spaces, where the robot needs to navigate intricate paths and avoid obstacles efficiently. Holonomic legged robots achieve motion through the coordinated control of their legs or limbs, enabling them to execute intricate maneuvers and adapt to dynamic environments effectively. This versatility in movement contributes to legged robots' suitability for applications that require agile and flexible locomotion.

11. The term that refers to the sensors and algorithms that enable a robot to perceive its environment is "perception." Perception encompasses the process through which a robot gathers, processes, and interprets sensory information from its surroundings to gain an understanding of its environment. Perception sensors include cameras, LiDAR, ultrasonic sensors, GPS, IMUs, and more, each providing different types of data such as images, distances, orientations, and positions. Perception algorithms analyze this sensor data to identify objects, obstacles, landmarks, and terrain features. By leveraging perception, robots can build maps, localize themselves, detect obstacles, and interact with their surroundings effectively, essential for enabling autonomous navigation and decision-making.

12. The navigation challenge that involves determining a robot's location within a known map is known as "localization." Localization refers to the process of estimating a robot's position and orientation within a predefined map or coordinate system. It requires integrating sensor measurements, such as GPS, LiDAR, and odometers, with

map data to accurately determine where the robot is situated within the environment. Successful localization is critical for enabling robots to follow planned paths, avoid obstacles, and interact with the environment effectively. It plays a key role in autonomous navigation, as a robot's ability to navigate depends on accurate self-localization within the reference map.

13. Localization systems play a fundamental role in mobile robotics by enabling robots to determine their precise position and orientation within their environment. These systems utilize sensor data, such as GPS, LiDAR, cameras, and IMUs, along with localization algorithms to estimate the robot's pose (position and orientation) relative to a known reference frame or map. Accurate localization is essential for enabling robots to navigate autonomously, plan paths, avoid obstacles, and interact with the environment effectively. Localization errors can lead to inaccurate decision-making, unintended collisions, and inefficient navigation. Therefore, robust and reliable localization systems are crucial for the success of mobile robotic applications.

14. Ultrasonic sensors are effective for measuring distances and detecting obstacles in close proximity to a robot. Ultrasonic sensors work based on the principle of sending out ultrasonic sound waves and measuring the time it takes for the waves to bounce off objects and return to the sensor. By analyzing this time-of-flight, the sensor can calculate the distance to the object. Ultrasonic sensors are commonly used for obstacle detection and avoidance in environments where precision at close ranges is essential, such as indoor navigation, robotic vacuum cleaners, and small-scale robotics applications.

15. Simultaneous Localization and Mapping (SLAM) addresses the simultaneous tasks of mapping and localization in mobile robotics through an integrated approach that utilizes sensor data and algorithms. SLAM algorithms process sensor measurements, such as laser range scans or camera images, to build a map of the environment's features, including landmarks and obstacles. Simultaneously, these algorithms estimate the robot's position and orientation relative to the map. By iteratively refining both the map and the pose estimates, SLAM enables a robot to create an accurate map of its environment while localizing itself within that map. This synchronization between mapping and localization is achieved through data fusion, probabilistic algorithms, and optimization techniques, ensuring that the robot's understanding of its surroundings evolves as it explores the environment.

16. The primary purpose of path planning algorithms in autonomous navigation is to determine the optimal route for a robot to reach its destination while avoiding obstacles and adhering to constraints. Path planning algorithms analyze the environment's map and the robot's localization information to generate a path that minimizes travel time, distance, or other defined criteria. These algorithms consider factors like the robot's dynamics, the presence of obstacles, terrain constraints, and safety margins. Path planning ensures that the robot can navigate from its current position to the desired goal while making informed decisions to avoid collisions and optimize its trajectory. This process is essential for enabling robots to move autonomously in complex and dynamic environments.

17. The acronym "LIDAR" stands for "Light Detection and Ranging." LIDAR technology contributes to mobile robot perception by emitting laser beams and measuring the time it takes for the light to reflect off objects and return to the sensor. By analyzing the time-of-flight and the angles of the returning laser beams, LIDAR sensors create detailed 3D point cloud representations of the environment. These point clouds provide accurate spatial information about the surroundings, including distances, shapes, and locations of objects. LIDAR is crucial for mapping, localization, obstacle detection, and terrain analysis in mobile robotics. Its ability to capture high-resolution environmental data enhances a robot's perception capabilities and enables it to navigate autonomously while making informed decisions based on accurate environmental information.

18. Reactive navigation refers to a navigation strategy employed by mobile robots that focuses on real-time responsiveness to the immediate environment. Unlike traditional planning approaches that involve creating complete maps and path plans, reactive navigation algorithms directly respond to sensor data without constructing a global map. These algorithms use local sensory information, such as proximity sensors, cameras, and obstacle detection systems, to make instantaneous decisions about motion and steering. Reactive navigation enables robots to navigate by reacting to the immediate sensory inputs, allowing them to adapt dynamically to changing conditions, obstacles, and unplanned events. This approach is particularly useful in environments where mapping and planning are challenging, and quick reactions are essential, such as navigating cluttered or unpredictable spaces.

19. Sensor fusion enhances the accuracy and reliability of mobile robot perception by integrating data from multiple sensors to create a more comprehensive and accurate understanding of the environment. Different sensors have varying strengths and weaknesses, and by combining their measurements, sensor fusion algorithms can compensate for individual sensor limitations. For example, by fusing data from cameras, LiDAR, GPS, IMUs, and other sensors, a robot can accurately estimate its position, map the environment, detect obstacles, and recognize objects. Techniques like Kalman filtering and particle filtering are used to fuse sensor data while considering their uncertainties. Sensor fusion ensures a holistic and robust perception system, improving a robot's ability to make informed decisions, navigate safely, and interact effectively with its surroundings.

20. An example of an industry where mobile robotics and SLAM technology are extensively used is autonomous agriculture. In precision agriculture, robots equipped with various sensors, including LiDAR, cameras, and GPS, are employed to perform tasks such as planting, seeding, fertilizing, and monitoring crops. SLAM technology enables these robots to navigate fields, create accurate maps of crop rows and features, and determine their position within the farm environment. By integrating SLAM data with information about soil quality, plant health, and weather conditions, these robots can make informed decisions to optimize planting and crop management processes. Autonomous agriculture enhances resource utilization, minimizes environmental impact, and increases crop yield, highlighting the significance of mobile robotics and SLAM in revolutionizing traditional farming practices.

Online Resources

Online Exam Papers

Link:https://forms.gle/VtAfcFdAB4RdzMiE6

QR Code:

YouTube Video Playlist Link

Link: https://www.youtube.com/playlist?list=PLzlh_31VWEFpi8Dxpm-Edeg7rQkydIjJ2

QR Code:

Chapter 7: Manipulators and Robot Arms

In the realm of robotics, the concept of extending human capabilities through mechanical devices has manifested itself most prominently in the form of manipulators and robot arms. These remarkable machines represent the embodiment of precision engineering and automation, enabling us to perform tasks with a level of accuracy, efficiency, and scale that would be unattainable by human hands alone. Manipulators and robot arms have evolved from humble beginnings into sophisticated instruments that traverse industries ranging from manufacturing and healthcare to space exploration and entertainment. At their core, these mechanical wonders are testament to the harmonious collaboration of mechanical design, control systems, and advanced materials. By replicating the intricate movements of human arms and hands, while also defying physical limitations, manipulators and robot arms have redefined how we approach complex tasks, from assembling intricate components to exploring the cosmos. This journey into the world of manipulators and robot arms uncovers a landscape where technology melds with imagination to create machines that are not just tools, but extensions of human ingenuity, laying the foundation for a future where precision and automation intertwine seamlessly.

In the intricate dance between human ingenuity and technological advancement, manipulators and robot arms take center stage as the embodiment of precision engineering and the epitome of automation. These mechanical marvels transcend mere machinery, weaving together the art of design, the finesse of control, and the power of materials to create instruments that emulate and extend the capabilities of the human arm. As we delve into the world of manipulators and robot arms, we uncover a narrative that spans centuries, shaped by the relentless pursuit of efficiency, accuracy, and innovation.

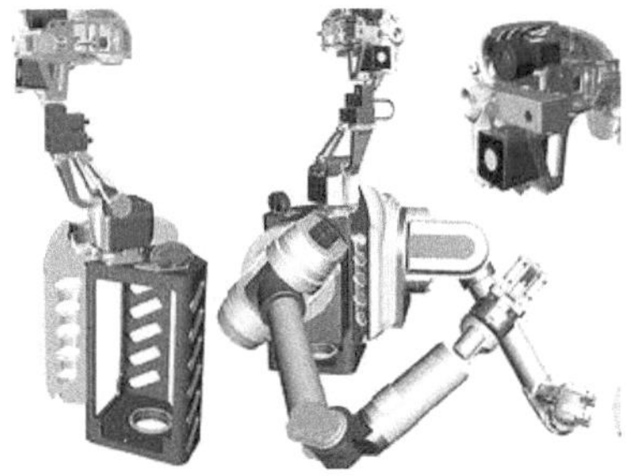

From the earliest designs inspired by clockwork mechanisms to the state-of-the-art robotic arms gracing modern factories and laboratories, the evolution of manipulators and robot arms tells a tale of human curiosity and the quest for enhancement. These machines, with their articulated joints and meticulously engineered linkages, mirror the dexterity and adaptability of the human hand while superseding its physical limitations. What began as mechanical aids for industrial tasks burgeoned into multifunctional systems that can delicately assemble microelectronics, execute complex surgical procedures, or weld heavy components with equal finesse?

Central to the allure of manipulators and robot arms is the synergy between mechanical design and intelligent control. The intricate choreography of joints, actuators, and sensors, seamlessly orchestrated by sophisticated control algorithms, gives rise to movements that are not just functional but graceful. This amalgamation of hardware and software propels the machines to navigate intricate paths, maintain steady trajectories, and precisely interact with their surroundings. These arms, often equipped with end effectors and grippers, become the extensions of human intent, carrying out tasks that range from the mundane to the extraordinary.

In the manufacturing sector, robot arms have revolutionized mass production, expediting processes, enhancing quality, and reducing human exposure to hazardous conditions. The concept of "lights-out manufacturing," where factories can operate autonomously with minimal human intervention, has transformed industries and economies. Beyond manufacturing, robot arms have found their place in medicine, assisting surgeons in intricate procedures and even enabling tele surgery across vast distances. In space exploration, these arms have delicately deployed satellites, repaired spacecraft, and roamed alien landscapes, expanding the horizons of human knowledge.

However, the realm of manipulators and robot arms is not without its challenges. Ensuring the safety of human interactions, fine-tuning control algorithms to achieve fluid motions, and optimizing the trade-off between precision and speed are all intricate puzzles that researchers and engineers tirelessly unravel. Moreover, the convergence of artificial intelligence, computer vision, and advanced materials has opened new vistas of possibilities, enabling robot arms to adapt to uncertain environments, learn from experiences, and collaborate seamlessly with humans.

As the symbiosis between human and machine deepens, manipulators and robot arms hold the potential to revolutionize how we interact with the physical world. In industries, they augment productivity and unlock new heights of efficiency. In healthcare, they redefine surgical precision and enable treatments that were once inconceivable. In exploration, they probe the far reaches of the cosmos, uncovering mysteries beyond our planet.

The convergence of these technologies creates a nexus where precision meets automation, craftsmanship meets computation, and the boundaries of human potential expand exponentially.

Robotic Arm Configurations: Robotic arms have evolved into indispensable tools across a diverse array of industries, from manufacturing and healthcare to space exploration and even domestic tasks. These mechanical wonders offer unparalleled precision, strength, and dexterity, allowing them to perform tasks that would be impossible or impractical for human operators alone. One critical aspect of robotic arms is their configuration, which determines their range of motion, flexibility, and suitability for specific applications. In this article, we delve into the realm of robotic arm configurations, focusing on the 7.1 most significant setups that have transformed industries and pushed the boundaries of what automation can achieve.

Cartesian Configuration: The Cartesian configuration is one of the most straightforward and commonly used robotic arm setups. It employs three linear joints along orthogonal axes – X, Y, and Z – which correspond to left-right, forward-backward, and up-down movements, respectively. This configuration is ideal for tasks requiring precise positioning within a three-dimensional workspace, such as pick-and-place operations in manufacturing lines and 3D printing.

SCARA Configuration: SCARA stands for Selective Compliance Assembly Robot Arm or Selective Compliance Articulated Robot Arm. This configuration features two parallel rotary joints, allowing for planar movement within a horizontal plane, combined with a linear vertical joint. The SCARA configuration excels in tasks requiring quick and precise lateral movements, such as assembling electronic components or performing small-scale packaging operations.

Articulated Configuration: The articulated robotic arm configuration closely mimics the structure of the human arm. Consisting of a series of interconnected rotary joints, it offers exceptional flexibility and dexterity. Each joint adds an additional degree of freedom, enabling complex movements and the manipulation of objects in intricate ways. This configuration is extensively utilized in applications ranging from surgical procedures to painting large surfaces with varying angles.

Delta Configuration: The delta configuration employs three or more arms connected to a common base, with the end-effectors attached to the arms' tips. This setup enables extremely fast and precise movement within a defined workspace. The delta configuration finds its niche in tasks requiring high-speed pick-and-place operations, such as packaging, sorting, and food processing.

Polar Configuration: The polar configuration, also known as the spherical configuration, employs a combination of rotary and linear joints to provide movement along radial and angular axes. This arrangement is particularly advantageous for applications demanding versatile movement within a spherical workspace. Welding, material deposition, and inspection tasks in complex shapes are among the areas where the polar configuration shines.

Parallel Configuration: Parallel robotic arms utilize multiple arms interconnected through joints and links to share the load and enhance stability. This configuration is revered for its strength and precision, making it well-suited for heavy-duty applications such as automotive assembly lines and aerospace manufacturing. The parallel setup offers the advantage of distributing forces among the arms, resulting in higher payload capacities and increased accuracy.

Snake or Continuum Configuration: Inspired by the flexibility of snakes, the snake or continuum configuration employs a series of interconnected segments that mimic vertebral joints. This arrangement grants the robotic arm the ability to navigate through confined spaces and manipulate objects in complex environments. Medical procedures, search and rescue missions, and hazardous material handling are some of the domains that benefit from the snake configuration's unique capabilities.

Soft Robotic Configuration: A relatively novel approach, the soft robotic configuration utilizes flexible and deformable materials to create robotic arms with inherent compliance. These arms can adapt to the shape of objects, allowing for gentle and safe interactions in unstructured environments. The soft robotic configuration finds applications in fields like medical assistance, where delicate interactions with patients are necessary, and in gripping fragile items in unpredictable settings.

In conclusion, the world of robotic arm configurations is a captivating landscape where engineering innovation meets real-world challenges head-on. From the foundational Cartesian setup to the revolutionary soft robotic configuration, each configuration brings its strengths and capabilities to the table, enabling robots to execute an astonishing array of tasks across various industries. As technology continues to advance, these configurations will likely evolve even further, opening up new avenues for automation, efficiency, and safety. As we move forward, understanding these configurations will be essential for engineers, designers, and innovators aiming to push the boundaries of robotics and redefine the possibilities of human-robot collaboration.

End Effectors and Grippers: In the realm of robotics, where machines emulate the dexterity and precision of human hands, end effectors and grippers stand as the unsung heroes of automation. These devices, often attached to the tip of robotic arms, serve as the "hands" of robots, enabling them to interact with the physical world in ways that mimic and even surpass human capabilities. From delicate tasks in manufacturing to intricate surgical procedures and even extraterrestrial exploration, end effectors and grippers have revolutionized industries and expanded the horizons of what robots can achieve. In this comprehensive exploration, we delve into the intricacies of these devices, unveiling their types, functions, and the transformative impact they have had on various sectors.

Types of End Effectors and Grippers:

Mechanical Grippers: Mechanical grippers are perhaps the most recognizable and straightforward type of end effector. These devices use mechanical jaws or fingers to grasp, hold, and manipulate objects. They come in various designs, including two-finger, three-finger, and multi-finger configurations, each tailored to different tasks and object geometries. Mechanical grippers find applications in pick-and-place tasks on assembly lines, material handling in warehouses, and even in simple household robots that can pick up objects.

Vacuum Grippers: Vacuum grippers rely on the principle of suction to secure objects. They create a pressure difference between the gripper's surface and the object, allowing atmospheric pressure to firmly hold the object in place. Vacuum grippers are ideal for handling items with flat surfaces, such as cardboard boxes, sheets of paper, and electronic components. These grippers are widely used in packaging, logistics, and industries where delicate handling of objects is crucial.

Pneumatic Grippers:Pneumatic grippers operate using compressed air to generate the gripping force. They provide a combination of speed and force, making them suitable for applications where rapid and robust gripping is essential. Pneumatic grippers are commonly used in industries like automotive manufacturing, where they can swiftly grip and release parts during assembly processes.

Hydraulic Grippers: Similar to pneumatic grippers, hydraulic grippers use hydraulic fluid to generate gripping force. They are known for their exceptional strength and gripping capabilities, making them suitable for heavy-duty applications. Hydraulic grippers are often employed in tasks involving large and weighty objects, such as in construction equipment and material handling at ports.

Electric Grippers: Electric grippers use motors and gears to provide precise and controlled gripping actions. These grippers offer programmability and the ability to adjust gripping force, making them versatile for various tasks. Electric grippers are frequently found in applications that require accurate positioning and force control, such as in electronics assembly and quality control processes.

Soft Grippers: Soft grippers are a relatively new addition to the realm of end effectors. Inspired by natural organisms, these grippers are made from soft and flexible materials that can conform to the shape of the object they are grasping. Soft grippers excel in tasks where delicate interaction with objects is necessary, such as handling fragile items, picking fruits without damaging them, and performing minimally invasive medical procedures.

Robotic Hands: Robotic hands represent the epitome of mimicking human capabilities. These advanced end effectors consist of multiple fingers with intricate joints and sensors, enabling them to grasp objects with a high degree of dexterity and adapt to various shapes. Robotic hands find applications in fields like research, space exploration, and industries where precise manipulation of objects is crucial.

Specialized End Effectors: Beyond the primary categories, specialized end effectors have emerged to cater to unique and niche tasks. Magnetic grippers, for instance, are employed in industries involving metallic objects. They use powerful magnets to hold and transport ferrous materials. In the realm of medical robotics, micro-grippers are designed for intricate surgeries, enabling minimally invasive procedures with minimal trauma to patients. Likewise, underwater grippers are engineered to operate in aquatic environments, contributing to underwater exploration and maintenance tasks.

Impact across Industries: The impact of end effectors and grippers extends across an array of industries, bringing unprecedented levels of efficiency, precision, and safety. In manufacturing, they accelerate production lines, enabling robots to repetitively assemble products with unparalleled accuracy. In the healthcare sector, they have revolutionized surgical procedures, allowing for minimally invasive surgeries that reduce patient trauma and recovery time. In space exploration, end effectors have enabled robots to repair and maintain satellites, as well as explore distant planets and moons.Furthermore, end effectors have played a pivotal role in addressing challenges that are too hazardous or cumbersome for humans to undertake. They contribute to disaster response by enabling robots to navigate through debris, search for survivors, and handle hazardous materials. Additionally, these devices are reshaping the landscape of industries such as agriculture, where robots equipped with specialized grippers can delicately harvest fragile crops without damaging them.

Future Directions:As technology continues to evolve, end effectors and grippers are poised to become even more sophisticated and versatile. Researchers are exploring novel materials and designs that enhance gripping efficiency and enable robots to handle objects with diverse properties. Additionally, advancements in sensors and artificial intelligence are paving the way for grippers that can adapt to object shapes and materials in real-time, further expanding their capabilities.

In conclusion, end effectors and grippers have transformed robotics from stationary machines to dynamic and adaptable systems capable of intricate interactions with the world around them. These devices, available in a multitude of types, are the linchpins that empower robots to perform tasks across a wide spectrum of industries. From the rigors of manufacturing to the precision of medical procedures and the challenges of space exploration, end effectors and grippers continue to push the boundaries of what automation can achieve. As technology marches forward, these devices will undoubtedly continue to define the future of robotics, driving innovation and expanding the possibilities of human-robot collaboration to new frontiers.

Pick-and-Place Applications: In the landscape of industrial automation, few developments have been as transformative as pick-and-place applications. At the heart of manufacturing and assembly processes, pick-and-place systems have revolutionized the way products are made, packaged, and even distributed. These systems, often driven by robotic arms equipped with advanced end effectors, excel at precisely selecting objects from one location and accurately placing them in another. The realm of pick-and-place applications is vast, spanning industries such as electronics, food and beverage, pharmaceuticals, logistics, and more. In this comprehensive exploration, we delve into the intricacies of pick-and-place systems, their diverse applications, and the significant role they play in shaping modern industrial automation.

The Mechanics of Pick-and-Place: At its core, a pick-and-place system is designed to replicate the fundamental human action of picking up an object and relocating it. However, these automated systems take this basic action to an entirely new level of efficiency and precision. The key components of a pick-and-place system typically include a robotic arm, an end effector or gripper, sensors, and a control system. The process begins with sensors detecting the presence and position of objects in the source location. The control system then calculates the optimal path and necessary force for the robotic arm to pick up the object, taking into account factors like object weight, shape, and orientation. Once picked up, the arm moves the object to the desired destination, where it is accurately placed.

Applications in Manufacturing: Pick-and-place applications have become the backbone of modern manufacturing processes, significantly enhancing efficiency, accuracy, and throughput. In electronics manufacturing, components as small as microchips are picked up and precisely placed onto circuit boards, enabling the creation of intricate devices like smartphones and computers. The pharmaceutical industry benefits from pick-and-place systems for assembling medical devices and packaging pharmaceutical products with the utmost precision. Automotive assembly lines rely on these systems to seamlessly integrate parts, ensuring that each vehicle is built to exact specifications. The cosmetic industry employs pick-and-place systems for the precise arrangement of packaging components, ensuring a consistent and aesthetically pleasing final product.

Food and Beverage Industry: The food and beverage industry is another domain where pick-and-place systems shine. These systems are adept at handling delicate food items, such as baked goods and confectionery, ensuring that the products remain intact and visually appealing. Pick-and-place applications in this industry extend to packaging products into boxes, trays, and containers, ready for distribution. The accuracy of these systems minimizes waste and optimizes packaging, contributing to both cost savings and improved customer satisfaction.

Logistics and Warehousing: In the realm of logistics and warehousing, pick-and-place systems play a pivotal role in streamlining operations. These systems efficiently sort and arrange items for storage or distribution, reducing human labor and potential errors. In e-commerce fulfillment centers, pick-and-place robots have revolutionized the picking process, enabling rapid and accurate retrieval of items from shelves to fulfill customer orders. The use of automation in these settings has dramatically increased order processing speeds and reduced fulfillment times.

Advantages and Impact: The advantages of pick-and-place applications are multifaceted and extend beyond increased speed and accuracy. By automating repetitive and labor-intensive tasks, these systems free human workers from mundane activities, allowing them to focus on higher-level tasks that require creativity and problem-solving. This not only improves job satisfaction but also contributes to the overall quality of the products being manufactured. Pick-and-place systems also significantly reduce the risk of errors, which can have far-reaching consequences in industries like electronics, where a single misplacement could result in a malfunctioning device.

Furthermore, the integration of pick-and-place systems enhances the overall efficiency of manufacturing processes. These systems can operate around the clock without fatigue, leading to continuous production cycles and shorter lead times. The consistency and precision they bring to manufacturing ensure that each product adheres to strict quality standards, reducing the likelihood of defects and recalls.

Challenges and Future Directions: While pick-and-place applications offer a multitude of benefits, they are not without challenges. The diversity of objects being handled, varying in size, shape, weight, and fragility, demands adaptable and intelligent gripping mechanisms. Innovations in soft grippers and advanced sensor technologies aim to address these challenges, allowing pick-and-place systems to handle a wider range of objects with greater flexibility.

As we look to the future, the evolution of pick-and-place applications is poised to continue. Advancements in machine learning and artificial intelligence will enable systems to recognize and handle objects with increasing complexity and variability. Collaborative robots, or cobots, which can work alongside human operators, will become more common, further blurring the line between manual and automated processes.

In conclusion, pick-and-place applications represent a cornerstone of industrial automation, shaping the efficiency and productivity of diverse industries. These systems, driven by robotic arms equipped with sophisticated end effectors, have revolutionized manufacturing, packaging, and distribution processes. The precision, speed, and accuracy they bring to these operations have not only increased output but also improved product quality and worker satisfaction. As technology continues to advance, pick-and-place systems will undoubtedly evolve to meet the challenges of an ever-changing manufacturing landscape, continuing to drive innovation and redefine the boundaries of automation.

Inverse Dynamics and Control: In the captivating realm of robotics, where machines emulate human movements and actions, the concepts of inverse dynamics and control stand as the linchpins that empower these mechanical wonders to perform complex tasks with astounding precision. Inverse dynamics is the technique that underlies the calculation of joint torques or forces required to achieve a desired motion, while control strategies dictate how these torques or forces are applied to guide the robotic system along a desired trajectory. This dynamic duo forms the backbone of robotics engineering, enabling machines to mimic human actions, respond to changing environments, and execute tasks with a finesse that was once relegated to the realm of science fiction. In this in-depth exploration, we dive into the intricacies of inverse dynamics and control, uncovering their significance, methodologies, and the profound impact they have on reshaping industries and advancing automation technology.

The Essence of Inverse Dynamics: At the core of inverse dynamics lies the quest to understand and replicate the underlying mechanics of motion. Just as humans instinctively adjust the forces applied to their muscles to move their limbs in a coordinated manner, robotic systems seek to achieve similar coordination. Inverse dynamics involves determining the forces or torques that need to be applied to the joints of a robotic system to achieve a specific trajectory or motion. This process requires a deep understanding of the robot's kinematics (geometry and motion) and dynamics (forces and torques) equations. By solving these equations in reverse, engineers can calculate the joint torques necessary for a robot to achieve a desired movement.

Control Strategies in Robotics: While inverse dynamics forms the theoretical foundation, control strategies provide the practical means to translate theoretical calculations into real-world actions. Control in robotics involves designing algorithms that regulate the system's behavior to achieve desired outcomes. These algorithms continuously adjust the joint torques or forces applied to the robot's actuators, ensuring that the robot follows the intended trajectory and maintains stability. Control strategies range from simple open-loop control, where predefined torques are applied, to complex closed-loop control, where feedback from sensors is used to adaptively adjust torques based on the robot's actual state and external conditions.

Applications and Significance: Inverse dynamics and control are pervasive across numerous industries, revolutionizing manufacturing, healthcare, space exploration, and more. In manufacturing, robots equipped with precise control systems execute intricate assembly tasks with impeccable accuracy. These systems navigate complex trajectories, assembling components with micron-level tolerances. In healthcare, robotic-assisted surgeries utilize inverse dynamics and control to enhance the precision of surgical procedures, allowing surgeons to perform delicate operations with minimal invasiveness and reduced patient trauma.

The field of space exploration benefits greatly from these principles. Robotic arms on spacecraft and rovers execute intricate movements on distant planets and moons, collecting samples and performing experiments with astonishing accuracy. In hazardous environments, such as nuclear power plants or disaster-stricken areas, robots controlled through inverse dynamics offer safer alternatives for inspection, maintenance, and recovery tasks.

Methodologies and Challenges: Inverse dynamics and control methodologies vary based on the complexity of the robotic system and the desired outcome. For simple robotic manipulators, analytical methods based on Newton-Euler equations are commonly employed to calculate joint torques. However, more complex systems, such as humanoid robots, may require numerical optimization techniques or machine learning algorithms to solve the inverse dynamics problem efficiently. The challenge lies not only in formulating accurate equations but also in managing uncertainties arising from sensor noise, external disturbances, and complex environments.Robotic control strategies are equally diverse, ranging from classical proportional-integral-derivative (PID) control to advanced model-based and adaptive control approaches. Model-based control leverages mathematical models of the robot's dynamics to compute control inputs. Adaptive control strategies continually adjust control parameters based on real-time measurements, adapting to changing conditions and uncertainties. Reinforcement learning and neural networks are also making their mark in control strategies, enabling robots to learn and optimize their actions through experience.

Future Directions: As technology continues to advance, the world of inverse dynamics and control is on the cusp of transformative changes. The integration of artificial intelligence and machine learning into control strategies holds the promise of enabling robots to adapt more effectively to dynamic and uncertain environments. Reinforcement learning algorithms allow robots to learn from trial and error, refining their control strategies over time.

Additionally, the development of soft robotics and compliant actuators is reshaping the landscape of robotics control. Soft robots, with their inherent flexibility and adaptability, require novel control strategies that can manage their complex deformation and movement. Research in bio-inspired control is paving the way for robots that can seamlessly interact with their environment, making them well-suited for tasks such as search and rescue missions.

In conclusion, the realm of inverse dynamics and control is the bedrock upon which the precision and versatility of modern robotics are built. These concepts enable robots to replicate human-like movements, respond to changing conditions, and execute tasks with remarkable accuracy. From manufacturing to healthcare, space exploration to disaster recovery, inverse dynamics and control are at the forefront of transforming industries and pushing the boundaries of what machines can achieve. As technology advances, these principles will undoubtedly continue to drive innovation, opening new avenues for human-robot collaboration and redefining the possibilities of automation technology.

Exercise Multiple-choice Questions

1. What is the primary function of a robot's end effector?
a) Generating power for the robot's movements
b) Providing stability to the robot's structure
c) Mimicking the human hand to interact with the environment
d) Controlling the robot's communication with other systems

2. Which robotic arm configuration closely resembles the structure of the human arm?
a) Cartesian configuration
b) SCARA configuration
c) Articulated configuration
d) Polar configuration

3. A robot arm with a delta configuration is best suited for which type of task?
a) Precise pick-and-place operations
b) Painting large surfaces
c) Heavy-duty lifting
d) Minimally invasive surgery

4. What distinguishes a vacuum gripper from other types of grippers?
a) It uses hydraulic fluid to generate gripping force.
b) It relies on magnetic attraction for gripping.
c) It utilizes suction to hold objects in place.
d) It employs a soft and deformable material for gripping.

5. In the context of robotics, what does "inverse dynamics" refer to?
a) Determining the forces or torques needed for a desired motion
b) Calculating the joint angles for a specific robot configuration
c) Analyzing the external forces acting on a robot
d) Estimating the weight of the robot's end effector

Review Questions:

1. What is the primary function of a robotic end effector, and how does it contribute to a robot's overall capabilities?
2. Compare and contrast the articulated and Cartesian configurations of robotic arms. Provide examples of industries where each configuration is advantageous.
3. How does the SCARA configuration differ from other robotic arm setups, and what types of tasks is it well-suited for?
4. Describe the delta configuration of robotic arms. What are its main characteristics, and in which applications does it excel?
5. Explain the working principle of a vacuum gripper. In what industries or scenarios is a vacuum gripper most commonly used?
6. What are soft grippers in robotics, and what advantages do they bring to tasks that require delicate object manipulation?
7. Define inverse dynamics and its significance in robotics. How does it relate to forward dynamics, and why is it essential for robotic control?
8. How do control strategies enhance the functionality of robotic systems? Provide examples of different types of control strategies and their applications.

9. Discuss the applications of pick-and-place systems in the manufacturing industry. How do these systems optimize production processes?
10. What challenges might arise when calculating inverse dynamics for complex robotic systems, and how can these challenges be addressed?
11. Explain the concept of closed-loop control in robotics. How does closed-loop control differ from open-loop control, and when is it preferred?
12. In what ways do robotic arms equipped with specialized end effectors impact the healthcare industry? Provide examples of medical applications.
13. How do robotic pick-and-place systems contribute to warehouse and logistics operations? What advantages do they offer in this context?
14. Describe the potential applications of robotic arms with snake or continuum configurations. Where are these systems particularly useful?
15. What are the main advantages of using robotic arms with parallel configurations? Provide examples of industries where these arms are commonly employed.
16. How does the integration of artificial intelligence impact the future of robotic control strategies? What benefits does AI bring to robotic systems?
17. Discuss the concept of collaborative robots (cobots). How do they differ from traditional robots, and what industries benefit from their use?
18. How does the choice of end effector impact a robot's ability to handle different types of objects and perform specific tasks?
19. Explain the concept of force feedback in robotic systems. How does force feedback enhance the robot's interaction with its environment?
20. In what ways are robotic arms with polar configurations used in industries beyond manufacturing? Provide examples of non-industrial applications.

Answers Multiple-choice Questions

Answers:
1. c)
2. c)
3. a)
4. c)
5. a)

Review Questions:

1. Robotic end effectors are devices attached to the tip of robotic arms that enable the robot to interact with its environment. They serve as the robot's "hands" and are crucial for performing tasks such as picking up objects, manipulating them, and placing them in desired locations. The design of the end effector depends on the specific application and the types of objects to be handled. For example, a vacuum gripper uses suction to hold objects, making it suitable for picking up flat and smooth surfaces like sheets of paper. On the other hand, a mechanical gripper with fingers can be used to grasp a wide range of objects, from cylindrical items to irregularly shaped components. Overall, the end effector determines the robot's versatility, adaptability, and efficiency in various tasks.
2. The articulated and Cartesian configurations are two distinct setups for robotic arms, each with its own advantages and applications. The articulated configuration, resembling the human arm with a series of interconnected joints, provides exceptional flexibility and dexterity. It is commonly used in industries requiring complex tasks, such as surgery and painting, where precision and a wide range of motion are crucial. In contrast, the Cartesian configuration employs linear joints along orthogonal axes, making it suitable for tasks that demand precise positioning within a three-dimensional workspace. Industries like manufacturing and 3D printing benefit from the Cartesian configuration's ability to follow precise paths while maintaining stability.
3. The SCARA configuration, which stands for Selective Compliance Articulated Robot Arm, features two parallel rotary joints and a linear vertical joint. This design allows the robot arm to move within a horizontal plane while maintaining stiffness in the vertical direction. This stiffness ensures accurate positioning and high repeatability, making the SCARA configuration ideal for applications that require quick and precise lateral movements. Industries such as electronics assembly and small-scale packaging benefit from SCARA robots, as they can swiftly manipulate components in tight spaces without sacrificing precision.
4. The delta configuration employs three or more arms connected to a common base, with the end-effectors attached to the arms' tips. The arms work in parallel, allowing the robot to achieve fast and precise movement within a specific workspace. Delta robots are particularly well-suited for high-speed pick-and-place operations, such as packaging, sorting, and food processing. Their rapid movement and efficient coordination make them valuable in

industries where tasks need to be completed with speed and accuracy, minimizing production cycle times and optimizing efficiency.

5. A vacuum gripper uses suction to secure objects by creating a pressure difference between the gripper's surface and the object being held. The gripper's surface is equipped with a vacuum source that generates negative pressure, causing the object to be drawn and held against the gripper. Vacuum grippers are particularly useful for handling objects with flat surfaces, such as cardboard boxes, sheets of paper, and electronic components. They are extensively used in packaging, logistics, and industries where gentle and precise handling is essential to prevent damage to delicate items during manipulation.

6. Soft grippers are a novel type of end effector that utilizes flexible and deformable materials to achieve grasping and manipulation tasks. Unlike traditional rigid grippers, soft grippers can adapt to the shape of the object they are handling. This adaptability is especially advantageous for tasks requiring delicate interaction with objects, such as picking and placing fragile items or gripping objects with irregular shapes. Soft grippers find applications in various fields, including medical assistance, where gentle interactions with patients are necessary, and in environments where objects have varying shapes and sizes that require a flexible grip for safe and efficient manipulation.

7. Inverse dynamics is a fundamental concept in robotics that involves calculating the forces or torques required at each joint of a robotic system to achieve a desired motion or trajectory. This concept is essential for robotic control, as it allows engineers to determine the control inputs necessary to guide the robot along specific paths while considering factors like the robot's dynamics, kinematics, and the desired motion profile. Inverse dynamics plays a critical role in tasks ranging from manufacturing to space exploration. For instance, in manufacturing, inverse dynamics calculations enable robots to execute precise and controlled movements during assembly processes, ensuring components are placed accurately and securely. In space exploration, inverse dynamics guides robotic arms on spacecraft and rovers to manipulate objects and collect samples on distant planets with high accuracy.

8. Control strategies in robotics refer to the methodologies and algorithms used to regulate the behavior and movements of robotic systems. These strategies ensure that the robot follows desired trajectories, maintains stability, and adapts to changes in its environment. A common classification of control strategies includes open-loop control and closed-loop control. Open-loop control involves predefined control inputs based on the desired motion profile, but it lacks the ability to respond to external disturbances or changes in the environment. Closed-loop control, on the other hand, uses feedback from sensors to continuously adjust control inputs, enabling the robot to adapt to uncertainties and disturbances in real-time. Advanced control strategies, such as model-based control and adaptive control, take advantage of mathematical models and sensor feedback to enhance the robot's performance and stability.

9. Pick-and-place applications have revolutionized the manufacturing industry by optimizing production processes, improving efficiency, and enhancing overall quality. These applications involve robots picking up objects from one location and accurately placing them in another. In the electronics manufacturing sector, pick-and-place robots assemble intricate electronic components onto circuit boards with high precision, contributing to the production of devices like smartphones and computers. The pharmaceutical industry benefits from pick-and-place systems for assembling medical devices and packaging pharmaceutical products with utmost accuracy. In the automotive sector, pick-and-place robots play a crucial role in assembling various vehicle components, ensuring that each vehicle is built to exact specifications. Overall, pick-and-place applications accelerate production cycles, reduce human error, and contribute to the advancement of industries by streamlining manufacturing processes.

10. Calculating inverse dynamics for complex robotic systems presents various challenges that stem from the complexity of the system's dynamics, the number of joints, and the nonlinearities involved. One challenge is the need for accurate and detailed information about the robot's physical properties, including mass distribution, inertia, and friction coefficients. Inaccurate parameter values can lead to inaccurate torque calculations, affecting the robot's performance. Another challenge is the potential for singularities, where the robot's kinematic chain becomes kinematically or dynamically ill-conditioned, making it difficult to calculate torques accurately. Additionally, external disturbances, sensor noise, and uncertainties in the environment can further complicate inverse dynamics calculations. To address these challenges, researchers are exploring machine learning techniques and numerical optimization algorithms to improve the accuracy and efficiency of inverse dynamics calculations for complex robotic systems.

11. Closed-loop control is a control strategy used in robotics that involves continuously adjusting control inputs based on feedback from sensors. This feedback allows the robot to adapt its behavior to changes in the environment, disturbances, and uncertainties. Closed-loop control differs from open-loop control, where control inputs are preplanned and executed without considering real-time feedback. Closed-loop control strategies improve the robot's performance by enabling it to respond to unexpected events and disturbances. For example, in a robotic pick-and-place application, closed-loop control allows the robot to adjust its gripper's force and position based on feedback

from force sensors, ensuring that delicate objects are handled gently and accurately placed. Closed-loop control contributes to the robot's robustness, accuracy, and adaptability, making it a valuable tool in applications where precision and reliability are paramount.

12. The healthcare industry has seen significant advancements due to the integration of robotic arms with specialized end effectors. Robotic-assisted surgeries, for instance, benefit from the precision and stability provided by these systems. Surgeons can perform minimally invasive procedures using robotic arms equipped with instruments and cameras, allowing them to operate with enhanced dexterity and precision. Robotic end effectors also play a vital role in rehabilitation and physical therapy. Exoskeletons and robotic limbs with force feedback mechanisms can assist patients in regaining motor functions after injuries or surgeries. Additionally, robotic arms with haptic feedback systems enable surgeons to feel and manipulate tissues remotely during teleoperated procedures, further enhancing surgical capabilities. The combination of robotic arms and specialized end effectors revolutionizes the medical field by enabling minimally invasive procedures, improving patient outcomes, and expanding the possibilities of surgical techniques.

13. Robotic pick-and-place systems have revolutionized warehouse and logistics operations by increasing efficiency, reducing labor costs, and improving order fulfillment speed. These systems use advanced algorithms to identify and pick items from shelves, conveyors, or bins and place them accurately into designated containers or shipping boxes. In e-commerce fulfillment centers, for instance, robotic pick-and-place systems work alongside human operators to process orders quickly and accurately. Robots equipped with cameras and sensors identify items, while control algorithms optimize picking paths and ensure items are packed securely. By automating the picking process, these systems minimize errors, reduce human fatigue, and significantly accelerate the order fulfillment process. This is particularly valuable in the modern age of online shopping, where rapid and accurate order processing is essential to meet customer expectations.

14. Robotic arms with snake or continuum configurations are designed to mimic the flexibility and movement patterns of natural organisms, such as snakes or elephants' trunks. These systems consist of multiple interconnected segments or modules that can bend, twist, and move in a highly flexible manner. Snake-like robotic arms excel in navigating complex and confined spaces where traditional rigid-arm robots might struggle. Applications range from search and rescue missions in disaster-stricken areas to inspecting tight and hazardous environments, such as pipelines and nuclear facilities. In medical applications, continuum robots can navigate through tortuous pathways in the body for minimally invasive surgeries. Their ability to adapt to various shapes and environments makes them invaluable for tasks that require intricate and flexible manipulation.

15. Robotic arms with parallel configurations are known for their exceptional stiffness, stability, and precision. These arms consist of multiple parallel branches, each with its set of joints and actuators. The parallel design provides enhanced rigidity, making these arms ideal for applications requiring high accuracy and force control. Industries such as aerospace engineering benefit from parallel robotic arms for tasks like drilling holes or assembling components that demand precision in high-stress environments. These arms are also employed in medical procedures, where they offer precise control for tasks such as bone drilling during orthopedic surgeries. Parallel configurations provide a unique combination of accuracy, stability, and robustness that caters to industries that require meticulous control and manipulation capabilities.

16. The integration of artificial intelligence (AI) into robotic control strategies is poised to bring transformative changes to the field of robotics. AI-enabled control strategies leverage machine learning algorithms to adapt and optimize the robot's behavior based on experience and real-time data. Reinforcement learning, a subset of machine learning, allows robots to learn from trial and error, optimizing their actions through repeated interactions with their environment. This enables robots to learn complex tasks and adapt to changes or uncertainties without relying solely on predefined control algorithms. Neural networks can provide robots with the ability to recognize objects, patterns, and even human gestures, enhancing their interaction capabilities. Furthermore, AI-driven control strategies enable robots to handle non-linear and complex dynamics, expanding their applications beyond traditional tasks. As AI technology continues to advance, robotic control strategies will become more adaptive, intelligent, and capable of tackling a broader range of tasks and challenges.

17. Collaborative robots, often referred to as cobots, are designed to work alongside human operators in a shared workspace. Unlike traditional industrial robots that are isolated in cages, cobots are equipped with advanced sensors, safety features, and control algorithms that enable them to operate safely in proximity to humans. Cobots offer several advantages, including enhanced human-robot interaction, increased productivity, and improved ergonomics. Industries such as manufacturing, where cobots can assist human operators in tasks like assembling components or lifting heavy objects, benefit from their collaborative nature. Cobots also find applications in healthcare, where they can work closely with healthcare professionals in tasks such as patient care and rehabilitation exercises. The ability

to work collaboratively with humans makes cobots a valuable tool for enhancing efficiency, safety, and flexibility in various industries.

18. The choice of end effector plays a crucial role in a robot's ability to handle different types of objects and perform specific tasks effectively. The end effector is the component of the robotic system that interacts directly with the objects being manipulated. It determines how the robot grasps, holds, and moves objects. For example, a vacuum gripper is suitable for flat and smooth objects, while a soft gripper can adapt to irregular shapes. Mechanical grippers with fingers are versatile and can grasp various objects, but they might not be suitable for delicate items. The choice of end effector is influenced by factors such as the object's size, shape, weight, material, and fragility, as well as the specific task requirements. Different end effectors can be swapped out based on the task at hand, allowing robots to perform a wide range of applications.

19. Force feedback in robotic systems refers to the ability of the robot to sense and respond to external forces applied to it or encountered during manipulation tasks. Force feedback enhances the robot's interaction with its environment by providing information about the forces and torques it experiences. This feedback allows the robot to adjust its grip, pressure, or movement based on the resistance encountered, preventing damage to objects or ensuring safe interactions with humans. In surgical applications, force feedback enables surgeons to feel the pressure they apply during tele operated procedures, providing a sense of touch and enhancing precision. In industries like manufacturing and assembly, force feedback assists robots in handling fragile objects with care and maintaining consistent force levels. Overall, force feedback improves the robot's perception, adaptability, and ability to perform tasks safely and accurately.

20. Robotic arms with polar configurations, also known as spherical robots, have joints arranged in a spherical geometry. This design allows the end effector to move in any direction within a three-dimensional space, resembling the movement of a human wrist. Beyond manufacturing, polar configurations find applications in various industries. In astronomy, robotic telescopes equipped with polar configurations can observe celestial objects at different angles and orientations. These telescopes can track stars, planets, and galaxies as they move across the sky. In research and development, polar robots can be used to simulate complex motions for testing and validation purposes.

Online Resources

Online Exam Papers

Link:https://forms.gle/CAWPwSW2prdLKaLP7

QR Code:

YouTube Video Playlist Link

Link:https://www.youtube.com/playlist?list=PLzlh_31VWEFofjdKDq7AHhiixR2I7EOGP

QR Code:

Chapter 8: Robot Programming and Software

In the intricate symphony of robotics, where mechanical precision meets the ethereal realm of artificial intelligence, lies the domain of robot programming and software. Here, lines of code transform inert metal and circuits into sentient agents capable of understanding, learning, and executing tasks with unparalleled finesse. The fusion of programming and software engineering infuses life into the mechanical bodies of robots, enabling them to navigate complex environments, interact with humans, and perform intricate operations across industries. From factory floors to hospitals, from outer space to our homes, the realm of robot programming shapes a new reality where machines become collaborators, assistants, and extensions of our capabilities. As we embark on this journey through the realms of algorithms, control systems, and sensor integration, we unravel the tapestry of creating intelligence in metal and silicone, crafting a future where the synergy between human ingenuity and computational prowess reshapes the way we perceive and interact with technology.

In the heart of the technological revolution, where the boundaries between the virtual and the physical blur, the realm of robot programming and software emerges as a pivotal force. It is here that the intangible realm of code converges with the tangible world of machinery, orchestrating the intricate dance of robotic intelligence. This dynamic interplay transforms lifeless mechanical structures into sentient entities, poised to navigate complex environments, comprehend human interactions, and execute tasks that were once the exclusive domain of human hands. The landscape of robot programming and software is a canvas where innovation, creativity, and computational prowess converge to redefine the capabilities of machines across sectors. From manufacturing floors to healthcare settings, from the depths of the ocean to the frontiers of outer space, the art of programming breathes life into robots, making them not just tools, but intelligent companions in our quest for progress. As we delve into this domain, we uncover the tapestry of algorithms, control strategies, and sensor integration that infuse robotics with autonomy and cognition, creating a future where the synergy between human intellect and digital acumen reshapes the boundaries of what is achievable in the realm of technology.

Robot Operating Systems (ROS): In the realm of robotics, the development and deployment of robotic systems have been transformed by the emergence of Robot Operating Systems (ROS). A comprehensive framework that provides a standardized and modular approach to building robotic applications, ROS has revolutionized the way researchers, engineers, and developers create, test, and deploy robotic technologies. This article delves into the intricacies of ROS, exploring its origins, key features, benefits, and the profound impact it has had on shaping the field of robotics.

Understanding ROS: Robot Operating System, commonly referred to as ROS, is not an operating system in the traditional sense; rather, it's an open-source middleware framework that acts as a middleware layer between the hardware and software components of a robotic system. ROS was initially developed in 2007 by the Stanford Artificial Intelligence Laboratory in collaboration with Willow Garage, a robotics research lab. It was designed to address the challenges faced by robotics researchers and developers, including the lack of a standardized platform for building, simulating, and testing robotic applications.

Key Features and Components: At its core, ROS is characterized by its modularity and flexibility. It consists of a collection of software libraries, tools, and frameworks that facilitate various aspects of robotics development. The key features and components of ROS include:

Node-based Architecture: ROS applications are structured as a collection of loosely coupled nodes. Each node is a standalone executable that performs a specific task, such as sensor data processing, path planning, or control. Nodes communicate with each other through a publish-subscribe messaging system, allowing for easy integration and communication between different components.

Communication Infrastructure: ROS employs a publish-subscribe communication model using topics, which are named buses that allow nodes to send and receive messages. This facilitates seamless communication between nodes, even when they are distributed across different computers or devices. Additionally, ROS supports service calls for synchronous communication and parameter servers for dynamic configuration.

Package Management: ROS organizes code and resources into packages, which encapsulate specific functionalities or algorithms. These packages can be easily shared, reused, and distributed, promoting collaboration and rapid development. The ROS community maintains a vast repository of open-source packages, offering a wide range of tools and capabilities.

Simulators: ROS provides simulators like Gazebo that enable developers to create virtual environments for testing and validating robotic applications. Simulators allow for safe and efficient testing of algorithms and control strategies before deploying them on physical robots, reducing the risk of damaging hardware during experimentation.

Visualization Tools: ROS offers visualization tools like RViz, which provides a graphical interface for visualizing sensor data, robot models, and trajectories. This tool aids in debugging, planning, and monitoring the behavior of robotic systems.

Diverse Libraries: ROS offers libraries for various functionalities, including perception (such as image and point cloud processing), motion planning, control algorithms, and machine learning. These libraries serve as building blocks for creating complex robotic applications.

Benefits of ROS: The adoption of ROS has brought about numerous benefits to the field of robotics, contributing to its rapid growth and innovation:

Collaboration and Community: ROS is an open-source project with a thriving community of researchers, developers, and enthusiasts. This collaborative ecosystem fosters knowledge sharing, code reusability, and the rapid advancement of robotic technologies.

Modularity and Reusability: ROS's modular architecture encourages the development of reusable and interchangeable components. Developers can create specialized nodes, sensors, and algorithms that can be easily integrated into different projects, saving time and effort.

Rapid Prototyping: ROS's simulation capabilities allow developers to prototype and test algorithms in virtual environments before deploying them on physical robots. This accelerates the development cycle and reduces the risk associated with real-world testing. Hardware Agnostic: ROS is designed to be hardware agnostic, meaning it can be used with a wide range of robotic platforms and sensors. This versatility allows developers to work with diverse hardware configurations and experiment with various setups.

Flexibility and Customizability: ROS's flexibility allows developers to choose the most suitable algorithms and libraries for their applications. This customization empowers developers to create robotic systems tailored to specific tasks and requirements. Education and Research: ROS has become a valuable tool for educational institutions and researchers. It provides a platform for teaching robotics concepts, conducting experiments, and validating research findings in a simulated or real-world environment.

Impact on Robotics and Future Directions: The impact of ROS on the field of robotics has been profound. It has democratized access to advanced robotics tools and technologies, enabling researchers and developers to create sophisticated robotic systems without starting from scratch. The collaborative nature of ROS has facilitated breakthroughs in various domains, from autonomous vehicles and industrial automation to healthcare and space exploration.

Looking to the future, ROS is expected to continue shaping the trajectory of robotics innovation. As technology advances, ROS will likely integrate more seamlessly with emerging technologies such as artificial intelligence, machine learning, and advanced sensors. Additionally, ROS is likely to evolve to accommodate the challenges posed by new application domains, such as human-robot collaboration, swarm robotics, and multi-agent systems.

In conclusion, Robot Operating Systems (ROS) have become a cornerstone of robotics development, revolutionizing the way robotic applications are created, tested, and deployed. Its modular and open-source nature fosters collaboration, accelerates innovation, and empowers developers to build complex robotic systems. ROS's impact spans industries, research, education, and beyond, and its influence is expected to continue shaping the future of robotics technology and applications. As robotics continues to advance, ROS remains a vital tool in the hands of those pushing the boundaries of what robots can achieve.

Motion Planning Algorithms: In the realm of robotics, the ability to navigate autonomously in complex and dynamic environments is a cornerstone of achieving true robotic autonomy. Motion planning algorithms play a pivotal role in enabling robots to move safely and efficiently, avoiding obstacles, following trajectories, and making informed decisions while navigating the world around them. From self-driving cars and drones to industrial robots and healthcare assistants, motion planning algorithms are at the forefront of shaping the future of robotics. This comprehensive exploration delves into the intricacies of motion planning algorithms, their classifications, methodologies, challenges, and the transformative impact they have on robotics.

Understanding Motion Planning: Motion planning refers to the process of determining a sequence of valid configurations or motions that guide a robotic system from an initial state to a desired goal state while avoiding obstacles and adhering to various constraints. These constraints can include physical limitations, environmental considerations, and optimization objectives. Motion planning algorithms enable robots to navigate through complex and cluttered environments, make real-time decisions, and adapt to changing conditions.

Classification of Motion Planning Algorithms: Motion planning algorithms can be broadly classified into several categories based on their underlying methodologies and approaches:

Sampling-Based Algorithms: These algorithms construct a discrete representation of the robot's configuration space by randomly sampling feasible configurations. Popular examples include Probabilistic Roadmaps (PRM) and

Rapidly-exploring Random Trees (RRT). PRM builds a roadmap of feasible paths through a connectivity graph, while RRT focuses on rapidly expanding a tree structure towards unexplored regions.

Grid-Based Algorithms: Grid-based methods discretize the environment into a grid of cells and create a two-dimensional or three-dimensional representation of the workspace. A* (A star) is a widely used grid-based search algorithm that finds the shortest path between two points in a grid, while D* (D star) dynamically updates paths as new information becomes available.

Potential Field Methods: Potential field algorithms treat obstacles as repulsive forces and the goal as an attractive force. The robot navigates by moving in the direction of decreasing potential, avoiding obstacles and reaching the goal. However, potential field methods can lead to local minima and get stuck in situations with complex obstacle configurations.

Optimization-Based Algorithms: These algorithms formulate motion planning as an optimization problem, where the goal is to find a trajectory that minimizes a certain cost function. Model Predictive Control (MPC) is an example of an optimization-based approach, where the robot's actions are optimized over a finite time horizon while considering dynamics and constraints.

Sampling-Based Heuristic Search: These algorithms combine elements of sampling-based methods and heuristic search algorithms. They generate a set of samples and iteratively improve them using search techniques like A* or D* to find an optimal path.

Methodologies and Challenges: Motion planning algorithms employ various methodologies to generate feasible and collision-free paths for robots. Regardless of the specific approach, several challenges persist across different algorithms:

High-Dimensional Configuration Spaces: In many scenarios, robots operate in high-dimensional configuration spaces. Exploring these spaces efficiently while avoiding exponential growth in computation complexity poses a significant challenge.

Real-Time Performance: Real-time execution is crucial, especially in dynamic environments. Algorithms must generate paths quickly to allow robots to react promptly to changes and obstacles.

Dynamic Environments: Handling dynamic obstacles and changing environments requires algorithms that can adapt on-the-fly, predicting the future positions of moving objects and updating plans accordingly.

Optimality vs. Computation Time: Many algorithms balance the trade-off between finding the optimal path and the time it takes to compute that path. Some scenarios prioritize finding a feasible path quickly over finding the globally optimal solution.

Local Minima and Constraints: Motion planning algorithms can get trapped in local minima or struggle with constraints imposed by the robot's kinematics, dynamics, or the environment. Strategies to escape local minima and efficiently handle constraints are essential.

Applications and Impact: The impact of motion planning algorithms reverberates across a multitude of industries and domains:

Autonomous Vehicles: Motion planning is crucial for self-driving cars to navigate safely through complex urban environments, make lane changes, avoid collisions, and respond to traffic conditions.

Drones and Aerial Robotics: Drones rely on motion planning algorithms to navigate airspace, avoid obstacles, and plan trajectories for tasks like surveillance, package delivery, and search and rescue missions.

Industrial Automation: Manufacturing robots use motion planning to optimize paths for assembly, welding, painting, and material handling, improving efficiency and reducing production cycle times.

Healthcare and Surgery: Surgical robots utilize motion planning to execute precise movements during minimally invasive procedures, enhancing surgical accuracy and reducing patient trauma.

Space Exploration: Robots on planetary rovers and spacecraft employ motion planning to traverse challenging terrains, collect samples, and execute scientific experiments.

Service Robotics: Robots in hospitality, healthcare, and assistance tasks use motion planning to navigate indoor environments, interact with humans, and perform tasks autonomously.

Future Directions: As technology continues to evolve, motion planning algorithms are poised to undergo further advancements and innovations:

Machine Learning Integration: Integrating machine learning techniques into motion planning algorithms can enhance adaptability, enabling robots to learn from experience and handle previously unseen scenarios.

Human-Robot Collaboration: Future algorithms will focus on enabling safe and intuitive collaboration between humans and robots, allowing robots to predict human intentions and adjust their paths accordingly.

Hybrid Approaches: Hybrid algorithms that combine the strengths of different motion planning methods are likely to emerge, addressing challenges such as high-dimensional spaces and dynamic environments.

Multi-Agent Systems: Motion planning for multi-robot systems will become more prominent, requiring algorithms that ensure efficient coordination, collision avoidance, and cooperation.

Certainty and Uncertainty Handling: Algorithms that can handle uncertainty in perception and sensor data will enable robots to navigate reliably in diverse and unpredictable environments.

In conclusion, motion planning algorithms are a cornerstone of robotics, enabling robots to navigate complex environments, make informed decisions, and achieve true autonomy. Their classifications, methodologies, challenges, and transformative impact highlight their importance across various industries and domains. As robotics continues to advance, motion planning algorithms will remain a driving force behind the realization of autonomous robots capable of safely and efficiently navigating the intricate world around them.

8.3Behavior-Based Programming: In the landscape of robotics, achieving true autonomy is a formidable challenge that requires intricate control strategies capable of guiding robots through complex and dynamic environments. Behavior-based programming has emerged as a powerful paradigm that empowers robots to exhibit intelligent behavior by reacting to sensory input and making decisions in real-time. This approach harnesses the principles of reactive intelligence, allowing robots to navigate, interact, and adapt to their surroundings without relying on pre-defined plans. In this comprehensive exploration, we delve into the nuances of behavior-based programming, its foundational concepts, methodologies, applications, challenges, and its transformative impact on shaping the trajectory of robotics.

Understanding Behavior-Based Programming: Behavior-based programming, often referred to as behavior-based robotics or behavior-based control, is a programming paradigm that emphasizes the design of robotic systems based on a collection of individual behaviors. These behaviors are simple, reactive modules that respond to sensory input from the environment, such as visual or tactile data. By composing these behaviors, complex robotic actions and decisions emerge as a result of the interaction between individual modules. Unlike traditional control architectures that rely on central planning and explicit instructions, behavior-based programming leverages the principles of emergence and decentralized control to achieve intelligent and adaptive behavior.

Key Concepts and Methodologies:

Behavior Modules: At the core of behavior-based programming are behavior modules, each responsible for a specific task or reaction. These modules are designed to be reactive, meaning they respond to specific stimuli or cues from the environment. For instance, a collision avoidance behavior might trigger when the robot senses an obstacle in its path, prompting it to change direction.

Layered Architectures: Behavior-based systems often employ layered architectures, where behaviors are organized into layers that handle different levels of abstraction and urgency. Lower layers might deal with basic survival behaviors, such as obstacle avoidance, while higher layers manage more complex tasks like goal pursuit or social interaction.

Arbitration and Blending: When multiple behaviors are active simultaneously, an arbitration mechanism decides which behavior takes precedence based on predefined rules or priorities. This allows robots to smoothly transition between behaviors and handle situations where conflicting behaviors might arise.

Sensor Fusion: Behavior-based programming relies heavily on sensor input to trigger behaviors. Sensor fusion techniques are employed to combine data from various sensors, enhancing the robot's perception and decision-making capabilities.

Learning and Adaptation: Some behavior-based systems incorporate learning mechanisms that enable robots to adapt their behaviors over time based on their experiences. This adaptability enhances the system's robustness and versatility.

Applications: Behavior-based programming finds applications across a wide spectrum of industries and domains:

Mobile Robotics: Mobile robots utilize behavior-based programming for tasks such as obstacle avoidance, path following, exploration, and navigation. These systems can navigate through dynamic environments and adapt to changes in real-time.

Robotic Exploration: Behavior-based approaches are essential for planetary rovers and drones, enabling them to autonomously explore remote and hazardous environments while responding to unexpected obstacles.

Industrial Automation: In manufacturing and logistics, behavior-based systems enhance the agility of robotic arms, allowing them to perform tasks like pick-and-place, sorting, and material handling without the need for precise pre-programming.

Search and Rescue: Behavior-based programming is crucial for search and rescue robots that must navigate through disaster-stricken areas, locate survivors, and overcome obstacles while avoiding dangerous terrain.

Swarm Robotics: In swarm robotics, behavior-based control enables a group of robots to collectively perform tasks like exploration, surveillance, and environmental monitoring.

Human-Robot Interaction: Behavior-based systems enhance human-robot interaction by enabling robots to respond to gestures, vocal commands, and other cues from humans in real-time.

Challenges: While behavior-based programming offers significant advantages, it also presents several challenges:

Complex Behavior Interactions: As the number of behaviors increases, managing the interactions and dependencies between behaviors can become complex. Arbitration mechanisms must be carefully designed to ensure coherent and desirable behavior.

Emergent Behavior: While emergent behavior can lead to innovative solutions, it can also result in unpredictable outcomes. Ensuring that emergent behavior aligns with desired goals and safety constraints is a critical challenge.

Robustness and Adaptation: Ensuring that behavior-based systems remain robust in the face of uncertainties, sensor noise, and changing environments requires robust sensor fusion and adaptation mechanisms.

Fine-Tuning and Calibration: Behavior-based systems often require extensive tuning of parameters and thresholds to achieve desired performance. This tuning can be time-consuming and challenging.

Lack of High-Level Planning: Behavior-based systems excel at reactive behavior but might lack the ability to engage in high-level planning, long-term decision-making, and goal reasoning.

Impact and Future Directions: Behavior-based programming has significantly impacted the field of robotics, offering a flexible and dynamic approach to achieving intelligent behavior. As technology advances, behavior-based systems are poised to undergo further evolution:

Hybrid Approaches: Hybrid systems that combine behavior-based programming with other paradigms, such as symbolic reasoning or learning-based approaches, can offer a balanced approach that integrates reactive behavior with high-level planning.

Learning-Enhanced Behaviors: Integrating machine learning techniques can enhance the adaptability of behavior-based systems, enabling them to learn from experience and improve their performance over time.

Ethical Considerations: As behavior-based systems become more autonomous, addressing ethical concerns such as safety, decision transparency, and accountability will become increasingly important.

Cognitive Architectures: The integration of cognitive architectures with behavior-based programming can enable robots to engage in more sophisticated reasoning, memory, and planning.

Neuromorphic Computing: Advances in neuromorphic computing might offer new ways to implement behavior-based systems that mimic biological neural networks, potentially leading to more biologically-inspired intelligent behavior.

Behavior-based programming stands as a cornerstone of robotics, offering an approach that emphasizes reactive intelligence and adaptability. By leveraging the principles of emergence and decentralized control, behavior-based systems empower robots to navigate complex environments, interact with their surroundings, and make decisions in real-time. Their applications span industries and domains, from mobile robotics and exploration to manufacturing and human-robot interaction. As technology advances, behavior-based programming is poised to evolve further, combining with other paradigms and incorporating learning mechanisms to unlock new levels of autonomy and intelligence in robotic systems.

Simulation and Virtual Environments: The journey to developing advanced robotic systems involves a plethora of challenges, ranging from hardware design and control algorithms to motion planning and real-world testing. Simulation and virtual environments have emerged as indispensable tools that bridge the gap between theoretical concepts and real-world applications. These technologies allow researchers, engineers, and developers to prototype, test, and refine robotic systems in a controlled, cost-effective, and safe digital realm before transitioning to physical implementations. This comprehensive exploration delves into the intricacies of simulation and virtual environments in the realm of robotics, exploring their significance, methodologies, applications, challenges, and their transformative impact on shaping the trajectory of robotic innovation.

Understanding Simulation and Virtual Environments: Simulation and virtual environments refer to computer-based systems that replicate real-world scenarios and interactions in a digital space. In the context of robotics, these technologies enable the creation of virtual models of robots, environments, sensors, and interactions. This virtual realm serves as a platform for designing, testing, and validating robotic systems without the need for physical hardware. Simulation can range from basic physics-based models to highly sophisticated representations that emulate sensor data, complex physics, and dynamic interactions.

Key Components and Methodologies:

Robot Models: Virtual environments require accurate representations of robotic systems, encompassing mechanical components, sensors, actuators, and control algorithms. These models serve as the foundation for simulating robot behavior.

Environment Models: Virtual environments recreate the physical spaces in which robots operate, ranging from indoor environments to outdoor landscapes. These models incorporate features like terrain, obstacles, lighting, and environmental dynamics.
Physics Simulation: Physics engines simulate the laws of physics, enabling virtual robots to interact with their environment realistically. These engines calculate forces, collisions, friction, and other physical properties that influence the robot's behavior.
Sensor Emulation: To mimic real-world sensor data, virtual environments emulate sensor outputs such as cameras, lidars, and proximity sensors. This enables testing of perception algorithms and sensor fusion in a controlled environment.
Control and Algorithms: Virtual environments enable the deployment and testing of control algorithms, motion planning strategies, and decision-making processes. These algorithms can be fine-tuned and validated in a digital space before implementation on physical robots.
Benefits and Significance: Simulation and virtual environments offer a multitude of benefits that drive their significance in the field of robotics:
Cost Savings: Developing and testing robots in the physical world can be expensive due to the cost of hardware, maintenance, and potential risks. Simulation reduces costs by minimizing the need for physical prototypes during early development stages.
Safety: Virtual environments provide a safe platform for testing and experimenting with robots, preventing damage to hardware and minimizing the risk of accidents.
Rapid Prototyping: Virtual environments enable rapid iteration and experimentation, allowing researchers to explore various design options and algorithms without the time constraints of physical testing.
Scalability: Simulation allows researchers to test their algorithms and systems at various scales, from individual robots to swarms, without the limitations of physical resources.
Risk Mitigation: Simulating extreme scenarios, adverse weather conditions, or dangerous environments allows researchers to evaluate robot performance without subjecting them to potential risks.
Education and Training: Simulation provides an educational platform for students to learn robotics concepts, algorithms, and principles in a hands-on manner without requiring physical hardware.
Applications: The applications of simulation and virtual environments in robotics are wide-ranging and impactful:
Autonomous Vehicles: Self-driving car algorithms are rigorously tested in virtual environments before real-world deployment. Simulations help assess navigation, collision avoidance, and decision-making in diverse scenarios.
Aerial Robotics: Drone simulations allow researchers to test flight dynamics, sensor integration, and path planning algorithms, optimizing drone performance and safety.
Industrial Robotics: Manufacturing robots are simulated to optimize production lines, identify bottlenecks, and refine robotic assembly processes.
Human-Robot Interaction: Virtual environments enable testing of human-robot interaction scenarios, fine-tuning robot behavior to ensure user comfort and safety.
Swarm Robotics: Simulation facilitates research on swarm behavior, coordination, and emergent properties in large groups of robots.
Space Exploration: Robotic systems for space exploration are tested in simulated space environments, ensuring their reliability in the harsh conditions of outer space.
Challenges and Limitations: Despite their numerous advantages, simulation and virtual environments pose challenges and limitations:
Realism vs. Simplification: Balancing the need for realism with computational efficiency is a challenge. Highly realistic simulations might be computationally intensive, affecting real-time performance.
Sensor Accuracy: Simulated sensor data might not perfectly reflect the uncertainties and noise present in real-world sensors. Ensuring accurate sensor emulation is crucial for meaningful results.
Dynamic Environments: Simulating dynamic environments with moving obstacles and changing conditions can be complex and computationally demanding.
Validation and Transferability: Ensuring that results obtained in simulation translate to real-world performance requires thorough validation and calibration.
Uncertainty Modeling: Capturing uncertainties arising from sensor noise, hardware imperfections, and unpredictable interactions remains a challenge in simulation.
Advancements and Future Directions: As technology evolves, simulation and virtual environments are poised to undergo further advancements, shaping the way robots are designed, tested, and integrated into various domains. The future holds a plethora of possibilities that can revolutionize the capabilities and impact of simulation technologies:

High-Fidelity Models: The continuous advancement of graphics and physics engines is expected to lead to simulations that closely resemble the real world. As computing power increases, simulations will be able to emulate complex interactions, environmental dynamics, and intricate robot behaviors with unparalleled realism.

Hardware-in-the-Loop Integration: The concept of hardware-in-the-loop (HIL) integration is gaining traction. This approach involves connecting physical robots or hardware components to the simulation environment, allowing for real-time testing of algorithms and control strategies. HIL integration bridges the gap between simulation and reality, enabling researchers to validate their algorithms on actual hardware while leveraging the safety and versatility of the virtual environment.

Data-Driven Simulation: Machine learning techniques are anticipated to play a pivotal role in enhancing the accuracy and realism of simulations. Data-driven simulation involves training machine learning models on real-world sensor data to improve the emulation of sensors in the virtual environment. This approach can lead to more accurate simulations that account for sensor noise, uncertainties, and complex interactions.

Multi-Modal Simulation: The integration of various simulation modalities is an exciting direction for the field. By combining physics-based simulations with machine learning models, simulations can provide a more holistic representation of robot-environment interactions. This multi-modal approach can lead to simulations that capture a wider range of scenarios and behaviors.

Cloud-Based Simulation: Cloud computing offers a scalable solution for simulation, enabling researchers and developers to access high-performance computing resources for running complex simulations. Cloud-based simulation platforms can provide the computational power required to simulate intricate scenarios and optimize algorithms, democratizing access to advanced simulation capabilities.

Real-Time Scenario Testing: Virtual environments can be used to recreate specific scenarios and challenges that robots might encounter in real-world applications. This enables researchers to rigorously test algorithms and control strategies under diverse conditions, preparing robots for a wide range of situations they might encounter.

Learning and Reinforcement: Simulation can serve as a platform for training robots using reinforcement learning techniques. Robots can learn complex tasks and behaviors in the virtual environment before being deployed in the real world. This approach accelerates the learning process and reduces the risk of damage to physical hardware during training.

Collaborative Simulation: Collaboration in virtual environments is another exciting avenue. Multiple researchers or teams can interact and collaborate within the same simulated environment, enabling cooperative testing, evaluation, and development of robotic systems.

Simulation and virtual environments have emerged as essential pillars of robotics research, development, and innovation. Their ability to replicate real-world scenarios, enable safe testing, and expedite the design process has revolutionized the way robots are conceptualized, developed, and integrated into various domains. As technology continues to advance, simulation technologies are poised to evolve further, offering increased realism, integration with physical hardware, and the incorporation of data-driven and machine learning approaches. The future of simulation holds the promise of unlocking new avenues for exploration, experimentation, and optimization in robotics, ultimately contributing to the creation of more capable, adaptable, and intelligent robotic systems. In the quest to push the boundaries of robotics, simulation and virtual environments stand as indispensable tools, propelling the field toward a future of unprecedented possibilities.

Exercise

Multiple-choice Questions

1. Which programming paradigm emphasizes the design of robotic systems based on a collection of simple, reactive modules that respond to sensory input from the environment?

a) Procedural programming
b) Object-oriented programming
c) Behavior-based programming
d) Functional programming

2. What are the key components of behavior-based programming?

a) Loops and conditional statements
b) Control flow and decision trees
c) Behavior modules, environment models, physics simulation
d) Data structures and algorithms

3. Which architecture organizes behavior-based systems into layers that handle different levels of abstraction and urgency?

a) Centralized architecture
b) Hierarchical architecture

c) Distributed architecture
d) Decentralized architecture
4. What is the primary benefit of behavior-based programming in robotics?
a) Minimizing the need for sensors
b) Reducing computational complexity
c) Enabling high-level planning
d) Reacting to real-time sensor input
5. What role do physics engines play in simulation and virtual environments?
a) Emulating sensor data
b) Simulating human interactions
c) Replicating physical laws and interactions
d) Generating random scenarios

Review Questions:

1. What is behavior-based programming in robotics, and how does it differ from traditional control architectures?
2. Explain the concept of behavior modules in behavior-based programming. Provide an example of a behavior module and its triggering condition.
3. How behavior-based systems are organized using layered architectures? Describe the role of different layers in controlling robot behavior.
4. What is the significance of arbitration mechanisms in behavior-based programming? How do they resolve conflicts between multiple active behaviors?
5. Describe the role of simulation in robot programming. What benefits does simulation offer in terms of cost, safety, and development speed?
6. How do physics engines contribute to the realism of simulations? Provide examples of physical interactions that can be simulated using physics engines.
7. Explain the concept of sensor emulation in virtual environments. Why is accurate sensor emulation crucial for meaningful simulation results?
8. What is hardware-in-the-loop (HIL) integration? How does HIL integration benefit the testing and development of robotic systems?
9. In the context of simulation, what is data-driven simulation? How can machine learning techniques enhance the accuracy of simulated sensor data?
10. How does multi-modal simulation enhance the capabilities of virtual environments? Provide examples of simulation modalities that can be integrated.
11. Describe the applications of simulation and virtual environments in the field of autonomous vehicles. How do simulations contribute to the development of self-driving car algorithms?
12. Discuss the concept of cloud-based simulation. How does cloud computing enhance the scalability and accessibility of simulation resources?
13. What challenges are associated with achieving a balance between realism and computational efficiency in simulations?
14. Explain the concept of uncertainty modeling in simulation. How can simulations account for uncertainties arising from sensor noise and dynamic environments?
15. How can simulation be used to recreate specific scenarios and challenges that robots might encounter in real-world applications?
16. What is the role of simulation in robotics education and training? How can students benefit from hands-on learning in virtual environments?
17. Discuss the potential limitations of simulation and virtual environments in accurately representing real-world interactions and dynamics.
18. How does behavior-based programming enhance a robot's ability to react to real-time sensor input? Provide an example of a reactive behavior in a robotic context.
19. What are the advantages of using behavior-based programming for swarm robotics applications? How can behavior modules be coordinated in swarm behavior?
20. In which robotic domains does behavior-based programming excel, and why? How does behavior-based programming adapt to various application requirements?

Answers Multiple-choice Questions

Answers:
1. c)
2. c)

3. b)
4. d)
5. c)

Review Questions:

1. Behavior-based programming in robotics is a paradigm that focuses on designing robotic systems by breaking down complex behaviors into simple, reactive modules. Unlike traditional control architectures that rely on centralized planning, behavior-based programming emphasizes real-time reactions to sensory input. This approach allows robots to exhibit intelligent behavior by combining these reactive modules. In contrast, traditional architectures often involve complex algorithms and pre-determined plans, making them less adaptable to dynamic environments. Behavior-based programming prioritizes responsiveness and adaptability, enabling robots to interact effectively with their surroundings and react to unforeseen situations.
2. Behavior modules are the building blocks of behavior-based programming. Each module represents a specific behavior or action that the robot can perform in response to sensory input. For instance, a collision avoidance behavior module might trigger when the robot senses an obstacle in its path. This module could instruct the robot to change its direction to avoid the obstacle. The triggering condition for the collision avoidance behavior could be based on sensor data indicating the proximity of the obstacle. Behavior modules are designed to be simple and reactive, ensuring that the robot can swiftly respond to changes in its environment.
3. Layered architectures are a common organizational structure in behavior-based programming. They arrange behavior modules into layers that handle different levels of abstraction and urgency. Lower layers often deal with fundamental survival behaviors, such as obstacle avoidance and basic navigation. As you move up the layers, behaviors become more complex and goal-oriented, such as path planning and task execution. This hierarchy enables robots to prioritize immediate safety and survival while also pursuing higher-level objectives. The layered approach allows for efficient arbitration between behaviors, ensuring that the most relevant behavior takes precedence in a given situation.
4. The primary benefit of behavior-based programming in robotics is its emphasis on real-time responsiveness and adaptability. Unlike traditional programming approaches, where robots follow pre-defined plans, behavior-based systems can react to sensory input on the fly. This means that robots can handle unexpected situations, adapt to changes in the environment, and interact more naturally with their surroundings. Behavior-based programming excels in dynamic and unpredictable environments where robots need to make split-second decisions based on their sensor data. It also simplifies the development process by breaking down complex behaviors into manageable modules, making it easier to design, test, and modify robot behavior.
5. Simulation is a vital component of robot programming that involves creating virtual environments to test and validate robotic systems. It provides a controlled and safe platform to experiment with robot designs, algorithms, and behaviors without the need for physical prototypes. One significant benefit of simulation is cost savings. Developing and testing robots in the physical world can be expensive due to the cost of hardware and potential risks. Simulation reduces these costs by minimizing the need for physical prototypes during the early stages of development. Additionally, simulations provide a safe environment to test robots in challenging scenarios that might be risky or unfeasible in real life.
6. Physics engines are essential components of simulations that replicate the laws of physics within a virtual environment. They simulate physical interactions, such as collisions, forces, friction, and object dynamics, to ensure that virtual robots and objects behave realistically. For example, a physics engine can calculate how a virtual robot's movements affect its surroundings and how objects respond to external forces. This enables researchers to test how robots interact with their environment in a digital space before implementing the behavior on physical hardware. Physics engines enhance the realism of simulations and provide valuable insights into how robots will perform in real-world scenarios.
7. Sensor emulation in virtual environments involves replicating the output of real-world sensors within the simulation. Accurate sensor emulation is crucial for ensuring that the behaviors and algorithms tested in simulation translate effectively to the physical robot. For instance, if a robot's sensors include cameras and lidars, the simulation must generate virtual sensor data that mimics the sensor's behavior in real-world conditions. Inaccurate sensor emulation can lead to misleading results, as the robot's behavior might not perform as expected when transferred to the physical world. Therefore, high-fidelity sensor emulation is essential for achieving meaningful and reliable simulation results.
8. Hardware-in-the-loop (HIL) integration is a concept that combines physical hardware with simulation environments. This approach allows researchers to test algorithms and control strategies on actual hardware while leveraging the safety and flexibility of a virtual environment. In an HIL setup, the physical robot or components are connected to the simulation, enabling real-time interaction between the hardware and the virtual environment. This

integration offers several benefits. It allows researchers to validate algorithms on the actual hardware they will be deployed on, ensuring that the behavior performs as intended. It also provides a safer testing environment, as failures or errors in the physical hardware are contained within the simulation.

9. Data-driven simulation involves using machine learning techniques to enhance the accuracy of simulated sensor data. Instead of relying solely on physics-based models, data-driven simulations incorporate real-world sensor data to improve the fidelity of sensor emulation. Machine learning algorithms analyze real sensor data to understand how sensors behave in various situations, including noise, uncertainty, and artifacts. The simulation then generates virtual sensor data that mirrors the patterns observed in the real data. This approach bridges the gap between the synthetic environment of simulation and the complexities of real-world sensor behavior, leading to more accurate and reliable simulation results.

10. Multi-modal simulation refers to the integration of various simulation modalities to create a holistic representation of robot-environment interactions. Instead of relying solely on physics-based simulations, multi-modal simulations combine different simulation techniques to provide a more comprehensive view. For example, a simulation might incorporate physics-based models for object interactions, machine learning models for realistic sensor data, and behavioral models for robot decision-making. This approach allows researchers to capture a wider range of scenarios and behaviors, enhancing the fidelity and versatility of the simulation environment.

11. Simulation and virtual environments play a pivotal role in the development of self-driving car algorithms. Autonomous vehicles require rigorous testing in a wide range of scenarios to ensure their safety and reliability. Simulations provide a controlled environment to replicate complex traffic situations, adverse weather conditions, and potential hazards that might be challenging to reproduce in the real world. By testing self-driving algorithms in simulations, researchers can evaluate how the algorithms react to various scenarios, fine-tune parameters, and optimize decision-making strategies. Simulations significantly accelerate the development process and reduce the risks associated with real-world testing.

12. Cloud-based simulation involves leveraging cloud computing resources to perform simulations. Instead of relying solely on local hardware, researchers can access high-performance computing resources provided by cloud service providers. This approach offers several advantages, including scalability, flexibility, and accessibility. Researchers can scale up or down based on the complexity of simulations, ensuring that they have the computational power required for accurate and detailed simulations. Cloud-based simulation also democratizes access to advanced simulation capabilities, as researchers around the world can access powerful computing resources without the need for specialized hardware.

13. Achieving a balance between realism and computational efficiency is a challenge in simulations. Highly realistic simulations that replicate complex interactions and environments often require significant computational resources. This can lead to slower simulation speeds, making real-time testing challenging. On the other hand, simulations that prioritize computational efficiency might sacrifice realism, potentially leading to inaccurate results. Striking the right balance involves optimizing the simulation's fidelity to match the specific goals of the research or development process, ensuring that the simulation provides meaningful insights while maintaining acceptable performance.

14. Uncertainty modeling in simulation involves capturing the uncertainties that arise in real-world robotic systems and emulating them within the virtual environment. Real-world sensors are subject to noise, inaccuracies, and other uncertainties that can impact robot behavior. Simulations must account for these uncertainties to ensure that the behaviors tested in the virtual environment reflect the challenges that robots face in reality. Uncertainty modeling might involve introducing noise to sensor data, simulating sensor errors, and replicating unpredictable events. By incorporating uncertainty modeling, simulations become more representative of real-world conditions, leading to more accurate and reliable results.

15. Simulation can be used to recreate specific scenarios and challenges that robots might encounter in real-world applications. For example, a search and rescue robot's behavior can be tested in a simulated disaster scenario, complete with collapsed buildings, debris, and hazardous conditions. By recreating these scenarios, researchers can assess how robots navigate through challenging environments, interact with obstacles, and respond to emergencies. Simulations allow researchers to fine-tune algorithms and strategies for specific use cases, ensuring that robots are well-prepared to handle complex situations once deployed in the real world.

16. Simulation has a significant role in robotics education and training. In educational settings, simulations provide a hands-on platform for students to learn robotics concepts, algorithms, and principles. Students can experiment with different behaviors, test algorithms, and observe the outcomes in a controlled environment. This practical experience enhances their understanding of robotic systems and their ability to apply theoretical knowledge to real-world challenges. Simulations also allow students to engage in collaborative learning, where they can work together to design and test robotic behaviors, fostering teamwork and problem-solving skills.

17. While simulations offer numerous advantages, they also come with limitations in accurately representing real-world interactions and dynamics. One limitation is the fidelity of physics simulations. Complex physical interactions might be challenging to model accurately in simulations, leading to discrepancies between virtual and real-world behaviors. Additionally, simulating dynamic environments with moving objects, fluid dynamics, and changing conditions can be computationally demanding. Capturing sensor accuracy and uncertainty accurately in simulations is another challenge, as real-world sensors often exhibit complex and nuanced behaviors that can be difficult to replicate.

18. Behavior-based programming enhances a robot's ability to react to real-time sensor input by prioritizing reactive behaviors over pre-defined plans. Reactive behaviors are triggered by specific sensory cues or events, allowing robots to respond immediately to changes in the environment. For example, a collision avoidance behavior might trigger when the robot detects an obstacle in its path, prompting it to change direction to avoid the obstacle. This responsiveness enables robots to handle dynamic and unpredictable situations, making them well-suited for applications where real-time decision-making is crucial, such as mobile robotics, exploration, and human-robot interaction.

19. Behavior-based programming is particularly well-suited for swarm robotics applications due to its decentralized and reactive nature. In swarm robotics, a group of robots collaboratively performs tasks without centralized control. Behavior modules in each robot can be designed to follow simple rules that allow them to coordinate with other robots. For instance, a swarm behavior module might instruct a robot to align with neighboring robots' directions. By combining such individual behaviors, emergent swarm behaviors like flocking, pattern formation, and exploration can arise. Behavior-based programming allows swarm robots to adapt to changes in the environment and collaborate effectively without the need for complex centralized coordination.

20. Behavior-based programming excels in various robotic domains due to its emphasis on responsiveness, adaptability, and simplicity. In domains where robots need to react to real-time events, such as mobile robotics and exploration, behavior-based systems allow robots to navigate dynamically changing environments effectively. In industrial automation, behavior-based programming enables robots to perform tasks like pick-and-place, assembly, and sorting without the need for precise pre-programming. In human-robot interaction, behavior-based systems allow robots to respond to human gestures and commands in real-time, creating natural and intuitive interactions. Overall, behavior-based programming's ability to handle real-time sensor input and adapt to diverse situations makes it a valuable approach across a wide range of robotic applications.

Online Resources

Online Exam Papers

Link:https://forms.gle/YVq3xYew6HWLoKTJ9

QR Code:

YouTube Video Playlist Link

Link:https://www.youtube.com/playlist?list=PLzlh_31VWEFraHESNxUEM9hyIaqjR_PTV

QR Code:

Chapter 9: Human-Robot Interaction

In the dynamic landscape of modern technology, a profound transformation is underway at the intersection of human and machine – a transformation encapsulated by the evolving field of Human-Robot Interaction (HRI). This domain transcends traditional boundaries, forging connections between human intellect and robotic capability. Human-Robot Interaction explores the intricacies of communication, cooperation, and coexistence between humans and machines, reshaping the way we perceive and interact with technology. As robots transition from being mere tools to intelligent companions and collaborators, the study of HRI becomes paramount. It encompasses a multidisciplinary tapestry woven with psychology, engineering, design, and ethics, as it seeks to decipher the nuances of gesture, expression, and intention that underpin effective collaboration. Through the lenses of empathy, trust, and usability, the exploration of HRI paves the way for machines that are not just functional entities, but true partners in our pursuit of progress. As we navigate the complex interplay between human emotion and artificial intelligence, we embark on a transformative journey into a world where the boundaries between human and machine blur, and a new era of harmonious coexistence emerges.

In the intricate landscape of modern technology, a paradigm shift is underway, where the boundaries between human and machine are dissolving to give rise to a harmonious coexistence. At the forefront of this evolution stands the captivating domain of Human-Robot Interaction (HRI). Far beyond a mere fusion of mechanics and algorithms, HRI delves into the profound intersection of human cognition, emotion, and artificial intelligence. It unveils the potential for machines not only to serve as tools but also as partners, companions, and collaborators. Through the intricate choreography of gestures, expressions, and dialogues, HRI explores the rich tapestry of communication that transcends language barriers and bridges the divide between man and machine.

HRI is an interdisciplinary voyage, weaving together psychology, engineering, design, and ethics, all in the pursuit of understanding and enhancing the interactions between humans and robots. It delves into the realms of user experience, adaptability, and cognitive load, unraveling the complexities that define seamless collaboration. From industrial settings to healthcare environments, from homes to outer space, the domain of HRI permeates various facets of our lives, revolutionizing how we work, learn, and even form emotional connections with machines.

Embedded within HRI lies the essence of empathy and trust – the building blocks of human relationships. Machines, guided by sophisticated algorithms, learn to interpret human emotions, respond with empathy, and adapt to individual preferences. From robotic assistants that anticipate our needs to therapy robots that offer solace, HRI transforms sterile machinery into companions that understand our desires and emotions. As we navigate the delicate landscape of coexistence, ethical considerations play an indispensable role. The exploration of questions surrounding autonomy, privacy, and accountability underscores the importance of responsible innovation in the realm of HRI.

The journey through HRI is a journey of transformation. It is a journey that molds the concept of interaction from one of mere functionality to one of genuine collaboration. The intricate dance between human curiosity and artificial intelligence creates a symphony where machines enhance human potential and humans infuse machines with understanding. As we navigate this realm, we step into a future where human emotion meets computational precision, where collaboration transcends language, and where the convergence of human intellect and artificial intelligence forms the foundation of a harmonious and technologically enriched world.

Interface Design and Usability: Introduction to Interface Design and Usability: In the modern digital age, where technology intertwines with every aspect of our lives, the concept of Interface Design and Usability has become paramount. It serves as the pivotal bridge between human cognition and technological functionality. This interdisciplinary realm combines art, psychology, and engineering to create interfaces that not only facilitate efficient interactions but also evoke emotions, stimulate engagement, and enhance user satisfaction. Interface Design and Usability delve into the delicate balance of aesthetics and functionality, unraveling the intricacies of human perception, behavior, and expectations in the realm of digital interfaces.

The Significance of Human-Centered Design: At the heart of Interface Design and Usability lies the principle of human-centered design. This approach places the user's needs, preferences, and experiences at the forefront of interface creation. By empathizing with users and understanding their goals, frustrations, and aspirations, designers can craft interfaces that seamlessly adapt to human behavior, minimizing cognitive load and maximizing efficiency. Human-centered design principles underscore the idea that technology should be a tool that augments human capabilities, rather than an obstacle that hinders them.

The Marriage of Aesthetics and Functionality: The design of digital interfaces is an intricate dance between aesthetics and functionality. Aesthetics encompass visual elements, typography, colors, and layouts that not only make interfaces visually appealing but also communicate information effectively. Functionality, on the other hand, involves creating interactions that are intuitive, consistent, and aligned with user expectations. The fusion of aesthetics and functionality creates interfaces that not only captivate users with their visual allure but also empower them with seamless interactions.

The Psychology of User Behavior: Understanding the psychology of user behavior is a cornerstone of effective interface design. Cognitive principles, such as Hick's Law and Fitts's Law, guide designers in optimizing the arrangement of elements and minimizing the effort required for users to achieve their goals. Gestalt principles delve into how users perceive and organize visual elements, aiding in the creation of interfaces that guide users' attention and comprehension. By embracing these psychological insights, designers can craft interfaces that feel intuitive and natural.

Usability Testing and Iterative Design: Usability testing is a pivotal phase in the Interface Design and Usability process. It involves observing users as they interact with prototypes, identifying pain points, and refining designs based on feedback. This iterative approach ensures that the interface evolves to meet user expectations, addressing potential pitfalls and enhancing user satisfaction. Usability testing not only validates design choices but also uncovers unforeseen challenges, leading to continuous improvement and refinement.

The Impact of Mobile and Responsive Design: In an era dominated by mobile devices and diverse screen sizes, the concept of responsive design has become paramount. Interfaces that adapt seamlessly to various devices and orientations ensure consistent user experiences across platforms. Mobile interfaces require a unique understanding of touch interactions, gestures, and the limitations of smaller screens. Responsive design not only caters to user convenience but also influences search engine rankings and overall user engagement.

The Role of Accessibility and Inclusivity: Interface Design and Usability extend their impact by fostering accessibility and inclusivity. Designing interfaces that are accessible to individuals with disabilities ensures that technology is available and usable for everyone. Implementing features such as screen readers, captions, and keyboard navigation empowers individuals with diverse needs to interact with digital interfaces on equal footing. Inclusivity in design enriches the user experience and reflects ethical considerations that transcend mere functionality.

Interface Design and Usability: In the dynamic landscape of technology, where human-machine interaction is at the forefront of innovation, the realm of Interface Design and Usability emerges as a linchpin that bridges the gap between complex systems and user experience. This domain encapsulates the art and science of creating intuitive pathways that empower users to effortlessly navigate digital interfaces, enabling them to harness the full potential of technology with minimal friction. Interface Design and Usability amalgamate creativity, psychology, and engineering to cultivate interactions that are not just functional but also aesthetically pleasing and emotionally resonant.

Aesthetic Elegance and User-Centered Focus: At the heart of Interface Design and Usability lies the marriage between aesthetic elegance and user-centered focus. Interfaces are no longer mere utilitarian tools; they are canvases where creativity meets functionality. By infusing interfaces with visually captivating elements, designers create experiences that engage users on an emotional level. Aesthetics, however, extend beyond superficial beauty; they encompass the arrangement of elements, the hierarchy of information, and the cohesiveness of the design, all of which contribute to a seamless and delightful user experience.

The Psychology of Perception and Interaction: Understanding the psychology of human perception and interaction forms the bedrock of effective interface design. Gestalt principles, cognitive load theory, and affordances guide designers in crafting interfaces that align with how users naturally perceive and interact with their environment. By embracing these psychological insights, designers create interfaces that intuitively guide users' attention, making interactions feel instinctive and fluid.

Navigation and Information Architecture: Navigating digital interfaces is akin to traversing a labyrinth of information. Effective navigation and information architecture become paramount in ensuring users can find what they seek without frustration. By organizing content logically, providing clear signposts, and minimizing cognitive effort, designers create pathways that empower users to explore interfaces with confidence. A well-structured information architecture enhances user engagement and satisfaction.

The Responsive Revolution: In a world dominated by a multitude of devices with varying screen sizes, the concept of responsive design has revolutionized interface development. Responsive design ensures that interfaces adapt seamlessly to different devices and orientations, providing a consistent experience across platforms. The challenge lies in preserving the essence of the interface while accommodating diverse contexts, making responsive design an art that balances flexibility and consistency.

Iterative Design and Usability Testing: The journey from concept to a polished interface involves a series of iterative design cycles, enriched by usability testing. Usability testing exposes design flaws, illuminates user frustrations, and provides invaluable feedback for refinement. This iterative approach ensures that interfaces evolve based on real-world interactions, resulting in designs that resonate with users and anticipate their needs.

Inclusivity and Ethical Considerations: The ethical dimension of Interface Design and Usability is underscored by the commitment to inclusivity. Designing interfaces that are accessible to individuals with disabilities is not just a legal obligation; it's a moral imperative. Accessibility features, such as screen readers and voice commands, transform interfaces into inclusive spaces, enabling everyone to participate in the digital realm.

Future Horizons: Interfaces in Virtual and Augmented Reality: As technology marches forward, the boundaries between the physical and digital worlds blur. Virtual and augmented reality (VR/AR) interfaces offer new dimensions of interaction, enabling users to immerse themselves in alternative realities. Designing for VR/AR demands a reimagining of spatial awareness, gesture-based interactions, and the fusion of the virtual and the real.

In conclusion, Interface Design and Usability are the conduits that empower users to navigate the digital landscape with confidence, grace, and engagement. The intricate dance between aesthetics, psychology, and engineering creates interfaces that transcend functionality, becoming conduits of memorable experiences. As designers strive to harmonize form and function, aesthetics and efficiency, they craft pathways that lead users to the heart of technology, where every click, swipe, and interaction is an invitation to engage with a world of endless possibilities.

Teleoperation: In the vast expanse of technological innovation, the concept of teleoperation emerges as a bridge that transcends physical distances, enabling humans to extend their presence and capabilities to remote environments through robotic proxies. Teleoperation, a multidimensional domain that melds human intent with robotic action, reshapes the boundaries of exploration, intervention, and interaction. As we delve into the intricacies of teleoperation, we navigate a landscape defined by remote control, haptic feedback, immersive interfaces, and the profound potential to engage with environments that would otherwise remain inaccessible. This exploration of teleoperation unearths a world where humans command machines from afar, enabling us to traverse hazardous terrains, delve into the depths of the oceans, venture into outer space, and interact with delicate materials or environments that demand precision beyond human capabilities. As the synergy between human intention and robotic execution propels us into uncharted territories, we unveil the nuances of teleoperation, its applications, challenges, and the ever-expanding horizon of human-machine collaboration.

In the intricate tapestry of technological advancement, the concept of teleoperation stands as a testament to the human desire to overcome physical limitations and traverse remote realms through the agency of robotic proxies. Teleoperation, a realm where human intention meets robotic action, heralds a new era of exploration, intervention, and interaction that defies geographical barriers. This multifaceted domain encompasses the intricacies of remote control, tactile feedback, immersive interfaces, and the symbiotic relationship between humans and machines, all culminating in the ability to wield influence and effect change in environments far removed from our physical presence.

The Essence of Teleoperation: At its core, teleoperation encapsulates the transmission of human commands across space to robotic entities that replicate those commands into tangible actions. It introduces a level of remote dexterity that transcends distance, enabling us to interact with environments ranging from the minuscule to the colossal. By channeling our intentions through tele operated systems, we navigate hazardous zones, survey uncharted territories, and manipulate intricate objects with a precision that belies the separation between human and machine.

Immersive Interfaces and Sensory Feedback: The evolution of teleoperation has witnessed the rise of immersive interfaces that enable operators to perceive and interact with remote environments as if they were present. Immersive teleoperation relies on an integration of visual, auditory, and tactile sensory feedback, allowing operators to perceive depth, distance, and even textures. By blurring the line between reality and remote presence, immersive teleoperation offers a heightened sense of immersion and control.

Robotic Avatars: Exploring Beyond Human Reach: Teleoperation extends its reach to domains that are inhospitable, hazardous, or beyond human physiological capabilities. It empowers robotic avatars to tread on terrains riddled with danger – from navigating nuclear disaster zones to exploring the abyssal depths of the oceans. This extension of human capabilities enables us to push the boundaries of exploration and gain insights into environments that were once shrouded in mystery.

Space Exploration: Virtually Venturing into the Cosmos: Teleoperation has manifested as a crucial tool in space exploration, allowing us to remotely maneuver rovers and spacecraft across the barren landscapes of distant planets. As we remotely control robotic extensions millions of miles away, the accomplishments of teleoperation open doors to unraveling the mysteries of the cosmos, all while operating from the comfort of our terrestrial abode.

Challenging Latency and Feedback Constraints: While teleoperation offers unprecedented capabilities, it also presents challenges inherent to the nature of remote control. The delay between human actions and robotic responses, known as latency, introduces complexities that demand anticipatory actions. Moreover, the fidelity of sensory feedback poses challenges in conveying the nuanced sensations necessary for delicate tasks. Overcoming these constraints demands advances in communication technologies and haptic interfaces.

The Ethical and Societal Dimensions: Teleoperation also raises questions of ethical responsibility, accountability, and the potential implications of exerting influence from a distance. The advent of remote warfare, medical interventions, and interventions in disaster-stricken regions highlights the need to consider the moral dimensions that arise when human decisions manifest through robotic actions.

The Essence of Teleoperation: At its core, teleoperation encapsulates the transmission of human commands across space to robotic entities that replicate those commands into tangible actions. It introduces a level of remote dexterity that transcends distance, enabling us to interact with environments ranging from the minuscule to the colossal. By channeling our intentions through tele operated systems, we navigate hazardous zones, survey uncharted territories, and manipulate intricate objects with a precision that belies the separation between human and machine.

Immersive Interfaces and Sensory Feedback: The evolution of teleoperation has witnessed the rise of immersive interfaces that enable operators to perceive and interact with remote environments as if they were present. Immersive teleoperation relies on an integration of visual, auditory, and tactile sensory feedback, allowing operators to perceive depth, distance, and even textures. By blurring the line between reality and remote presence, immersive teleoperation offers a heightened sense of immersion and control.

Robotic Avatars: Exploring Beyond Human Reach: Teleoperation extends its reach to domains that are inhospitable, hazardous, or beyond human physiological capabilities. It empowers robotic avatars to tread on terrains riddled with danger – from navigating nuclear disaster zones to exploring the abyssal depths of the oceans. This extension of human capabilities enables us to push the boundaries of exploration and gain insights into environments that were once shrouded in mystery.

Space Exploration: Virtually Venturing into the Cosmos: Teleoperation has manifested as a crucial tool in space exploration, allowing us to remotely maneuver rovers and spacecraft across the barren landscapes of distant planets. As we remotely control robotic extensions millions of miles away, the accomplishments of teleoperation open doors to unraveling the mysteries of the cosmos, all while operating from the comfort of our terrestrial abode.

Challenging Latency and Feedback Constraints: While teleoperation offers unprecedented capabilities, it also presents challenges inherent to the nature of remote control. The delay between human actions and robotic responses, known as latency, introduces complexities that demand anticipatory actions. Moreover, the fidelity of sensory feedback poses challenges in conveying the nuanced sensations necessary for delicate tasks. Overcoming these constraints demands advances in communication technologies and haptic interfaces.

The Ethical and Societal Dimensions: Teleoperation also raises questions of ethical responsibility, accountability, and the potential implications of exerting influence from a distance. The advent of remote warfare, medical interventions, and interventions in disaster-stricken regions highlights the need to consider the moral dimensions that arise when human decisions manifest through robotic actions.

In conclusion, teleoperation stands as a testament to humanity's indomitable spirit of exploration and intervention. As we navigate the terrain of remote interaction, we unlock the ability to engage with environments that were once beyond our grasp. Teleoperation converges human intent and robotic capability, enabling us to traverse the hazardous, delve into the unknown, and chart the cosmos. The journey through teleoperation is a journey through technological empowerment – a journey that propels us beyond our physical confines, forging connections between worlds that were once distant, and expanding the horizons of human possibility.

Collaborative Robotics: In the tapestry of technological advancement, the emergence of Collaborative Robotics stands as a testament to humanity's pursuit of harmonious interaction between humans and machines. Collaborative Robotics, often referred to as "cobots," represents a paradigm shift in the realm of automation, where robots transcend their role as mere tools and evolve into collaborative partners that work alongside humans. This domain navigates the delicate balance between machine precision and human intuition, reshaping industries and applications that range from manufacturing and healthcare to logistics and beyond. As we delve into the intricacies of Collaborative Robotics, we embark on a journey that explores the fusion of advanced technology, cognitive empathy, safety protocols, and the promise of enhanced efficiency and productivity. This exploration reveals a world where robots are not just instruments of automation but facilitators of human potential, ushering in a new era of collaboration that enriches our industries, propels innovation, and redefines the boundaries of what machines and humans can achieve together. In the ever-evolving landscape of robotics, the emergence of Collaborative Robotics signifies a profound shift in how humans and machines interact and collaborate. Often referred to as "cobots," Collaborative Robotics redefines the conventional boundaries of automation by fostering a symbiotic partnership between human workers and robotic entities. This paradigm shift transcends the traditional concept of robots as isolated tools by creating a dynamic ecosystem where machines and humans coexist, share tasks, and amplify each other's capabilities. This domain delves into the intricacies of creating a harmonious blend of advanced technology, safety protocols, human intuition, and adaptable systems. As we navigate the landscape of Collaborative Robotics, we uncover a realm where precision meets empathy, where efficiency is magnified through collaboration, and where innovation thrives on the fusion of human ingenuity and robotic precision.

Evolution beyond Automation: Collaborative Robotics redefines the very essence of automation. While traditional industrial robots operate within fenced environments, cobots share workspace with humans, amplifying human capabilities while retaining the precision and efficiency of machines. This evolution signifies a shift from automated processes to human-assisted, adaptable systems that thrive on cooperation and partnership.

Synergy of Human and Machine Expertise: The heart of Collaborative Robotics lies in the seamless blend of human and machine expertise. By leveraging the dexterity and adaptability of humans alongside the repeatability and precision of robots, cobots amplify productivity. This partnership capitalizes on the strengths of both entities, leading to an environment where creativity meets consistency.

Ensuring Safety through Sensing and Intelligence: Central to the success of Collaborative Robotics is the assurance of human safety. Cobots are equipped with advanced sensors and intelligent algorithms that enable them to detect the presence of humans and respond dynamically to changing situations. This fusion of sensing and intelligence ensures that humans and cobots can work side by side without compromising safety.

Human-Centric Design Philosophy: Collaborative Robotics is rooted in a human-centric design philosophy. Cobots are designed to be intuitive, easy to program, and adaptable to changing tasks. The user-friendly interfaces and programming methods allow even non-experts to collaborate with cobots, democratizing access to automation and augmenting human skills.

Versatility Across Industries: The influence of Collaborative Robotics is felt across diverse industries. From manufacturing lines where cobots assist assembly tasks to healthcare settings where they aid in patient care and logistics, cobots demonstrate their versatility. This adaptability reflects the potential to revolutionize industries and streamline processes.

Enhancing Innovation and Creativity: By delegating repetitive, mundane tasks to cobots, Collaborative Robotics liberates human workers to focus on tasks that require creativity, critical thinking, and problem-solving. This empowerment fosters an environment where innovation flourishes, as human workers can channel their cognitive abilities toward higher-value activities.

Addressing the Skills Gap: Collaborative Robotics addresses the growing skills gap by augmenting human expertise. As industries evolve, the demand for specific skills increases. Cobots can be quickly reprogrammed to perform new tasks, enabling companies to pivot their operations in response to changing market dynamics without the need for extensive retraining.

Ethical Considerations and Socioeconomic Impact: The advent of Collaborative Robotics raises ethical considerations related to job displacement and the evolving nature of work. While cobots enhance productivity, they also reshape job roles, necessitating conversations about reskilling and upskilling the workforce. Balancing automation with workforce development becomes pivotal in ensuring the socioeconomic impact is positive.

Evolution of Human-Robot Interaction: Collaborative Robotics represents a pivotal evolution in the trajectory of human-robot interaction. While conventional industrial robots operated behind safety barriers, cobots integrate seamlessly into human workspaces, enabling cooperative tasks that benefit from the unique strengths of both humans and machines. This evolution is not just about automating processes but about reimagining how humans and robots can collaborate to achieve shared objectives.

Synergy in Skills and Capabilities: At the core of Collaborative Robotics lies the idea of synergy – a harmonious blend of human skills and robotic capabilities. Cobots leverage the flexibility, creativity, and adaptability of human workers alongside the precision, repeatability, and endurance of machines. This collaborative synergy creates an environment where efficiency is heightened, tasks are completed with accuracy, and human workers are liberated from mundane tasks to focus on higher-value activities.

Safety and Trust through Advanced Sensing: One of the cornerstones of Collaborative Robotics is ensuring the safety of human workers in close proximity to cobots. Advanced sensors, including vision systems and force sensors, equip cobots with the ability to detect the presence of humans and respond instantly to changes in the environment. This fusion of technology and safety protocols instills a sense of trust and confidence in working alongside machines.

Human-Centric Design Philosophy: Collaborative Robotics places a strong emphasis on human-centric design. Cobots are designed to be user-friendly, intuitive to program, and adaptable to changing tasks. This design philosophy democratizes access to automation, enabling workers of varying skill levels to collaborate seamlessly with cobots, thereby bridging the gap between machine complexity and user-friendliness.

Transforming Industries and Processes: The influence of Collaborative Robotics transcends industries, infusing innovation into sectors such as manufacturing, healthcare, logistics, and more. In manufacturing, cobots collaborate with humans on assembly lines, optimizing efficiency and product quality. In healthcare, they assist in tasks ranging from patient care to surgical procedures, amplifying the skills of medical professionals.

Fostering Innovation and Creativity: By relieving humans from repetitive and mundane tasks, Collaborative Robotics paves the way for increased innovation and creativity. Human workers are empowered to engage in tasks that require critical thinking, problem-solving, and complex decision-making. This dynamic allows industries to tap into the full cognitive potential of their workforce.

Addressing Workforce Challenges: As industries evolve, the Collaborative Robotics paradigm offers a solution to the growing skills gap. Cobots can be rapidly reprogrammed to adapt to new tasks, enabling companies to swiftly pivot their operations without extensive retraining. This agility in task allocation addresses workforce challenges by providing a versatile and adaptive solution.

Ethical and Societal Implications: The proliferation of Collaborative Robotics brings forth ethical considerations regarding the future of work, job displacement, and the role of humans in an increasingly automated landscape. Balancing automation with the preservation of job roles and skill development becomes paramount in ensuring a positive societal impact.

In conclusion, Collaborative Robotics ushers in an era where humans and machines coexist as partners, collaborators, and co-contributors. By seamlessly integrating technology with human skills, cobots create an environment where innovation thrives, efficiency soars, and safety remains a paramount concern. This journey into Collaborative Robotics demonstrates our ability to harness technology not as a replacement for human involvement but as an amplifier of human potential. As industries and sectors continue to integrate cobots into their operations, the collaborative synergy between humans and machines serves as a testament to our capacity for innovation, adaptation, and collaboration on the path to a technologically enriched future.

9.3 Ethical Considerations in Human-Robot Interaction: As the realm of robotics and artificial intelligence continues to advance at an unprecedented pace, the integration of machines into our everyday lives raises a host of profound ethical considerations. In the intricate tapestry of human-robot interaction, the domain of Ethical Considerations holds a pivotal place, urging us to navigate the moral compass that guides our interactions with intelligent machines. From collaborative robots working side by side with humans to autonomous systems making consequential decisions, the ethical dimensions woven into these interactions touch on questions of accountability, privacy, bias, trust, and the very essence of what it means to be human. As we delve into the nuances of Ethical Considerations in Human-Robot Interaction, we are confronted with a landscape that demands careful contemplation, multidisciplinary collaboration, and a forward-thinking approach to shaping the technological future. This exploration unravels a world where the lines between human and machine blur, prompting us to ponder the moral implications that arise when technology and ethics converge. In the dynamic intersection of human ingenuity and technological prowess, the field of robotics stands as a testament to our capacity to create machines that mirror human cognition and actions. As we embark on this journey of human-robot interaction, we are confronted with a profound moral terrain that demands careful navigation and ethical introspection. The domain of Ethical Considerations in Human-Robot Interaction delves deep into the ethical complexities that arise when intelligent machines become integral to our lives, shaping our actions, decisions, and relationships. This domain encapsulates the ethical dilemmas that range from the autonomy of machines to accountability, from the distribution of resources to issues of bias and fairness. The exploration of ethical considerations in this context extends beyond technological advancement.

The Moral Fabric of Human-Robot Interactions: Ethical Considerations in Human-Robot Interaction delve into the fabric of moral and philosophical questions that arise when technology intersects with human values and principles. The introduction of intelligent machines prompts us to assess how our actions, intentions, and responsibilities are reshaped when we engage with entities that mirror human cognition and decision-making.

Accountability and Responsibility in Autonomous Systems: As machines become more autonomous and capable of making decisions that impact human lives, the question of accountability becomes paramount. Ethical considerations dictate that while machines may carry out actions, the ultimate responsibility and consequences rest with humans. This intersection of accountability and autonomy brings to light the need for clear lines of responsibility.

Addressing Bias and Fairness: Ethical considerations extend to addressing bias and fairness in machine learning algorithms. The potential for algorithms to perpetuate and amplify biases inherent in training data calls for vigilant scrutiny and mitigation strategies. Ensuring equitable and just outcomes in human-robot interactions requires proactive measures to eliminate discriminatory practices.

Privacy in an Era of Ubiquitous Sensors: The proliferation of sensors and data-driven technologies in human-robot interactions poses challenges to individual privacy. From surveillance drones to smart homes, the collection and utilization of personal data by intelligent machines necessitate careful contemplation of privacy rights, data ownership, and consent.

Establishing Trust in Human-Robot Relationships: The foundation of any interaction, including between humans and robots, is trust. Ethical considerations underscore the importance of establishing and maintaining trust in human-robot relationships. Transparency, predictability, and reliability are key components that contribute to building trust and confidence in intelligent machines.

Ethical Dilemmas in Decision-Making: Autonomous systems equipped with decision-making capabilities raise ethical dilemmas that parallel human ethical quandaries. When machines are entrusted with making choices that impact human lives, their decisions must align with ethical principles. Navigating scenarios where machines must make difficult choices poses philosophical and moral challenges.

Human Identity and Transhumanism: The integration of machines into our lives prompts contemplation of human identity and the potential for transhumanism – a movement that seeks to enhance human abilities through technology. Ethical considerations delve into the implications of blurring the lines between humans and machines, raising questions about authenticity, autonomy, and the essence of being human.

Multidisciplinary Collaboration and Ethical Frameworks: Addressing the ethical dimensions of human-robot interaction demands a multidisciplinary approach. Collaborations between ethicists, technologists, policymakers, and social scientists are essential to create ethical frameworks that guide the development and deployment of intelligent machines.

The Ethical Imperative in Human-Robot Collaboration: The foundation of Ethical Considerations in Human-Robot Interaction rests on the realization that the integration of intelligent machines into our lives is not merely a technical feat but an ethical obligation. As machines become more integrated into our workplaces, homes, and public spaces, we are compelled to evaluate how their actions align with ethical principles and human values.

Autonomy and Accountability: The rise of autonomous machines introduces the challenge of assigning responsibility. Ethical considerations dictate that while machines may possess a degree of autonomy, humans remain the ultimate moral agents accountable for their actions. This intersection of autonomy and accountability sparks discussions about the extent of human control over machine decisions.

Bias, Fairness, and Equity: The ethical dimensions of bias and fairness loom large as algorithms and machine learning systems are integrated into decision-making processes. Ensuring that machines make decisions that are free from discriminatory biases and uphold principles of fairness and equity becomes a moral imperative, necessitating transparency and algorithmic audits.

Human Dignity and Relationships: The integration of robots into various aspects of life invites questions about the dignity of human existence and the impact on human relationships. Ethical considerations compel us to assess whether robotic companionship enhances or diminishes our humanity, and how it reshapes our interactions with both machines and fellow humans.

Privacy and Data Ethics: The pervasive nature of intelligent machines equipped with sensors and data collection capabilities raises ethical concerns surrounding privacy. As machines gather personal data, the ethical landscape demands robust frameworks for data ownership, consent, and protection, safeguarding individuals' right to privacy.

Ethical Decision-Making in Machines: The advent of machines capable of autonomous decision-making introduces complex ethical dilemmas. As machines are entrusted with decisions that carry moral consequences, discussions emerge about the ethical principles that guide their choices and the parameters within which their decisions align with human values.

Shaping Technological Futures: Ethical considerations extend beyond the immediate present to the long-term implications of our technological choices. By recognizing the profound societal impact of intelligent machines, we are challenged to envision ethical trajectories for their development, deployment, and integration into our lives.

Cultivating Ethical Awareness: Navigating the ethical dimensions of human-robot interaction necessitates raising ethical awareness among technologists, policymakers, and the broader public. An informed understanding of the ethical considerations associated with intelligent machines is essential to shape policies, regulations, and standards that align with societal values.

In conclusion, Ethical Considerations in Human-Robot Interaction form an integral pillar in the ever-evolving landscape of technology. The ethical discourse around intelligent machines calls for a holistic approach that transcends technical prowess and engages with the core values that define us as humans. As we venture into the territory of human-robot synergy, ethical considerations guide our path, reminding us that technology is not merely a tool but an extension of our ethical values. By embracing these considerations, we cultivate a future where machines and humans coexist in harmony, where technological advancement aligns with human dignity, and where ethical introspection shapes the trajectory of our shared journey into the realms of innovation and collaboration.

Exercise

Multiple-choice Questions

1. What does the term "cobots" refer to in the context of robotics?
a) Robots designed for combat purposes
b) Robots with advanced AI capabilities
c) Collaborative robots that work alongside humans
d) Robots used in underwater exploration
2. Why is trust important in human-robot interaction?
a) It ensures that robots always make the correct decisions
b) It helps robots become more autonomous
c) It establishes a sense of confidence and reliability
d) It eliminates the need for human intervention
3. What does the ethical dilemma of bias in human-robot interaction refer to?
a) The tendency of humans to treat robots unfairly
b) The presence of preconceived notions in robotic decision-making
c) Robots being overly biased in favor of human decisions
d) The difficulty of programming robots to make unbiased decisions
4. Which aspect of human-robot interaction raises concerns about individual privacy?
a) The use of robots for entertainment purposes
b) The sharing of emotions between humans and robots
c) The integration of sensors and data-driven technologies
d) The ability of robots to perform complex tasks
5. What is the primary ethical concern when it comes to assigning accountability in autonomous robots?
a) Ensuring that robots always make the correct decisions
b) Defining who is responsible for the actions of autonomous robots
c) Encouraging robots to act independently without human intervention
d) Preventing robots from taking any actions that could be considered harmful

Review Questions:

1. What is the significance of the term "cobots" in the context of robotics, and how do they differ from traditional industrial robots?
2. Why is the establishment of trust an essential factor in effective human-robot interaction? Provide examples to illustrate your answer.
3. Explain the ethical concerns related to bias in machine learning algorithms used in human-robot interaction. How can these concerns be mitigated?
4. How does the integration of sensors and data-driven technologies in robots raise ethical questions about privacy? Discuss the balance between data collection and individual rights.
5. In the context of autonomous robots, what challenges arise when assigning accountability and responsibility for actions? How can these challenges be addressed?
6. Define the concept of "human dignity" as it relates to human-robot interaction. How can the integration of robots impact human dignity in various scenarios?
7. Describe the concept of transhumanism and its ethical implications. How might the integration of robotic technologies into the human body raise ethical concerns?
8. How do ethical considerations intersect with decision-making by autonomous robots? Provide examples of scenarios where machine decisions have ethical implications.
9. Discuss the importance of establishing ethical frameworks for human-robot interaction. What are the key components that should be included in these frameworks?
10. How can the principles of transparency and explainability enhance the ethical aspects of human-robot interaction? Provide examples of how transparency can be implemented.
11. Compare and contrast the ethical implications of using robots in environments such as healthcare and military applications.
12. In what ways can bias in machine learning algorithms impact the fairness and equity of decisions made by robots? How can algorithmic bias be mitigated?
13. Explain the concept of "human-centric design" in the context of ethical considerations in human-robot interaction. How does this design philosophy ensure a positive user experience?

14. How does the integration of robots into various aspects of life raise questions about human relationships and companionship? Discuss the ethical dimensions of developing emotional connections with robots.
15. Describe the challenges associated with ensuring transparency and accountability when robots are involved in making high-stakes decisions, such as in self-driving cars or medical diagnoses.
16. How does the notion of "shared responsibility" play a role in the ethical considerations of human-robot interaction? Provide examples of how humans and robots can collaboratively share responsibilities.
17. Discuss the concept of "algorithmic auditing" and its importance in addressing bias and discrimination in machine learning algorithms used in human-robot interaction.
18. Explore the ethical implications of using robots for entertainment purposes, such as humanoid robots in theme parks or interactive AI companions.
19. How can interdisciplinary collaboration between ethicists, technologists, policymakers, and social scientists contribute to the development of ethical frameworks for human-robot interaction?
20. Reflect on the long-term societal impact of intelligent machines and autonomous robots. How can ethical considerations help shape a future that benefits humanity as a whole?

Answers Multiple-choice Questions

Answers:
1. c)
2. c)
3. a)
4. c)
5. b)

Review Questions:

1. Collaborative robots, or "cobots," are robots designed to work alongside humans in shared workspaces, performing tasks collaboratively. Unlike traditional industrial robots that operate in isolated environments, cobots are designed to interact with humans, assisting them in various tasks. The significance of cobots lies in their potential to enhance productivity, efficiency, and safety in workplaces by combining human capabilities with robotic precision. Cobots differ from traditional robots as they are equipped with advanced sensors and technologies that allow them to detect human presence and react accordingly, ensuring safe and effective collaboration.
2. Trust is essential in human-robot interaction as it establishes a foundation of confidence and reliability in the interactions between humans and robots. When humans interact with robots, they need to trust that the robots will perform their tasks accurately and safely. Trust influences how humans perceive the decisions and actions of robots, affecting their willingness to work alongside robots, follow their recommendations, or delegate tasks. Establishing trust in human-robot interaction is crucial for fostering successful collaboration and ensuring that humans feel comfortable and confident while working with robots.
3. Bias in machine learning algorithms used in human-robot interaction refers to the presence of unfair or discriminatory outcomes in the decisions made by robots due to biased training data or algorithmic design. This bias can perpetuate societal inequalities and lead to unjust treatment of individuals. To mitigate these concerns, developers and researchers must ensure diverse and representative training datasets, identify potential bias during algorithm development, and implement strategies to correct and mitigate bias. Transparency in algorithmic decision-making and continuous monitoring can help identify and address bias, promoting fairness and equity in human-robot interaction.
4. The integration of sensors and data-driven technologies in robots enables them to collect and process vast amounts of information about their environment and human interactions. This raises ethical questions about privacy, as the data collected may include personal or sensitive information. The challenge lies in balancing the benefits of data collection for improved functionality with respecting individuals' rights to privacy. Striking this balance involves implementing robust data protection measures, obtaining informed consent from users, anonymizing data, and ensuring that data collection is proportionate to the intended purpose. Respecting individual rights while harnessing the potential of data-driven technologies is crucial for ethical human-robot interaction.
5. Assigning accountability and responsibility for actions in autonomous robots is challenging due to the decentralized decision-making process and the absence of human agency in the actions of robots. These challenges intensify when robots operate in complex, unstructured environments where it is difficult to predict their behavior. Addressing these challenges involves establishing clear legal frameworks that define liability and responsibility, developing standardized procedures for investigating incidents involving robots, and implementing advanced technologies that enable traceability of robot actions. Collaborative efforts between legal experts, technologists, and policymakers are crucial to navigating these challenges and ensuring ethical accountability.

6. Human dignity refers to the inherent value and worth of every individual, encompassing their autonomy, rights, and respect. In the context of human-robot interaction, maintaining human dignity involves ensuring that the integration of robots respects human rights, autonomy, and individuality. The impact of robots on human dignity varies depending on the context. For instance, using robots in caregiving scenarios can enhance human dignity by providing assistance and support to individuals with disabilities. However, the uncritical substitution of human caregivers with robots may undermine human dignity by neglecting the emotional and relational aspects of caregiving. It is essential to consider the broader societal implications and potential impacts on human dignity when integrating robots into various roles.

7. Transhumanism is a philosophical movement that advocates for the use of advanced technologies, including robotics and AI, to enhance human physical and cognitive capabilities beyond their natural limits. Ethical implications arise in transhumanism due to concerns about altering human nature, creating disparities between enhanced and non-enhanced individuals, and the potential for unintended consequences. The integration of robotic technologies into the human body, known as robo-augmentation, raises ethical concerns about consent, long-term health effects, and the potential for inequality based on access to enhancement technologies. Ethical discussions surrounding transhumanism focus on striking a balance between technological advancements and preserving human identity, autonomy, and social equality.

8. Ethical considerations intersect with decision-making by autonomous robots when robots are entrusted with choices that carry moral consequences. For example, in autonomous vehicles, robots must make split-second decisions in situations where human lives are at stake, such as deciding between saving passengers or pedestrians. These scenarios raise ethical dilemmas about the value of human life and the moral framework that guides machine decisions. Similarly, in healthcare, robots may be involved in triage decisions, allocation of medical resources, or even end-of-life care. Ethical discussions surrounding these scenarios involve programming robots to adhere to principles like minimizing harm, respecting individual autonomy, and upholding societal values.

9. Establishing ethical frameworks for human-robot interaction is essential to ensure responsible and morally aligned development, deployment, and use of robots. These frameworks guide developers, policymakers, and users in making informed decisions that uphold ethical values. Key components of ethical frameworks include guidelines for transparency in robot behavior and decision-making, protocols for accountability and liability, principles for ensuring fairness and non-discrimination, strategies for addressing bias, guidelines for safeguarding individual privacy, and mechanisms for informed consent and user education. By integrating these components, ethical frameworks provide a roadmap for navigating complex ethical challenges and promoting the responsible integration of robots into society.

10. Transparency and explain ability are essential principles in enhancing the ethical aspects of human-robot interaction by ensuring that the actions and decisions of robots are understandable and justifiable to humans. Transparent systems provide users with insights into how robots arrive at their decisions, fostering trust and accountability. For example, in healthcare, if a robot provides medical recommendations, it should explain the reasoning behind its suggestions, enabling healthcare professionals and patients to make informed choices. Implementing transparency can involve designing user interfaces that visualize robot decision-making processes, providing access to algorithms' internal logic, and generating human-readable explanations for complex decisions. Ultimately, transparency and explain ability promote responsible and ethically aligned human-robot interactions.

11. Using robots in healthcare and military applications raises distinct ethical implications. In healthcare, robots assist in patient care, surgery, and diagnostics. Ethical considerations include ensuring patient safety, respecting patient autonomy, and maintaining human oversight in critical decisions. In contrast, military robots involve considerations related to the use of force, adherence to international laws, and potential dehumanization of warfare. In both cases, ethical frameworks must address the balance between technological benefits and potential risks, as well as the overarching principles of beneficence, non-maleficence, justice, and autonomy.

12. Bias in machine learning algorithms can lead to unfair or discriminatory outcomes, disproportionately affecting certain groups. For instance, biased algorithms might lead to incorrect medical diagnoses or unjust hiring decisions. Algorithmic bias can be mitigated by using diverse and representative training datasets, regularly auditing algorithms for bias, and employing techniques such as re-sampling, re-weighting, or fairness-aware machine learning. Developing guidelines for handling bias and involving diverse teams in algorithm development can help ensure that algorithms are fair, equitable, and uphold human rights.

13. Human-centric design prioritizes the needs, preferences, and well-being of users in the development of robotic systems. This design philosophy ensures that robots are designed to be intuitive, user-friendly, and aligned with human values and ethical principles. By involving users in the design process, considering cultural sensitivities, and focusing on usability and accessibility, human-centric design enhances the overall user experience. For example,

designing a social robot that respects personal space and cultural norms demonstrates human-centric design, fostering positive human-robot interactions and minimizing potential discomfort.

14. The integration of robots into various roles, such as caregivers, companions, or even romantic partners, raises questions about the boundaries and nature of human-robot relationships. Developing emotional connections with robots carries ethical dimensions, as individuals might form deep attachments to machines that lack human emotions and consciousness. Ethical considerations include ensuring that such connections do not replace meaningful human relationships, addressing potential exploitation, and ensuring that individuals fully understand the nature of their interactions. Striking a balance between promoting positive well-being and avoiding harmful emotional dependence is crucial in this context.

15. Ensuring transparency and accountability in high-stakes robotic decisions involves challenges such as the complexity of algorithms, difficulty in explaining machine reasoning to humans, and the risk of over-reliance on automation. In self-driving cars, decisions about avoiding collisions can be challenging to explain to passengers. In medical diagnoses, understanding how AI arrived at a certain diagnosis can be crucial for medical professionals. Solutions include developing interpretable machine learning models, creating visualization tools to explain decisions, and providing clear guidelines for human oversight and intervention in critical situations.

16. Interdisciplinary collaboration brings together diverse perspectives to address the complex ethical challenges of human-robot interaction. Ethicists contribute moral insights, technologists provide technical feasibility, policymakers offer legal and regulatory guidance, and social scientists provide insights into societal impact. Collaboratively, these experts can develop comprehensive ethical frameworks that consider a wide range of perspectives, align with societal values, and address the multifaceted dimensions of robotics. This collaboration ensures that ethical frameworks are robust, informed, and capable of guiding responsible human-robot interaction.

17. The primary goal of ethical considerations in human-robot interaction is to ensure that the development, deployment, and use of robots align with human values, promote well-being, and uphold human rights. Ethical considerations impact robotics by guiding technological choices, shaping design principles, and influencing policy decisions. Ethical guidelines influence the integration of safety mechanisms, transparency features, and accountability frameworks into robotic systems. By addressing concerns related to fairness, privacy, bias, accountability, and human dignity, ethical considerations ensure that robots enhance society while minimizing potential harm and negative consequences.

18. Transparency and explain ability enhance ethical aspects by fostering trust, accountability, and human understanding of robot behavior. For example, in AI-assisted healthcare, transparency can be achieved by providing patients with explanations of how the AI made medical recommendations. Similarly, in financial services, transparent algorithms can explain lending decisions to borrowers. Transparency can be implemented through visualizations, decision trees, and natural language explanations, allowing users to comprehend and trust the rationale behind robotic decisions.

19. The integration of robots into various roles, such as caregivers, companions, or even romantic partners, raises questions about the boundaries and nature of human-robot relationships. Developing emotional connections with robots carries ethical dimensions, as individuals might form deep attachments to machines that lack human emotions and consciousness. Ethical considerations include ensuring that such connections do not replace meaningful human relationships, addressing potential exploitation, and ensuring that individuals fully understand the nature of their interactions. Striking a balance between promoting positive well-being and avoiding harmful emotional dependence is crucial in this context.

20. Intelligent machines and autonomous robots are poised to transform industries, economies, and daily life. Ethical considerations play a pivotal role in shaping this future by guiding technological advancement in alignment with human values. Ethical frameworks ensure that AI and robotics uphold fairness, avoid biases, respect human rights, and promote transparency. By considering the societal implications of automation, ethics drives discussions about retraining the workforce, mitigating job displacement, and ensuring access to benefits for all. Ultimately, ethical considerations help create a future where robots serve as tools for human progress, enhancing quality of life, rather than diminishing it.

Online Resources

Online Exam Papers

Link:https://forms.gle/Zdz3fAtNrBuwGmdc6

QR Code:

YouTube Video Playlist Link

Link: https://www.youtube.com/playlist?list=PLzlh_31VWEFqMQn6mcgtPWiK0Z_VTMpmR

QR Code:

Chapter 10: Advanced Topics in Robotics

In the realm of robotics, as technology continues to advance at an unprecedented pace, the field has delved into an array of complex and cutting-edge topics that push the boundaries of traditional robotic systems. These advanced topics in robotics are a testament to the ever-evolving nature of the discipline, exploring areas that extend beyond the fundamental concepts. From intricate algorithms that drive autonomous decision-making to the integration of AI and machine learning for enhanced capabilities, these advanced topics not only showcase the potential of robots to revolutionize industries but also pose new challenges that demand innovative solutions. As we navigate through the intricacies of teleoperation, human-robot collaboration, ethics, and beyond, the exploration of these advanced topics becomes pivotal in shaping the future landscape of robotics and its profound impact on society.

In the ever-evolving landscape of robotics, the exploration of advanced topics stands as a testament to the remarkable progress made in this field. These topics delve into the frontiers of innovation, where robotic systems transcend their conventional boundaries and delve into realms that were once considered the realm of science fiction. As technology accelerates, concepts like machine learning, artificial intelligence, and complex motion planning have become integral components of the robotic toolkit. The exploration of advanced topics in robotics not only highlights the field's potential to revolutionize industries but also underscores the necessity of addressing intricate challenges that arise in tandem. From the ethical considerations surrounding robotic autonomy to the intricacies of controlling complex multi-armed systems, these advanced topics pave the way for groundbreaking advancements that have the potential to reshape how we interact with, utilize, and benefit from robotic technologies.

As the field of robotics continues to evolve, embracing advanced topics has become essential to unraveling the full potential of robotic systems. These topics delve into the intricacies that lie beyond the fundamental principles, exploring the frontiers where innovation and complexity intersect. With the integration of artificial intelligence, machine learning, and sophisticated algorithms, robots are now capable of autonomous decision-making, adaptability, and even learning from their experiences. The exploration of advanced topics in robotics not only showcases the impressive capabilities of modern technology but also confronts us with ethical dilemmas and challenges that demand thoughtful consideration. From human-robot interaction in intricate environments to the fusion of robotics with virtual reality and simulations, these advanced topics illuminate the dynamic nature of the field and its profound impact on industries ranging from healthcare and manufacturing to entertainment and exploration. As we navigate these uncharted territories, it becomes evident that understanding and harnessing these advanced concepts is crucial for shaping the future of robotics in a rapidly evolving world.

In the fascinating realm of robotics, the journey from foundational principles to advanced topics is a captivating trajectory that mirrors the rapid pace of technological advancement. Advanced topics in robotics encapsulate a realm where innovation pushes the boundaries of what was once deemed possible. These topics delve into intricate algorithms, intricate mechanisms, and complex interactions that propel robotic systems into realms of autonomy, intelligence, and adaptability. From intricate motion planning and optimization algorithms that enable robots to navigate complex environments to the seamless integration of robots into virtual environments and simulations, these advanced topics reflect the symbiotic relationship between technology and human aspirations. As we explore the nuances of swarm robotics, soft robotics, and the ethical dimensions of AI, it becomes evident that these advanced inquiries have the power to reshape industries, enhance human capabilities, and provoke ethical discussions that guide responsible innovation. These explorations symbolize not just the evolution of robotics, but a testament to human ingenuity and our insatiable curiosity to unravel the mysteries of the universe through intelligent machines.

Swarm Robotics: Swarm robotics, encompassed within the realm of advanced topics, represents a paradigm shift in the way robotic systems operate by drawing inspiration from the collective behaviors of social insects like ants, bees, and termites. At its core, swarm robotics involves the coordination and collaboration of a large number of relatively simple robots, referred to as "agents," to accomplish complex tasks collectively. Unlike traditional single-robot approaches, swarm robotics capitalizes on the power of decentralized decision-making and emergent behavior. Each agent follows a set of simple rules, often influenced by local interactions and sensory information, without relying on central control. This decentralized nature grants swarm robots the ability to adapt swiftly to dynamic environments, recover from individual failures, and collectively achieve objectives that are beyond the capabilities of individual robots. By harnessing the principles of self-organization and cooperation, swarm robotics opens avenues for applications such as environmental monitoring, disaster response, exploration, and even artistic performances. However, the intricacies of designing effective communication mechanisms, ensuring robustness in the face of agent heterogeneity, and addressing ethical considerations related to the potential for unintended emergent behaviors challenge researchers in this burgeoning field. Swarm robotics, thus, not only showcases the potential for scalable, adaptable, and resilient systems but also demands a profound understanding of complex interactions and emergent phenomena to unlock their full capabilities.

Certainly, let's delve deeper into the advanced topic of "Swarm Robotics" across various facets:

Emergence of Collective Intelligence: Swarm robotics embodies the concept of "collective intelligence," where a group of relatively simple robots, collectively termed a "swarm," can achieve tasks that are complex and challenging for individual robots. This phenomenon is inspired by the behaviors observed in social insects, where the whole colony exhibits behaviors beyond the capabilities of individual insects. In swarm robotics, agents interact locally with their neighbors, resulting in emergent global behaviors that lead to efficient problem-solving, pattern formation, and navigation.

Decentralized Decision-Making: One of the hallmarks of swarm robotics is its reliance on decentralized decision-making. Instead of relying on a central control system, each agent operates based on local information and interactions with nearby agents. This approach enhances adaptability, as individual robots can respond to changes in the environment or the status of other robots without needing constant communication with a central entity.

Robustness and Scalability: Swarm robotics offers inherent robustness and scalability. As the number of agents' increases, the collective behavior tends to become more robust and fault-tolerant. If a few agents malfunction or are lost, the swarm can often continue its operation. This makes swarm robotics suitable for scenarios where redundancy and reliability are crucial, such as search and rescue missions in hazardous environments.

Applications across Domains: Swarm robotics finds applications in a wide range of domains. In environmental monitoring, swarms of robots can efficiently cover large areas to collect data on pollution levels or natural phenomena. In agriculture, they can collaborate to optimize crop yield and minimize resource usage. Disaster response scenarios benefit from swarm robots that can explore unstable terrains and locate survivors. Furthermore, swarm robotics has artistic applications, where robots coordinate to create intricate choreographies or patterns.

Challenges and Research Directions: Despite its potential, swarm robotics presents challenges. Designing effective communication mechanisms among agents while maintaining efficiency can be complex. Ensuring robustness and adaptability in heterogeneous swarms—where robots may differ in capabilities—requires innovative approaches. Ethical considerations arise regarding unintended emergent behaviors and the impact on natural ecosystems if swarm robots are deployed in the wild. Balancing individual behaviors with collective objectives is another challenge that researchers strive to address.

Future Prospects: Swarm robotics continues to evolve as researchers explore novel algorithms, communication protocols, and strategies. The development of bio-inspired techniques, such as mimicking the pheromone trails of ants, holds promise for more efficient swarm coordination. Additionally, advances in artificial intelligence and machine learning contribute to enhancing swarm behavior prediction and optimization.

Collective Behavior and Emergence: At the heart of swarm robotics lies the concept of collective behavior and emergence. Individual robots, equipped with limited intelligence, cooperate through local interactions to create emergent group behaviors. This phenomenon mirrors the behavior observed in natural systems, such as bird flocks or fish schools, where coordinated movement emerges from the actions of individual entities. In swarm robotics, emergent behaviors enable the accomplishment of tasks that are beyond the scope of a single robot's capabilities.

Decentralized Control and Self-Organization: Swarm robotics challenges the conventional centralized control paradigm. Each robot in a swarm operates independently, making decisions based on its local environment and interactions with nearby robots. This decentralized control mechanism grants swarm robots the ability to adapt rapidly to changes, navigate complex environments, and recover from unforeseen situations. The principle of self-organization, where global behaviors arise from simple local interactions, is central to achieving efficient and robust swarm performance.

Cooperation and Task Allocation: Swarm robotics encompasses various models of cooperation and task allocation. Cooperative behaviors can range from foraging in a dynamic environment to exploring unknown territories collectively. Task allocation involves distributing different tasks among swarm members based on their individual capabilities and the demands of the mission. These mechanisms ensure that the swarm optimally utilizes its resources to achieve common objectives.

Communication and Information Sharing: Effective communication among swarm members is pivotal for coordinated actions. Robots exchange information, such as position, status, and sensed data, to maintain situational awareness and align their actions. Communication methods range from simple local signaling to complex information dissemination strategies. Innovations in communication protocols enhance the swarm's ability to synchronize behaviors and respond to environmental changes.

Applications in Real-World Scenarios: Swarm robotics finds applications in various real-world scenarios, spanning diverse domains. In environmental monitoring, swarms can be deployed to map pollution levels, study wildlife behaviors, or track weather patterns. Disaster response benefits from swarms that search for survivors in hazardous environments, where human intervention may be risky. Industrial applications include tasks like warehouse automation, where swarms of robots collaborate to optimize logistics and improve efficiency.

Designing Robust and Scalable Algorithms: Developing algorithms that facilitate effective swarm coordination, while accounting for factors like scalability, heterogeneity, and dynamic environments, remains a significant research challenge. Balancing the trade-off between exploration and exploitation, ensuring fairness in task allocation, and managing potential conflicts among swarm members are ongoing areas of exploration.

Ethical Considerations and Social Impact: Deploying swarm robots introduces ethical considerations. Unintended emergent behaviors, privacy concerns, and potential effects on local ecosystems must be carefully evaluated. Additionally, questions regarding accountability and control in situations where robots make autonomous decisions arise. Addressing these ethical aspects ensures responsible and beneficial deployment of swarm robotics technologies.

In essence, swarm robotics stands as a fascinating realm that bridges natural behavior with technological innovation. By harnessing the power of collective intelligence, decentralized control, and emergent behaviors, swarm robotics offers solutions to intricate challenges across industries. As researchers continue to unravel the complexities of swarm dynamics and refine their algorithms, the potential of this advanced topic to reshape our interaction with robotics and redefine problem-solving methodologies remains an exciting and impactful journey.

Soft Robotics: In the landscape of advanced robotics, the emergence of "Soft Robotics" marks a revolutionary departure from the traditional rigid structures and mechanisms that have defined the field for decades. Soft robotics is an interdisciplinary domain that draws inspiration from biology and materials science to create robots composed of flexible, deformable materials. Unlike their traditional counterparts, which rely on rigid components, soft robots emulate the pliability and compliance found in natural organisms. This departure from rigid structures enables soft robots to interact with their environments in ways that were previously challenging or impossible. Soft robotics finds applications in areas such as medical devices, human-robot interaction, and exploration of fragile environments. By employing materials like elastomers and textiles, soft robots can adapt to complex surfaces, handle delicate objects, and navigate constrained spaces. This groundbreaking paradigm also brings forth challenges in control, modeling, and fabrication techniques. The integration of advanced sensors and actuators into soft robotic systems further enhances their dexterity and adaptability. As researchers delve deeper into this fascinating domain, the potential of soft robotics to revolutionize industries and redefine the relationship between robots and their surroundings becomes increasingly evident.

Biologically-Inspired Flexibility: Soft robotics draws inspiration from nature, mimicking the pliable characteristics of biological organisms. Unlike traditional rigid robots, soft robots are constructed from materials that can deform, bend, and stretch. This biologically-inspired flexibility enables them to navigate complex and unpredictable environments, opening doors to applications that demand delicate interactions or adaptation to variable terrains.

Deformable Structures and Actuators: Soft robotics leverages a range of deformable structures and actuators to achieve movement and manipulation. Pneumatic networks, fluidic channels, and shape-memory alloys are some examples of components that enable controlled deformation. By inflating or deflating chambers, these robots can achieve versatile movements and transformations, making them suitable for tasks that require intricate motion patterns.

Human-Centric Applications: The pliability and safety of soft robots make them well-suited for applications where close human interaction is required. In medical contexts, soft robots can navigate through intricate body structures with minimal risk of injury. Rehabilitation devices employing soft robotics facilitate natural and comfortable movement, aiding patients in their recovery journey.

Unconventional Locomotion: Soft robotics explores unconventional methods of locomotion. For instance, some soft robots utilize crawling, undulating, or slithering movements that mimic the locomotion of snakes or caterpillars. These modes of motion allow soft robots to traverse challenging terrains, such as rubble or tight spaces, where traditional robots might struggle.

Soft Grippers and Manipulators: Soft robotics excels in the design of adaptable grippers and manipulators. Soft grippers can conform to objects of varying shapes and sizes, making them suitable for tasks like picking up fragile items or handling irregular objects. The inherent compliance of soft manipulators reduces the risk of damage during interactions.

Challenges and Fabrication Techniques: Developing soft robots presents unique challenges. Modeling and controlling deformable structures require sophisticated approaches that account for the complex interaction between materials and forces. Fabrication techniques involve innovative methods such as 3D printing, molding, and textile integration, as traditional manufacturing methods may not be applicable.

Hybrid Systems: Hybrid systems that combine rigid and soft components are gaining traction. These systems capitalize on the strengths of both rigid and soft robotics, enabling precise control and manipulation while benefiting from the adaptability and compliance of soft structures. This integration broadens the scope of applications and potential robotic capabilities.

Future Directions: Soft robotics continues to evolve as researchers explore new materials, actuation methods, and control strategies. Developments in smart materials, such as those that change shape in response to external stimuli, hold promise for expanding the capabilities of soft robots. Additionally, advances in AI and machine learning enhance their autonomy and adaptability.

In summary, biohybrid and bio-inspired robotics push the boundaries of traditional robotics by integrating biological principles with artificial technology. This fascinating convergence not only broadens the scope of robotic capabilities but also raises questions about ethics, autonomy, and the potential for new applications. By drawing inspiration from nature and harnessing the power of living systems, this advanced topic exemplifies the synergy of biology and technology in shaping the future of robotics.

Bio-inspired Robotics: In the captivating realm of advanced robotics, "Bio-inspired Robotics" emerges as a field that draws inspiration from the intricate designs and adaptive mechanisms found in the natural world. It seeks to replicate the remarkable solutions that evolution has sculpted over millions of years, imbuing robots with capabilities that mirror the efficiency and elegance of biological organisms. Bio-inspired robotics delves into the exploration of locomotion, perception, sensing, and interaction patterns that have evolved in animals and plants. By mimicking nature's blueprints, researchers aim to create robots that are not only more adept at navigating diverse environments but also exhibit a higher degree of adaptability and resilience. From the study of avian flight for the development of drones to the analysis of insect locomotion for agile robotic platforms, bio-inspired robotics pioneers solutions that surpass traditional engineering methods. This field sparks interdisciplinary collaboration, as engineers, biologists, and computer scientists collaborate to translate the nuances of nature into innovative robotic systems. The potential applications span a wide spectrum, including exploration of challenging terrains, disaster response, and even medical advancements such as prosthetics that emulate natural limb movements. As the boundary between technology and biology blurs, bio-inspired robotics ignites the imagination, offering the promise of a future where robots and nature's ingenuity intertwine to redefine the capabilities of both worlds.

Nature's Design Inspiration: Bio-inspired robotics draws inspiration from nature's remarkable designs. From the wings of birds to the locomotion of insects, researchers dissect biological mechanisms to uncover principles that can be translated into robotic systems. By mimicking nature's solutions, robots gain the ability to navigate complex environments and perform tasks with greater efficiency and adaptability.

Leveraging Biomimicry: Biomimicry involves imitating biological functions to solve engineering challenges. Bio-inspired robots leverage biomimicry to replicate functionalities such as propulsion, sensing, and manipulation. For instance, robots designed with flexible and compliant structures imitate the locomotion of animals, enabling them to traverse challenging terrains with ease.

Locomotion Strategies: Locomotion is a key area of exploration in bio-inspired robotics. Robots borrow locomotion strategies from various creatures, including crawling, slithering, and flying. These adaptations enable robots to navigate environments that were previously inaccessible to rigid robots, such as disaster zones or remote landscapes.

Sensory Systems: Nature's finely tuned sensory systems serve as models for bio-inspired robots. Robots equipped with visual, auditory, or tactile sensors emulate the way animals perceive their surroundings. This enhances robots' ability to gather information, detect obstacles, and make informed decisions in dynamic environments.

Soft Robotics and Muscle-Like Actuators: Soft robotics, inspired by the flexibility of biological tissues, employs pliable materials and muscle-like actuators. These components enable robots to achieve fluid movements and manipulate objects with dexterity. Soft grippers, for example, replicate the way organisms interact with their environment.

Adaptation and Resilience: Bio-inspired robots embrace the adaptability and resilience seen in nature. These robots can alter their behaviors based on environmental changes, akin to how animals adjust to evolving circumstances. This adaptability ensures that robots can accomplish tasks efficiently and reliably in unpredictable conditions.

Applications in Exploration: Bio-inspired robots find applications in exploration scenarios. Robots modeled after animals can navigate environments that are challenging for traditional robots, such as underwater caves or complex terrains. These robots aid in collecting data, conducting surveys, and uncovering insights in remote and hazardous locations.

Interdisciplinary Collaboration: Bio-inspired robotics thrives at the intersection of diverse disciplines. Collaboration between biologists, engineers, materials scientists, and computer scientists fosters the integration of biological insights with robotic innovations. This interdisciplinary approach enriches the field and accelerates the development of novel solutions.

Biological Ethics and Sustainability: While bio-inspired robots offer tremendous potential, ethical considerations arise when living organisms are integrated into robotic systems. Ensuring the welfare of biological components and addressing ethical concerns surrounding animal use in research is crucial. Additionally, bio-inspired robotics promotes sustainability by leveraging nature's efficiency and minimizing the environmental impact of robots.

Future Horizons: The field of bio-inspired robotics continues to evolve with advancements in materials science, sensors, and AI. Researchers strive to create robots that closely emulate the versatility and elegance of biological organisms. This pursuit holds the promise of revolutionizing industries, from healthcare to exploration, by offering robots that can seamlessly integrate into natural ecosystems and perform tasks with a level of intricacy previously unseen in traditional robotics.

Neuromorphic Computing: Neuromorphic computing involves creating hardware and algorithms that mimic the neural networks of the brain. Bio-inspired robots integrate neuromorphic components to process sensory data and make decisions more akin to biological organisms. These robots can exhibit cognitive capabilities, learn from experiences, and adapt to dynamic environments.

Evolutionary Algorithms: Evolutionary algorithms imitate the process of natural selection to optimize robot designs and behaviors. By subjecting virtual robot populations to simulated environmental challenges, researchers can identify traits and behaviors that lead to improved performance. This approach accelerates the development of robots that are better suited for specific tasks or environments.

Swarm Intelligence: Bio-inspired robotics explores the behaviors of social organisms, such as ants and bees, to develop swarm robotics. Swarm robots operate collectively, leveraging local interactions to achieve global objectives. These systems excel in tasks such as search and rescue, environmental monitoring, and exploration of inaccessible areas.

Biorobotics and Biomechanics: Biorobotics involves replicating the biomechanics of living organisms to create robots with similar movement and functionality. Researchers study the mechanics of animal locomotion, from fish swimming to bird flight, to engineer robots with biomimetic motions. These robots find applications in fields like underwater exploration and aerospace.

Plant-Inspired Robotics: Bio-inspired robotics expands beyond animals to include plants as sources of inspiration. Plant-inspired robots replicate mechanisms like growth, bending towards light, or even photosynthesis. These robots have potential applications in environmental monitoring, agriculture, and biomimetic design.

Morphological Computation: Morphological computation focuses on utilizing the physical properties of a robot's body to simplify control and computation. This approach emphasizes the interaction between the robot's body and the environment, reducing the need for complex control algorithms. Robots designed with morphological computation can adapt their shapes to perform specific tasks efficiently.

Biohybrid Symbiosis: Biohybrid robots combine biological components with artificial systems. Muscle cells, neurons, or even living organisms can be integrated into robotic platforms to enhance their capabilities. Biohybrid robots enable the exploration of human-robot interfaces, neural control, and the potential for creating robots that harness the power of living cells.

Artificial Evolution: Artificial evolution simulates the process of genetic evolution in software to evolve robot designs or behaviors. Through successive generations, robots with improved traits are selected, leading to the emergence of more efficient solutions. This technique facilitates the rapid development of robots suited for specific tasks and environments.

Biological Feedback and Learning: Bio-inspired robots incorporate feedback mechanisms inspired by biological organisms. These robots can learn from their experiences, adjust their behaviors based on outcomes, and improve over time. This learning capability enables robots to adapt to changing conditions and optimize their performance.

Bioethics and Regulation: The integration of living components into robots raises ethical concerns about the treatment of living organisms. The field of bio-inspired robotics prompts discussions about the ethical boundaries of using biological materials in robots and the implications for animal welfare. Regulation and guidelines are crucial to ensure responsible research and development.

In essence, the world of bio-inspired robotics encompasses a spectrum of topics that explore nature's solutions for engineering challenges. From neuromorphic computing to evolutionary algorithms and beyond, these topics converge to create robots that mirror the elegance, adaptability, and complexity of the natural world. By embracing the wisdom of evolution, researchers in bio-inspired robotics pioneer groundbreaking solutions that redefine the potential of robotic systems across industries.

Neural Networks and Deep Learning in Robotics: Neural networks, inspired by the human brain's interconnected neurons, form the foundation of deep learning. In robotics, neural networks offer a versatile tool to process data, recognize patterns, and make informed decisions. These networks consist of layers of interconnected nodes, each emulating a neuron's activation. By adjusting the connections' strengths through training, neural networks can learn complex relationships and extract meaningful features from raw data.

Convolutional Neural Networks (CNNs): CNNs specialize in processing grid-like data, such as images and videos. They consist of convolutional layers that automatically learn and detect patterns in visual data, enabling object recognition, image segmentation, and even depth perception. In robotics, CNNs are pivotal for tasks like object manipulation and navigation, as they provide robots with the ability to perceive and interpret their environment visually.

Recurrent Neural Networks (RNNs): RNNs are designed to work with sequential data, making them suitable for time-series or sequential tasks. They possess memory cells that retain information from previous time steps, allowing them to process dynamic sequences. In robotics, RNNs excel in tasks like natural language understanding, gesture recognition, and even motion planning, where temporal dependencies play a crucial role.

Long Short-Term Memory (LSTM) Networks: LSTMs are a type of RNN designed to overcome the "vanishing gradient" problem by maintaining long-term dependencies in data. These networks are particularly effective at capturing context and patterns in sequential data. In robotics, LSTMs find application in tasks requiring memory-based decision-making, such as grasping objects with varying shapes and sizes.

Generative Adversarial Networks (GANs): GANs consist of two neural networks—a generator and a discriminator—competing against each other. The generator creates data instances, while the discriminator evaluates their authenticity. Through this adversarial process, GANs generate data that closely resembles real-world samples. In robotics, GANs have been used to generate realistic synthetic training data for robotic vision tasks, facilitating robust learning.

Transfer Learning and Fine-Tuning: Transfer learning leverages pre-trained neural network models on large datasets to kick start training on specific tasks or environments with limited data. This approach saves time and resources, enabling robots to learn new tasks more efficiently. Fine-tuning adapts the pre-trained model to the robot's specific requirements, enhancing its performance in the target domain.

Reinforcement Learning with Neural Networks: Reinforcement learning involves training agents to make decisions based on interactions with an environment. Neural networks serve as function approximates to model an agent's policy or value function. In robotics, reinforcement learning with neural networks has enabled robots to learn complex tasks through trial and error, achieving impressive feats like autonomous grasping and walking.

Sensor Fusion and Perception: Neural networks are pivotal in sensor fusion, where data from various sensors are combined to create a holistic perception of the environment. By processing inputs from cameras, LiDAR, and other sensors, neural networks enable robots to create accurate and comprehensive representations of their surroundings, enhancing navigation and decision-making.

Challenges and Ethics: Despite their power, neural networks also present challenges such as overfitting, interpretability, and bias in decision-making. Ethical considerations arise when deploying robots with deep learning capabilities, particularly regarding transparency in decision-making and potential biases in training data.

Interdisciplinary Collaborations: Neural networks and deep learning thrive at the intersection of robotics, computer science, and machine learning. Collaboration between experts in these fields accelerates the development of advanced robotic systems, enabling robots to learn, adapt, and perform tasks with a level of complexity previously unattainable.

Attention Mechanisms: Attention mechanisms enable neural networks to focus on specific parts of input data, enhancing their ability to process and extract relevant features. In robotics, attention mechanisms aid in tasks like object detection, where the network learns to selectively attend to important regions in an image, improving accuracy and efficiency.

Self-Supervised Learning: Self-supervised learning leverages unlabeled data to pretrain neural networks, which are then fine-tuned on labeled data for specific tasks. This approach reduces the need for extensive labeled datasets and benefits robotics applications where data collection is challenging or expensive.

Capsule Networks: Capsule networks, an innovative approach, aim to address the limitations of traditional neural networks in handling hierarchical relationships between features. These networks can represent complex spatial arrangements, making them suitable for tasks like pose estimation, where capturing spatial hierarchies is crucial.

Explainable AI and Interpretability: As neural networks become more complex, understanding their decision-making processes becomes vital. Research focuses on creating models that provide explanations for their predictions, enhancing transparency and trust in robotics applications, such as medical diagnosis and autonomous driving.

Meta-Learning: Meta-learning involves training neural networks to learn how to learn. These networks acquire knowledge from multiple tasks, enabling rapid adaptation to new tasks with limited data. In robotics, meta-learning facilitates quick adaptation to changing environments and tasks, reducing the need for extensive retraining.

One-Shot Learning: One-shot learning addresses scenarios where a neural network must learn from only one or a few examples of each class. This is particularly valuable in robotics applications where acquiring large amounts of labeled data is impractical, such as customized object recognition for manipulation tasks.

Quantum Machine Learning: Quantum machine learning explores the intersection of quantum computing and neural networks. Quantum neural networks leverage quantum states to perform computations, potentially accelerating deep learning tasks and enabling robots to process complex data more efficiently.

Neuromorphic Hardware: Neuromorphic hardware aims to create specialized hardware architectures that mimic the brain's computation principles. These hardware accelerators enable real-time processing of neural networks, making them suitable for applications where low-latency decision-making is essential, such as robotics for autonomous vehicles.

Continual Learning and Catastrophic Forgetting: Continual learning addresses the challenge of neural networks forgetting previously learned tasks when trained on new ones. Catastrophic forgetting occurs when learning a new task erases knowledge of previous tasks. Research focuses on developing algorithms that allow robots to learn new tasks without losing proficiency in previous ones.

Ethical Considerations: The adoption of neural networks in robotics raises ethical questions surrounding accountability, transparency, and the potential biases embedded in learned models. Ensuring fairness, preventing bias, and establishing regulations are crucial to deploying neural network-powered robots responsibly.

Real-World Robotic Deployments: Neural networks have enabled robots to excel in real-world environments, from autonomous driving to warehouse automation. The deployment of deep learning models in robotics highlights the necessity of robustness, adaptability, and continuous learning to ensure safe and reliable operation.

In summary, the integration of neural networks and deep learning into robotics propels the field toward unprecedented levels of intelligence and adaptability. These professional topics delve into the cutting-edge advancements, challenges, and ethical considerations that shape the trajectory of neural network-powered robotics, ultimately redefining the capabilities and potential of robotic systems across industries.

Exercise

Multiple-choice Questions

1. What is the primary goal of swarm robotics?
a) Individual robot optimization
b) Centralized control
c) Mimicking human behavior
d) Collective behavior through decentralized control

2. Soft robotics focuses on:
a) Rigid structures for precise movements
b) Mimicking human anatomy
c)Using traditional actuators for increased strength
d) Flexible and compliant structures for versatile movements

3. Bio-inspired robotics draws inspiration from:
a) Artificial intelligence only
b) Human-made technologies
c) Biological organisms and natural processes
d) Theoretical concepts in physics

4. Neural networks are inspired by:
a) Quantum computing
b) The human brain's interconnected neurons
c) Traditional computer programming languages
d) Mechanical engineering principles

5. Convolutional Neural Networks (CNNs) are well-suited for processing:
a) Text data
b) Tabular data
c) Sequential data
d) Image data

6. What type of neural network is designed to work with sequential data?
a) Convolutional Neural Networks (CNNs)
b) Recurrent Neural Networks (RNNs)
c) Long Short-Term Memory (LSTM) Networks
d) Generative Adversarial Networks (GANs)

7. What does Transfer Learning involve?
a) Training neural networks without any pre-existing knowledge
b) Learning from scratch for each new task
c) Using pre-trained models to kickstart training on specific tasks
d) Using only labeled data for training
8. What is the primary focus of Reinforcement Learning?
a) Unsupervised learning
b) Supervised learning
c) Learning from sequential data
d) Learning through interactions with an environment
9. Generative Adversarial Networks (GANs) consist of:
a) A single neural network
b) Two neural networks competing against each other
c) Multiple interconnected neural networks
d) A combination of convolutional and recurrent networks
10. What is the purpose of Attention Mechanisms in neural networks?
a) To minimize overfitting
b) To focus on specific parts of input data
c) To speed up training
d) To reduce the number of parameters

Review Questions:

1. What is the main objective of swarm robotics, and how does it differ from traditional robotics approaches?
2. How does soft robotics differ from traditional rigid robotics, and what advantages does it offer in certain applications?
3. Explain the concept of bio-inspired robotics and provide examples of robotic systems inspired by biological organisms or processes.
4. What is the fundamental principle behind neural networks, and how do they contribute to the field of deep learning?
5. How does a Convolutional Neural Network (CNN) differ from a Recurrent Neural Network (RNN) in terms of data processing and applications?
6. Describe the concept of transfer learning in neural networks. How does it benefit robotics applications with limited data?
7. What is the significance of Generative Adversarial Networks (GANs) in robotics, and how do they generate synthetic data?
8. How do attention mechanisms enhance the capabilities of neural networks? Provide an example of a robotics application where attention mechanisms are beneficial.
9. Explain the concept of self-supervised learning and its role in training neural networks using unlabeled data.
10. How do Capsule Networks address limitations of traditional neural networks in handling spatial hierarchies? Give an example of a task where Capsule Networks are advantageous.
11. What is Reinforcement Learning, and how does it differ from other learning paradigms in the context of neural networks?
12. Describe the concept of Continual Learning and its relevance in enabling robots to learn new tasks without forgetting previously acquired knowledge.
13. How does ethical consideration play a role in the deployment of neural networks in robotics applications? Provide examples of ethical challenges in this domain.
14. What role does neuromorphic hardware play in enhancing the processing capabilities of neural networks in real-time robotic applications?
15. Discuss the interdisciplinary nature of neural networks and deep learning in robotics, highlighting the collaboration between robotics, computer science, and machine learning.
16. How does one-shot learning address the challenge of limited labeled data in robotics applications? Provide an example scenario where one-shot learning is advantageous.
17. What is the concept of quantum machine learning, and how does it leverage neural networks to process data more efficiently?
18. Explain the importance of explainable AI and interpretability in neural network-powered robotics. How does this impact real-world applications?

19. What is the significance of meta-learning in robotics? How does it enable robots to adapt quickly to new tasks and environments?
20. In what ways can neural networks contribute to sensor fusion and perception in robotics? Provide examples of tasks where sensor fusion enhances robotic perception.
21. Describe the concept of catastrophic forgetting in neural networks. How does continual learning aim to mitigate this issue in robotic applications?
22. How can bio-inspired robotics benefit fields beyond traditional robotics? Provide examples of industries or applications that can leverage bio-inspired concepts.
23. How do advancements in neural networks contribute to enhancing the safety and reliability of autonomous vehicles in the field of robotics?
24. Discuss the potential societal impacts of deploying robotics systems powered by neural networks. What considerations should be taken into account to ensure responsible deployment?
25. How might the integration of advanced neural networks into robotics reshape industries and job roles in the future? What opportunities and challenges could arise?

Answers Multiple-choice Questions

Answers:
1. d)
2. d)
3. c)
4. b)
5. d)
6. b)
7. c)
8. d)
9. b)
10. b)

Review Questions:

1. The main objective of swarm robotics is to harness the collective power of a group of simple robots, known as swarm agents, to accomplish tasks that are often complex, distributed, or difficult for a single robot to achieve. Unlike traditional robotics approaches that often rely on centralized control and intricate programming for individual robots, swarm robotics emphasizes decentralized control and emergent behavior arising from interactions among the swarm agents. This distinction lies in the fact that while traditional robotics focuses on optimizing the capabilities of a single robot to perform specific tasks, swarm robotics focuses on leveraging the interactions and cooperation of multiple robots to achieve tasks collectively. This difference highlights the shift from relying on complex individual robots to harnessing the power of numbers, leading to robustness, adaptability, and scalability in swarm robotics applications.
2. Soft robotics represents a departure from the traditional rigid robotics paradigm by utilizing materials and mechanisms that are compliant, flexible, and deformable. Unlike traditional robots that are constructed using rigid materials like metal and plastic, soft robots are constructed using elastomers and flexible materials that allow them to deform and adapt to their environment. This approach offers several advantages in specific applications. Soft robots can interact safely with humans and delicate objects, making them suitable for tasks like medical procedures, where precision and safety are crucial. The inherent flexibility of soft robots enables them to navigate through confined spaces and interact with complex and unstructured environments, such as disaster recovery scenarios or exploration in tight spaces. Moreover, their deformable nature enables them to absorb impacts, increasing their resilience in unpredictable environments.
3. Bio-inspired robotics involves designing and developing robotic systems that draw inspiration from biological organisms or processes found in nature. These systems emulate the behavior, morphology, and adaptability of living organisms to achieve specific functionalities. An example of bio-inspired robotics is the development of robotic fish that replicate the movement and hydrodynamics of real fish, enabling them to maneuver efficiently in water environments. Another example is the creation of robotic limbs that mimic the movement and dexterity of human limbs, allowing amputees to regain mobility and perform intricate tasks. Additionally, biomimetic flying robots imitate the flight dynamics of birds and insects, enabling them to navigate complex environments with agility. These examples demonstrate how bio-inspired robotics leverages insights from nature to design robots that excel in specific tasks or environments.

4. The fundamental principle behind neural networks lies in their attempt to simulate the functioning of the human brain's interconnected neurons. Neural networks consist of interconnected nodes (neurons) organized in layers. Each neuron processes inputs, applies weights and biases, and produces an output. The network learns from data by adjusting these weights and biases through a process called training. Neural networks contribute to the field of deep learning by enabling the creation of complex models capable of learning representations from data. Deep learning leverages the power of neural networks with multiple hidden layers to automatically extract hierarchical features from data, enabling tasks such as image recognition, natural language processing, and robotics perception.

5. Convolutional Neural Networks (CNNs) and Recurrent Neural Networks (RNNs) have distinct architectures and applications. CNNs are designed to process grid-like data, such as images, by using convolutional layers that capture local patterns. They excel in tasks like image classification, object detection, and image generation. RNNs, on the other hand, are specialized for sequential data processing, making them suitable for tasks involving sequences, such as time series analysis, speech recognition, and language modeling. RNNs maintain an internal memory to capture temporal dependencies within sequential data. This differentiation underscores that CNNs are optimized for spatial data, while RNNs are tailored for sequential patterns.

6. Transfer learning is a strategy in which a pre-trained neural network model is adapted to perform a new task, even when only limited task-specific data is available. The idea is to leverage knowledge acquired during the training on a source task (usually a large dataset) to enhance performance on a target task. This is particularly advantageous in robotics applications where collecting labeled data can be challenging or expensive. For instance, a neural network pre-trained on a large dataset of real-world images (source task) can be fine-tuned on a smaller dataset of specific robotic images (target task). Transfer learning allows the model to generalize its understanding from the source domain to the target domain, improving learning efficiency and performance.

7. Generative Adversarial Networks (GANs) hold significant importance in robotics as they facilitate the generation of synthetic data that closely resembles real data distributions. GANs consist of two networks: a generator and a discriminator. The generator creates synthetic data samples, while the discriminator differentiates between real and synthetic data. Through an adversarial training process, the generator continually improves its ability to create data that is indistinguishable from real data. This is valuable in robotics, where acquiring sufficient real-world data can be expensive or impractical. GANs enable the generation of synthetic training data for tasks such as object recognition, simulation, and sensor calibration, improving the robustness and generalization of robotic systems.

8. Attention mechanisms enhance neural networks by allowing them to focus on specific parts of input data that are relevant to the task. These mechanisms enable networks to learn to allocate more attention to important features and less attention to less relevant ones. In a robotics application, consider a robotic vision system for object manipulation. By applying attention mechanisms, the network can dynamically focus on relevant regions of an image, such as the object to be manipulated, while filtering out background noise. This attention-based selection improves the accuracy and efficiency of the network's predictions and actions, making it a valuable tool for tasks requiring selective perception.

9. Self-supervised learning is a paradigm in which neural networks learn from unlabeled data by generating auxiliary tasks based on the data itself. In robotics, self-supervised learning leverages the inherent structure of the data to create learning objectives. For instance, a robot can predict the next frame in a video sequence or estimate the relative position of objects in an image. By creating such auxiliary tasks, the network learns to capture meaningful representations of the data, which can then be transferred to other tasks. Self-supervised learning is particularly useful when labeled data is scarce or expensive to obtain, enabling robots to utilize abundant unlabeled data for training.

10. Capsule Networks address a limitation of traditional neural networks in handling spatial hierarchies and relationships between parts of an object. Traditional networks struggle to recognize objects when parts are arranged differently, as their hierarchical structure is not explicitly encoded. Capsule Networks introduce "capsules," which are groups of neurons that represent specific features or parts and their relationships within an object. This allows the network to recognize objects based on the spatial arrangement of their parts, improving generalization and robustness. An example application is image classification, where Capsule Networks can recognize objects from different viewpoints and orientations.

11. Reinforcement Learning is a learning paradigm that involves an agent interacting with an environment to learn optimal actions that maximize a reward signal. Unlike supervised learning where the model is trained on labeled data, reinforcement learning agents learn through trial and error, exploring the environment to find the best actions that lead to the highest cumulative rewards. This learning paradigm is particularly well-suited for robotic applications where an agent needs to learn from interactions in dynamic and uncertain environments. Reinforcement Learning differs from other learning paradigms in that it focuses on sequential decision-making and aims to find a balance between exploration (trying new actions) and exploitation (choosing actions with known high rewards). It

has found applications in robotics tasks such as robotic control, navigation, and manipulation, enabling robots to adapt and learn in real-time scenarios.

12. Continual Learning, also known as lifelong learning or incremental learning, refers to the ability of a neural network to learn new tasks without forgetting previously learned tasks. In the context of robotics, continual learning is crucial for robots that need to adapt to new tasks, environments, or situations without losing their previously acquired knowledge. Traditional neural networks tend to suffer from catastrophic forgetting, where learning new tasks erases knowledge of previous tasks. Continual learning techniques aim to mitigate this by selectively retraining the network on new data while preserving knowledge from previous tasks. This enables robots to accumulate skills over time and gradually build a diverse set of capabilities, making them more versatile and adaptive in dynamic scenarios.

13. Ethical considerations play a vital role in the deployment of neural networks in robotics applications. As neural networks become increasingly integrated into various aspects of society, they raise ethical challenges related to bias, transparency, accountability, and safety. For example, a neural network-powered autonomous vehicle must make split-second decisions that impact human lives. Ensuring that these decisions are fair, unbiased, and transparent becomes critical. Additionally, the data used to train neural networks can inadvertently reflect human biases, leading to biased predictions and actions. It's essential to implement safeguards and rigorous testing to identify and rectify biases. Ethical challenges also arise in the context of deepfakes, where neural networks can create realistic but misleading content. Responsible deployment of neural networks in robotics demands a thorough understanding of these ethical challenges and the development of guidelines and regulations to address them.

14. Neuromorphic hardware refers to specialized hardware architectures designed to mimic the structure and functioning of the human brain's neural networks. These hardware systems leverage the principles of neuromorphic engineering to process information in ways that emulate the brain's synaptic connections and spiking behavior. In robotics, neuromorphic hardware plays a significant role in enhancing the processing capabilities of neural networks for real-time applications. Traditional neural networks often require substantial computational resources, limiting their deployment in time-critical tasks. Neuromorphic hardware offers energy-efficient and parallel processing that accelerates the execution of neural network algorithms. This advancement is crucial for applications like real-time perception, decision-making, and control in robotics, enabling robots to operate swiftly and effectively in dynamic environments.

15. The interdisciplinary nature of neural networks and deep learning in robotics underscores the collaboration between robotics, computer science, and machine learning. Robotics leverages neural networks and deep learning techniques to enhance perception, decision-making, and control, enabling robots to perform complex tasks in various domains. Computer science contributes by developing efficient algorithms, architectures, and tools to train and deploy neural networks. Machine learning provides the theoretical foundation and methodologies for training and optimizing neural networks. The synergy between these disciplines has propelled innovations in robotics, ranging from autonomous vehicles to medical robots. This collaborative effort continues to push the boundaries of what robots can achieve, driving advancements in fields like healthcare, manufacturing, exploration, and entertainment.

16. One-shot learning addresses the challenge of limited labeled data by training a model to recognize new objects or classes with only a single or a few examples. This is particularly relevant in robotics applications where collecting large amounts of labeled data for every possible object or scenario is impractical. One-shot learning leverages techniques such as Siamese networks or prototypical networks. Siamese networks learn to distinguish between pairs of data points, making them suitable for tasks like object verification. Prototypical networks create a prototype representation for each class based on a few examples, enabling classification with minimal data. An example scenario is training a robot to recognize new tools in a workshop environment using only a few images of each tool. One-shot learning enables robots to quickly adapt to new objects without the need for extensive training data.

17. Quantum machine learning is an emerging field that combines quantum computing principles with machine learning algorithms to perform data processing more efficiently than classical computers. In the context of neural networks, quantum machine learning aims to leverage the unique properties of quantum systems, such as superposition and entanglement, to enhance the processing capabilities of neural networks. Quantum neural networks have the potential to solve complex optimization problems faster, which is relevant in robotics for tasks such as path planning and optimization. Quantum machine learning also offers the potential for improved data representation and pattern recognition. However, this field is still in its early stages, and practical implementations and applications in robotics are ongoing areas of research.

18. Explainable AI (XAI) and interpretability are crucial aspects of deploying neural network-powered robotics systems responsibly. XAI refers to the ability of a model to provide understandable explanations for its decisions and predictions. Interpretability, on the other hand, involves understanding the inner workings of neural networks to

identify factors influencing their decisions. In robotics, XAI is essential to ensure that robots' actions can be explained and justified, especially in safety-critical applications like autonomous vehicles and medical robots. Interpretable models help users, developers, and regulators understand how decisions are made, enabling accountability and trust. The integration of XAI techniques allows stakeholders to comprehend complex neural network behaviors, uncover biases, and identify potential sources of errors, making neural network-powered robotics more transparent and reliable.

19. Meta-learning is a concept where neural networks are trained to learn how to learn. This involves training a model on a variety of tasks, enabling it to acquire knowledge and strategies for adapting to new tasks more effectively. In robotics, meta-learning allows robots to quickly adapt to new environments, tasks, or scenarios by leveraging knowledge learned from previous experiences. For example, a robot trained through meta-learning might be able to navigate different terrains by generalizing knowledge about movement dynamics from previous experiences. This accelerates the learning process and reduces the need for extensive task-specific training. Meta-learning is crucial for robotics applications where robots need to continuously acquire new skills and adapt to changing conditions.

20. Neural networks contribute to sensor fusion and perception in robotics by enabling the integration of information from various sensors to create a comprehensive understanding of the environment. For instance, in autonomous vehicles, neural networks can fuse data from LiDAR, cameras, radar, and other sensors to accurately perceive the surroundings, detect objects, and make informed decisions. Similarly, in industrial automation, neural networks can fuse data from multiple sensors to monitor production lines, detect defects, and optimize processes. This fusion enhances perception accuracy and robustness, allowing robots to make more informed and reliable decisions in complex and dynamic environments.

21. Catastrophic forgetting occurs when a neural network trained on multiple tasks forgets previously learned tasks as it learns new ones. Continual learning addresses this issue by developing algorithms that allow neural networks to learn new tasks while preserving knowledge from previous tasks. For example, a robot trained to perform various tasks, such as grasping objects and navigating, should retain its proficiency in earlier tasks even as it learns new ones. Continual learning algorithms prioritize important knowledge and incorporate strategies like regularization and replay to mitigate catastrophic forgetting. This is crucial for robots that need to accumulate skills over time and adapt to new tasks without sacrificing previously acquired capabilities.

22. Bio-inspired robotics holds the potential to benefit various fields beyond traditional robotics. In healthcare, bio-inspired robots can mimic the behavior of biological organisms to develop prosthetics and assistive devices that closely replicate human mobility and function. In agriculture, robots inspired by the movement patterns of insects can be used for pollination and pest control. In aerospace, drones inspired by the flight mechanics of birds and insects can navigate complex environments. Additionally, biomimetic robots can contribute to environmental monitoring, disaster relief, and space exploration. The principles of adaptability, efficiency, and versatility found in nature inspire innovative solutions that can address real-world challenges across diverse industries.

23. Advancements in neural networks have a profound impact on enhancing the safety and reliability of autonomous vehicles. Neural networks enable vehicles to perceive their surroundings accurately through sensor fusion, making informed decisions to avoid obstacles, navigate complex traffic scenarios, and respond to dynamic environments. Additionally, neural networks facilitate advanced driver assistance systems (ADAS) that enhance vehicle safety by assisting with tasks like lane keeping, adaptive cruise control, and emergency braking. Deep learning models enable vehicles to interpret and predict the behavior of other road users, improving the overall safety of autonomous and human-driven vehicles alike.

24. The deployment of neural network-powered robotics systems raises significant societal impacts and ethical considerations. As these systems become more integrated into daily life, issues related to job displacement, privacy, bias, and security emerge. Ethical challenges involve ensuring that robots' decisions align with human values, addressing biases in data and algorithms, and determining liability in case of accidents. Responsible deployment requires collaboration between roboticists, policymakers, ethicists, and the public to establish guidelines, regulations, and frameworks that prioritize safety, fairness, and transparency. It is essential to anticipate these impacts and foster a multidisciplinary dialogue to shape the ethical landscape of neural network-powered robotics.

25. The integration of advanced neural networks into robotics has the potential to reshape industries and job roles in profound ways. Industries like manufacturing, logistics, healthcare, and agriculture can benefit from increased automation and efficiency through robotics. As robots equipped with neural networks take on complex tasks, human workers can shift to more strategic and creative roles, focusing on oversight, maintenance, and collaboration with robots. New job roles in robotics design, programming, and management are likely to emerge, requiring expertise in machine learning, computer science, and robotics. This evolution presents both opportunities and challenges,

requiring education, training, and policy adaptation to ensure a harmonious transition into a future where neural network-powered robots play a central role.

Online Resources

Online Exam Papers

Link: https://forms.gle/YGtxoaGyKYkmjLJ8A

QR Code:

YouTube Video Playlist Link

Link: https://www.youtube.com/playlist?list=PLzlh_31VWEFrDBMI3dTPgIWlUh97DWWQf

QR Code:

Chapter 11: Future Trends in Robotics

The field of robotics is on an exciting trajectory of continuous evolution, with future trends poised to revolutionize the way we interact with and benefit from robotic systems. As technology advances, new avenues are opening up, allowing us to explore innovative applications and solutions that were once confined to the realm of science fiction. The convergence of various disciplines, such as artificial intelligence, materials science, and bioengineering, is driving the development of robotic systems with unprecedented capabilities and adaptability. From swarming nabobs to human-robot collaboration in diverse industries, the future of robotics promises to reshape our world in ways that were previously unimaginable. In this era of rapid technological progress, it's crucial to delve into the potential trends that will shape the landscape of robotics and influence the way we live, work, and interact with machines.

Advancements in robotics have already transformed industries and daily life, but the future holds even more remarkable possibilities. As we stand on the cusp of a new era, several emerging trends are set to propel robotics into uncharted territories. One prominent trend is the fusion of AI and robotics, leading to machines that possess not just automation skills but also the ability to learn, adapt, and make decisions autonomously. Additionally, the integration of soft and flexible materials into robot design is revolutionizing their physical interactions with the environment, enabling safer and more versatile applications. Furthermore, the evolution of human-robot interaction is ushering in an era of collaboration, where robots become trusted partners in sectors like healthcare, manufacturing, and customer service. Embracing sustainability, robotics is also pivoting towards eco-friendly designs, addressing environmental concerns. With the potential to reshape industries, redefine job roles, and improve quality of life, the future trends in robotics are captivating, raising questions about ethics, regulation, and the role of humans in an increasingly robotic world.

The future of robotics is characterized by a convergence of cutting-edge technologies and innovative approaches, paving the way for transformative trends that will shape our world in profound ways. As we peer into the horizon of possibilities, several key trends stand out. One of the most compelling trends is the rise of autonomous robotics, where machines equipped with advanced sensors, AI algorithms, and precise control mechanisms will navigate and interact with their environment without constant human intervention. This shift towards autonomy has the potential to revolutionize industries such as logistics, transportation, and agriculture, where robots can perform intricate tasks with efficiency and accuracy.

Another trend that holds immense promise is the emergence of biohybrid robots, which seamlessly blend biological and artificial components to create machines that mimic the behavior and functionality of living organisms. These biohybrids could find applications in healthcare, environmental monitoring, and exploration, offering a new dimension of adaptability and efficiency. Additionally, robotics is set to explore the frontiers of nanorobotics, where minuscule machines operate on the nanoscale, enabling precise medical procedures, targeted drug delivery, and manipulation of matter at the molecular level.

Ethical considerations are also playing a pivotal role in shaping the future of robotics. As robots become more integrated into our lives, discussions around the ethical use of AI, privacy concerns, and potential job displacement are gaining prominence. Striking a balance between technological advancement and responsible deployment is crucial to ensure that robotics contributes positively to society.

Moreover, the fusion of robotics with virtual reality (VR) and augmented reality (AR) is opening up new dimensions in human-robot interaction and training. These technologies allow users to immerse themselves in virtual environments to control robots remotely, perform complex simulations, and enhance training for various applications.

In the realm of medical robotics, advancements are poised to revolutionize surgical procedures, diagnostics, and patient care. Robotic-assisted surgeries are becoming more precise, reducing invasiveness and recovery times. Furthermore, soft robotics is gaining traction, offering robots with flexible, adaptable structures that can navigate complex environments and interact more naturally with humans.

As we explore these trends, it's important to recognize the interdisciplinary nature of robotics, where fields like AI, materials science, neuroscience, and ethics converge to create machines that transcend conventional limitations. The future of robotics is one of boundless innovation and limitless potential, as we strive to harness technology to enhance our lives, push boundaries, and unlock new frontiers of human capability.

AI Integration in Robotics: The integration of Artificial Intelligence (AI) into robotics marks a significant milestone in the evolution of both fields, unlocking a realm of possibilities that promises to reshape the way robots perceive, interact, and make decisions within their environment. AI, with its ability to simulate human-like cognitive processes and learning capabilities, is breathing new life into robotics by enabling machines to transcend their traditional roles of automation and perform tasks that demand complex reasoning, adaptability, and interaction with dynamic surroundings. This synergy between AI and robotics holds the potential to usher in a new era of intelligent machines that not only augment human capabilities but also pave the way for advancements in industries ranging from healthcare and manufacturing to exploration and entertainment. As AI becomes an intrinsic part of the robotic landscape, the boundaries of what machines can achieve continue to expand, promising a future where robots seamlessly navigate the complexities of the real world while adapting to an ever-changing landscape of challenges and opportunities. The convergence of Artificial Intelligence (AI) and robotics has given rise to a new era of technological possibilities, where machines are not only capable of performing predefined tasks but also possess the ability to learn, reason, and adapt. The integration of AI into robotics is a pivotal development that transcends traditional notions of automation, imbuing robots with cognitive abilities akin to human intelligence. This chapter delves into the profound impact of AI integration in robotics, exploring how AI algorithms, machine learning techniques, and advanced perception systems empower robots to perceive, process, and respond to their environment in ways that were once considered the realm of science fiction. From enabling robots to recognize objects and navigate complex terrains to fostering human-robot collaboration and enhancing decision-making, the infusion of AI redefines the capabilities and potential applications of robotic systems. As we navigate this chapter, we will uncover the intricacies of how AI is reshaping the landscape of robotics, paving the way for a future where intelligent machines coexist seamlessly with humans and contribute to solving a diverse array of challenges across industries.

The fusion of Artificial Intelligence (AI) and robotics has brought about a paradigm shift in the capabilities and possibilities of robotic systems. This integration goes beyond the conventional realm of robotics, where machines were primarily programmed to execute predefined tasks. Instead, it empowers robots with cognitive abilities, enabling them to perceive, understand, and navigate complex real-world environments with a degree of autonomy and adaptability previously thought unattainable.

At its core, AI integration in robotics involves equipping robots with AI algorithms and techniques that mimic human cognitive functions. Machine learning, a subset of AI, plays a pivotal role by allowing robots to learn from data and experiences, enabling them to refine their actions over time. This learning ability is particularly transformative as it enables robots to adapt to unforeseen situations and evolving environments, a trait essential for real-world applications.

One of the significant areas where AI integration shines is perception. Robots can now utilize advanced sensors, such as cameras, LiDAR, and depth sensors, to gather vast amounts of data from their surroundings. AI algorithms process this data to identify objects, people, obstacles, and patterns, enabling robots to make informed decisions based on their environment. This enhanced perception is instrumental in applications ranging from autonomous vehicles navigating busy streets to drones surveying disaster-stricken areas.

Furthermore, AI integration enables robots to excel in tasks involving pattern recognition, language processing, and decision-making. Natural language processing (NLP) allows robots to understand and respond to human commands and queries, facilitating intuitive human-robot interaction. Decision-making processes benefit from AI algorithms that analyze data, assess potential outcomes, and select the optimal course of action. In industrial settings, robots equipped with AI-driven decision-making capabilities enhance efficiency and adapt to variations in production processes.

Human-robot collaboration is another compelling facet of AI integration in robotics. Advanced AI algorithms enable robots to understand human actions, intentions, and emotions, fostering a symbiotic relationship. Collaborative robots, also known as cobots, work alongside humans on tasks that require precision, strength, or monotonous actions. AI ensures that cobots can adjust their behavior in real-time to accommodate human actions, ensuring safety and efficiency in shared workspaces.

The integration of AI in robotics has transcended individual machines, leading to the development of swarm robotics. In swarm robotics, a group of relatively simple robots collaborates to achieve complex tasks. AI algorithms guide the collective behavior of these robots, enabling them to self-organize, communicate, and adapt to changing conditions. Applications of swarm robotics range from exploring hazardous environments to environmental monitoring and search-and-rescue missions.

In conclusion, AI integration in robotics has ushered in a new era of intelligent machines capable of perceiving, learning, adapting, and interacting with the world in unprecedented ways. The marriage of AI and robotics is reshaping industries, advancing automation, enhancing decision-making, and creating opportunities in fields as diverse as healthcare, manufacturing, agriculture, and exploration. As AI continues to evolve, its integration into robotics holds the promise of pushing the boundaries of what robots can achieve, making them indispensable partners in addressing the challenges and opportunities of the modern world.

Robotic Ethics and Regulations: The integration of robotics and artificial intelligence (AI) into various aspects of human life raises profound ethical and regulatory considerations. As robots become increasingly autonomous, capable of decision-making, and integrated into critical domains such as healthcare, transportation, and public safety, it becomes imperative to establish a framework that guides their behavior, ensures accountability, and safeguards human interests. This chapter delves into the intricate landscape of robotic ethics and regulations, exploring the challenges, principles, and initiatives that shape the responsible development and deployment of robotic systems.

The rapid advancements in robotics and AI bring forth complex ethical challenges that demand thoughtful examination. One of the fundamental concerns is the potential displacement of human jobs by robots, particularly in industries susceptible to automation. Striking a balance between increased efficiency and preserving livelihoods necessitates a robust ethical framework that ensures responsible technological disruption.

Another ethical consideration revolves around the accountability and transparency of AI-driven decisions. As robots make decisions based on algorithms, the lack of transparency in these decision-making processes raises concerns about fairness, bias, and discrimination. Ensuring that AI decisions are explainable and comprehensible to humans is crucial to avoid instances where machines reinforce societal biases.

Furthermore, the development of autonomous systems that can inflict harm, such as military drones or self-driving vehicles, poses ethical dilemmas surrounding their deployment and potential consequences. Establishing rules and guidelines for the use of such systems to prevent unintended harm or malicious intent is essential for maintaining human safety and well-being.

Principles of Robotic Ethics: The establishment of ethical principles is a cornerstone in addressing the ethical challenges posed by robotic systems. The widely recognized principles include:

Beneficence: Robots should strive to maximize benefits to humanity, minimize harm, and prioritize human well-being.

Non-Maleficence: Robots should not harm humans, either physically or mentally, and should not compromise their dignity.

Autonomy: Robots should respect human autonomy and act in ways that allow individuals to make informed choices and maintain control over their decisions.

Justice: Robots should be designed and deployed in ways that ensure fair distribution of benefits, avoid bias, and uphold social and economic justice.

Transparency and Accountability: Developers and operators of robots should ensure transparency in decision-making processes and be accountable for their creations' actions.

Regulations in Robotic Systems:The fast-paced evolution of robotics has prompted policymakers and regulatory bodies to address the challenges posed by these technologies. Governments and international organizations are working to establish regulatory frameworks that ensure the ethical and safe use of robots while fostering innovation.

The European Union's General Data Protection Regulation (GDPR) serves as an example of regulations that impact robotics. GDPR addresses concerns about the collection, storage, and use of personal data, including data collected by robots. Ensuring that robots handle sensitive information responsibly is crucial for preserving privacy and preventing misuse.

Similarly, the International Organization for Standardization (ISO) has developed standards for robotics to address issues such as safety, interoperability, and ethical considerations. These standards provide guidelines for developers and operators to ensure that robots are designed, manufactured, and operated in a manner that aligns with ethical principles and prioritizes safety.

Emerging Initiatives: As robotics continues to evolve, efforts are underway to establish ethical guidelines specific to AI and robotics. Organizations like the IEEE Global Initiative on Ethics of Autonomous and Intelligent Systems are developing standards and recommendations to ensure that AI and robotics technologies are developed in ways that prioritize ethical considerations.

In the realm of autonomous vehicles, the development of ethical decision-making algorithms is a critical initiative. These algorithms must navigate complex scenarios, such as choosing between minimizing harm to passengers or pedestrians in potential accidents. Establishing universally accepted ethical frameworks for such decisions is a complex but necessary endeavor.

Conclusion: The integration of robotics and AI into society comes with a responsibility to address ethical challenges and establish regulations that ensure their positive impact. Balancing innovation with ethical considerations and human safety is paramount for a harmonious coexistence between humans and robots. As the field of robotics continues to advance, the collaboration between ethicists, policymakers, developers, and society at large is crucial in shaping a future where robots enhance human well-being while upholding ethical principles and regulatory standards.

Emerging Applications (Medical, Space, Industry): The rapid evolution of robotics and artificial intelligence (AI) has ushered in a new era of possibilities across a spectrum of industries. This chapter delves into the burgeoning landscape of emerging applications in three critical sectors: medical, space exploration, and industrial automation.

These domains are witnessing transformative advancements as robots and AI technologies redefine the boundaries of what is achievable, offering solutions that address complex challenges, enhance human capabilities, and unlock unprecedented opportunities.

Medical Robotics: The convergence of robotics and healthcare is on the brink of revolutionizing medical practices and patient outcomes. Robotic surgical systems, a prime example, epitomize precision and innovation. These systems allow surgeons to perform intricate procedures with unprecedented accuracy, leading to reduced trauma for patients, faster recovery times, and improved clinical outcomes. The da Vinci Surgical System, for instance, employs a combination of robotic arms and advanced imaging technology to enable minimally invasive surgeries. This approach has proven particularly effective in procedures like prostatectomies and gynecological surgeries. In diagnostics, robotic technologies are streamlining processes and enhancing accuracy. Robotic-assisted imaging systems allow for more precise targeting of tumors during interventions, reducing the risk of collateral damage. Beyond diagnostics, rehabilitation is benefitting from robotic assistance. Exoskeletons and rehabilitation robots aid patients with mobility impairments by providing tailored physical therapy. These devices facilitate repetitive and consistent movements, which are essential for effective recovery. The amalgamation of robotics and the medical field holds immense promise in revolutionizing healthcare practices and elevating patient outcomes. One of the most significant advancements lies in robotic-assisted surgeries, where precision and dexterity are enhanced by robotic systems. These systems, exemplified by the da Vinci Surgical System, enable surgeons to perform intricate procedures with heightened accuracy, translating into reduced trauma for patients and quicker postoperative recovery. In cases such as prostatectomies and hysterectomies, robotic assistance results in smaller incisions, less blood loss, and shorter hospital stays.

Beyond surgical interventions, medical robotics has diversified its applications to diagnostics, rehabilitation, and telemedicine. Robotic imaging systems, through their capacity for meticulous targeting, enhance the precision of interventions such as biopsies and radiation therapies. Rehabilitation robots, like exoskeletons, have emerged as transformative tools in aiding patients with mobility impairments. These wearable devices facilitate therapeutic movements, enabling patients to regain strength and mobility with personalized support.

Space Exploration Robotics: The realm of space exploration is characterized by its inhospitable and uncharted terrains. Robotic systems are vital companions in this endeavor, venturing into places humans cannot yet reach. Robotic rovers, exemplified by NASA's Mars rovers, traverse the Martian landscape, conducting experiments, analyzing soil samples, and capturing breathtaking images. These robots enable scientists to study distant worlds, search for signs of past or present life, and better understand the universe's origins. Moreover, robotic spacecraft are instrumental in gathering data from celestial bodies beyond Earth. Missions like the Voyager probes, the Hubble Space Telescope, and the upcoming James Webb Space Telescope leverage robotic technology to transmit images and data that shape our understanding of the cosmos. Autonomous navigation, AI-driven data analysis, and real-time communication are just a few of the advancements that make these missions possible. The intersection of robotics and space exploration is redefining humanity's reach beyond Earth's boundaries. Robotic spacecraft and rovers are invaluable in the pursuit of scientific discovery and understanding distant celestial bodies. Mars rovers, exemplified by the Opportunity and Curiosity rovers, have traversed the Red Planet's surface, conducting experiments and capturing images that have deepened our knowledge of Martian geology and potential habitability. These robots function autonomously, navigating treacherous terrain and transmitting vital data back to Earth.

Furthermore, robotic systems contribute to constructing and maintaining structures in space. The assembly of the International Space Station (ISS) heavily relies on robotic arms and systems that enable astronauts to connect modules, perform repairs, and conduct experiments. The significance of robotics in space exploration extends to ambitious plans such as crewed missions to Mars, where robots are pivotal in conducting preliminary investigations, resource prospecting, and habitat preparation.

Industrial Automation: The manufacturing landscape is undergoing a profound transformation, driven by the integration of robotics and AI. Collaborative robots, or cobots, have emerged as game-changers in industrial automation. These robots work harmoniously alongside human workers, complementing their skills and handling repetitive or hazardous tasks. The automotive industry, for instance, employs cobots to assist in assembling intricate components with precision and speed. The concept of Industry 4.0, often referred to as the fourth industrial revolution, is reshaping factories and production lines. Smart factories are characterized by interconnected machines, sensors, and AI-powered analytics. These elements work in synergy to optimize production processes, predict maintenance needs, and reduce downtime. For instance, robots equipped with sensors collect data on machine performance, and AI algorithms analyze this data to identify patterns, anticipate issues, and ensure smooth operations. The integration of robotics and artificial intelligence in the industrial sector is steering manufacturing into a new era of efficiency and adaptability. Collaborative robots, known as cobots, are designed to work alongside human operators, amplifying their capabilities and tackling monotonous or physically taxing tasks. In automobile manufacturing, cobots collaborate with workers to streamline production lines, enhance precision, and minimize errors.

The concept of Industry 4.0, underpinned by the fusion of robotics and AI, heralds smart factories characterized by interconnectivity, data-driven insights, and predictive analytics. Robotic systems equipped with sensors gather real-time data, offering insights into machine performance, production rates, and potential bottlenecks. AI algorithms process this data, allowing factories to predict maintenance needs, optimize resource allocation, and respond dynamically to changing demands.

Interplay of Robotics and AI: The fusion of robotics and AI amplifies the capabilities of each technology. AI equips robots with the ability to process and interpret data, adapt to changing environments, and make informed decisions. In medical robotics, AI-driven algorithms aid in diagnosis by analyzing medical images and identifying patterns that might elude the human eye. In space exploration, AI enables robotic systems to autonomously navigate obstacles and make real-time decisions, optimizing mission success. In industrial automation, AI-powered analytics enable predictive maintenance, resource optimization, and continuous process improvement. The symbiotic relationship between robotics and AI lies at the heart of these emerging applications. AI equips robots with cognitive capabilities, enabling them to process vast amounts of data, learn from experience, and make informed decisions. In medical robotics, AI-driven algorithms analyze medical images to detect anomalies, supporting radiologists in diagnosing diseases like cancer. In space exploration, AI empowers robots to autonomously navigate complex environments, adapt to unexpected obstacles, and conduct experiments with minimal human intervention. In industrial settings, AI-powered analytics optimize production processes, identify inefficiencies, and facilitate predictive maintenance.

Challenges and Considerations: As these emerging applications advance, challenges and ethical considerations arise. In medical robotics, safety and accuracy are paramount to prevent unintended consequences in surgical procedures or patient care. In space exploration, the reliability of robotic systems is crucial due to the remote and harsh conditions of outer space. Ensuring that robots can withstand the challenges of radiation, extreme temperatures, and communication delays is imperative for mission success. In industrial automation, questions about job displacement, data privacy, and cybersecurity must be addressed to ensure a smooth transition to smart factories.

While the potential of these emerging applications is undeniable, several challenges and ethical considerations demand attention. In medical robotics, ensuring patient safety and the accuracy of surgical procedures remains a critical concern. Proper training of medical personnel and rigorous testing of robotic systems are essential to minimize errors and maximize positive outcomes. In space exploration, the reliability of robotic systems is paramount due to the harsh conditions of space. Ensuring robustness, resilience, and redundancy in these systems is crucial for mission success. In industrial automation, addressing workforce concerns about job displacement and addressing data privacy and security are vital considerations.

Conclusion: The emerging applications of robotics and AI in medical, space exploration, and industrial sectors exemplify the transformative power of technology to reshape industries, push boundaries, and drive innovation. These applications demonstrate the potential to enhance human capabilities, tackle complex challenges, and open new horizons for exploration and discovery. As the realms of robotics and AI continue to evolve, collaborative efforts across research, engineering, policy, and industry will be pivotal in harnessing the full potential of these technologies across diverse sectors and shaping a future marked by unprecedented advancements.

Challenges and Opportunities in Robotics Research: The field of robotics is at the forefront of technological innovation, with far-reaching implications for various industries and aspects of human life. As robotics continues to evolve, it faces a plethora of challenges and opportunities that shape the trajectory of research, development, and implementation. From technical hurdles to ethical considerations, the landscape of robotics research is marked by complexity and potential. This exploration delves deep into the challenges and opportunities that define the realm of robotics research, highlighting its transformative potential while acknowledging the hurdles that lie ahead.

Challenges:

Autonomy and Intelligence: Developing robots that exhibit true autonomy and intelligence remains a daunting challenge. While significant strides have been made in AI and machine learning, achieving human-like cognitive abilities in robots is still a distant goal. Robots must navigate unpredictable environments, adapt to unforeseen scenarios, and make complex decisions, all while ensuring safety and ethical considerations.

Perception and Sensing: Accurate perception of the environment is fundamental for robotic systems. However, the challenges of sensory integration, dealing with uncertainty, and managing sensor noise persist. Creating algorithms that enable robots to process sensor data in real-time and make informed decisions amidst ambiguous situations remains a key research frontier.

Manipulation and Dexterity: Endowing robots with dexterity and fine motor skills comparable to humans is an intricate task. Grasping and manipulating objects of varying shapes, textures, and sizes in unstructured environments necessitates breakthroughs in robotic mechanics and control algorithms.

Human-Robot Interaction: Enabling seamless and natural interaction between humans and robots is crucial for their widespread adoption. Overcoming challenges such as understanding human gestures, emotions, and social cues while ensuring user comfort and safety is an ongoing endeavor.

Safety and Ethical Considerations: As robots become more integrated into daily life, ensuring their safety and addressing ethical concerns becomes paramount. The potential for robots to cause harm, invade privacy, or perpetuate bias requires a comprehensive framework for regulations and standards.

Robustness and Reliability: Robotics systems must demonstrate robustness in real-world scenarios, overcoming challenges like uneven terrains, changing lighting conditions, and unexpected disturbances. Ensuring reliable performance and minimizing failures are essential to building trust in robotic technology.

Energy Efficiency: Many robotic applications require energy-intensive computations and power sources. Developing energy-efficient robots that can operate for extended periods without frequent recharging or refueling is crucial for sustainable deployment.

Cost and Scalability: Creating affordable robotic systems that can be scaled for mass production remains a challenge. Balancing advanced technology with cost-effectiveness is necessary for widespread adoption across industries.

Opportunities:

Innovation and Advancements: Robotics research offers boundless opportunities for technological innovation. From soft robotics that mimic biological structures to advanced AI algorithms that enable robots to learn and adapt, researchers have the chance to push the boundaries of what robots can achieve.

Healthcare Transformation: Medical robotics presents immense potential for enhancing diagnosis, treatment, and patient care. Robotic surgical systems, personalized rehabilitation robots, and AI-driven diagnostics have the potential to revolutionize healthcare practices and improve patient outcomes.

Space Exploration: As space exploration endeavors become more ambitious, robots are indispensable tools for venturing into uncharted territories. The development of autonomous rovers and robotic systems capable of withstanding extreme conditions expands our understanding of the universe.

Environmental Conservation: Robotics research offers avenues for addressing environmental challenges. Autonomous drones can monitor deforestation, pollution levels, and wildlife populations, providing valuable data for conservation efforts.

Manufacturing and Automation: In the industrial sector, robotics research drives advancements in automation, streamlining manufacturing processes, enhancing precision, and reducing human exposure to hazardous tasks. Collaborative robots are redefining human-robot collaboration in factory settings.

Agriculture and Food Production: Robots have the potential to transform agriculture by automating tasks like planting, harvesting, and monitoring crop health. Precision agriculture driven by robotics can optimize resource utilization and yield while minimizing environmental impact.

Disaster Response and Search-and-Rescue: Robots equipped with sensors and AI can navigate hazardous environments, aiding in disaster response and search-and-rescue missions. These robots can enter spaces too dangerous for humans, increasing the chances of saving lives.

Education and Skill Development: The integration of robotics in education offers opportunities for students to engage with hands-on STEM learning. Robotic kits, competitions, and educational platforms foster creativity and technical skills among the next generation of researchers and engineers.

Social and Assistive Robotics: The development of socially adept robots holds potential for addressing challenges in elderly care, child education, and companionship for those in isolation. These robots can provide companionship, cognitive support, and aid in daily tasks.

Conclusion:Robotics research is a dynamic and multifaceted domain that holds both challenges and opportunities that are shaping the future of technology and its impact on society. As researchers continue to explore new frontiers in autonomy, interaction, and intelligence, they are unlocking possibilities that were once relegated to science fiction. However, the journey forward is not without hurdles. The delicate balance between technological advancement, ethical considerations, and societal integration must be navigated with care. Ultimately, the fusion of human ingenuity and robotic capabilities has the potential to usher in an era of innovation, transformation, and improved quality of life across a spectrum of industries and applications.

Exercise

Multiple-choice Questions

1. What is the primary challenge in achieving true autonomy and intelligence in robots?
a) Energy efficiency
b) Safety regulations
c) Environmental adaptation
d) Cognitive abilities
2. Soft robotics offers advantages over traditional rigid robotics in terms of:
a) Speed and precision
b) Durability and strength
c) Adaptability and interaction
d) Efficiency and control
3. Which field of robotics takes inspiration from biological organisms or processes?
a) Quantum robotics
b) Bio-inspired robotics
c) Virtual robotics
d) Nano robotics
4. The fundamental principle behind neural networks is:
a) Feedback loops
b) Biochemical reactions
c) Parallel processing
d) Artificial intelligence
5. How does a Convolutional Neural Network (CNN) differ from a Recurrent Neural Network (RNN)?
a) CNN processes sequential data
b) CNN specializes in image recognition
c) RNN is better for spatial analysis
d) RNN lacks deep learning capabilities

Review Questions:

1. What are some challenges associated with achieving true autonomy and intelligence in robotic systems?
2. How does soft robotics differ from traditional rigid robotics? What advantages does it offer in specific applications?
3. Provide examples of robotic systems that are inspired by biological organisms or processes. How do these systems leverage bio-inspired concepts?
4. Explain the fundamental principle behind neural networks and their contribution to the field of deep learning.
5. Compare and contrast Convolutional Neural Networks (CNNs) and Recurrent Neural Networks (RNNs) in terms of their data processing and application domains.
6. Describe the concept of transfer learning in neural networks. How does it address challenges related to limited data availability in robotics applications?
7. What role do Generative Adversarial Networks (GANs) play in robotics? How do they contribute to generating synthetic data?
8. How do attention mechanisms enhance the capabilities of neural networks? Provide an example of a robotics application where attention mechanisms are beneficial.
9. Explain the concept of self-supervised learning and its significance in training neural networks using unlabeled data.
10. Describe how Capsule Networks address limitations of traditional neural networks, particularly in handling spatial hierarchies.
11. What is Reinforcement Learning, and how does it differ from other learning paradigms in the context of neural networks?
12. How does Continual Learning enable robots to learn new tasks without forgetting previously acquired knowledge?
13. Discuss the role of ethics and regulations in the deployment of AI-integrated robotics systems. Provide examples of ethical considerations in this context.
14. How can AI-powered robotics contribute to emerging applications in the medical field? Provide specific examples of medical robotics advancements.

15. Describe the potential applications of robotics in space exploration. How can robots enhance space missions and exploration?
16. In what ways can robotics revolutionize industrial processes and automation? Provide examples of industries benefiting from robotic advancements.
17. Explain the challenges and opportunities in robotics research. How can researchers address these challenges while leveraging emerging technologies?
18. How might the interdisciplinary nature of robotics, involving AI, computer science, and machine learning, lead to new breakthroughs in the field?
19. Discuss the potential societal impacts of widespread deployment of robotics systems powered by AI and advanced technologies.
20. How do advancements in neural networks contribute to enhancing the safety and reliability of autonomous vehicles in the field of robotics?

Answers Multiple-choice Questions

Answers:
1. d)
2. c)
3. b)
4. c)
5. b)

Review Questions:

1. Achieving true autonomy and intelligence in robotic systems presents a complex challenge due to the need for robots to exhibit human-like decision-making, adaptability, and learning capabilities. Unlike traditional robotics, where robots follow pre-programmed instructions, achieving autonomy involves integrating AI and deep learning techniques. True autonomy requires robots to perceive and interpret their environment, make context-aware decisions, and adapt to dynamic situations. This challenge is rooted in the need to develop algorithms that can mimic human cognitive abilities, handle uncertainty, and learn from experience. It involves advancements in machine learning, neural networks, and sensor technologies to enable robots to learn, reason, and act in diverse and unstructured environments.
2. Soft robotics is a paradigm shift from traditional rigid robotics, where robots are constructed from flexible materials that can change shape, deform, and interact with their surroundings. Unlike rigid robots, which have predefined configurations, soft robots offer inherent adaptability and compliance. This enables them to navigate complex environments, interact with delicate objects, and provide safer interactions with humans. The advantages of soft robotics include improved safety in human-robot interactions, enhanced dexterity, and the ability to operate in constrained spaces. These characteristics are particularly useful in applications such as medical devices, exploration of unstructured environments, and human-assistive technology.
3. Bio-inspired robotics draws inspiration from biological organisms and processes to design robotic systems that replicate natural behaviors. Examples of bio-inspired robotics include the development of robotic limbs based on the movement of animals, drones inspired by bird flight, and snake-like robots for search and rescue missions. The concept behind bio-inspired robotics is to leverage millions of years of evolution to create efficient and adaptive robotic systems. This approach has led to innovative designs that exhibit capabilities such as agility, energy efficiency, and adaptability, which are not easily achievable through traditional engineering methods.
4. Neural networks are computational models inspired by the human brain's structure and function. They consist of interconnected nodes (neurons) organized into layers that process and analyze data. Neural networks contribute to the field of deep learning by enabling machines to recognize patterns, learn from data, and make decisions. This is achieved through the process of training, where neural networks learn to adjust their internal parameters based on input data and desired outputs. Neural networks have revolutionized various fields, including robotics, by enabling machines to process complex sensory data, perform object recognition, and make informed decisions in real-time.
5. Convolutional Neural Networks (CNNs) and Recurrent Neural Networks (RNNs) are two distinct architectures within the realm of neural networks. CNNs are primarily designed for image analysis tasks, leveraging convolutional layers to extract spatial features and patterns from images. In contrast, RNNs specialize in processing sequential data by incorporating feedback loops that allow information to be retained and utilized across time steps. While CNNs excel at tasks like image recognition, RNNs are suitable for tasks involving temporal dependencies, such as natural language processing and speech recognition.
6. Transfer learning in neural networks involves leveraging knowledge gained from one task and applying it to a different but related task. This benefits robotics applications with limited data by allowing models trained on a large dataset to be fine-tuned on a smaller, task-specific dataset. Transfer learning reduces the need for extensive data

collection and training, making it particularly valuable in robotics where data acquisition can be resource-intensive. For example, a neural network trained to recognize objects in one environment can be adapted to recognize objects in a different environment with minimal retraining, saving time and resources.

7. Generative Adversarial Networks (GANs) are significant in robotics for their ability to generate synthetic data that closely resembles real-world data. GANs consist of a generator network that produces synthetic data and a discriminator network that distinguishes between real and synthetic data. The generator and discriminator engage in a competitive process, refining the generator's output over iterations. In robotics, GANs are used to create diverse datasets for training, enabling robots to learn from a broader range of scenarios and improving their generalization capabilities. This synthetic data generation is particularly useful when real-world data is scarce or when simulating environments that are costly to create physically.

8. Attention mechanisms enhance neural network capabilities by allowing models to focus on specific parts of input data that are most relevant to the task at hand. In robotics, attention mechanisms are valuable for tasks such as object detection, image captioning, and visual tracking. For instance, in an autonomous driving scenario, attention mechanisms can enable a robot to focus on relevant objects, such as pedestrians or traffic signs, while filtering out less important details. This selective focus improves the efficiency of data processing and decision-making, ultimately enhancing the robot's performance.

9. Self-supervised learning is a training technique in neural networks where models learn from unlabeled data by generating their own labels from the input data. In robotics, self-supervised learning addresses challenges related to the availability of labeled data, which is often costly and time-consuming to obtain. By utilizing the inherent structure and relationships in the data, self-supervised learning enables robots to learn useful representations and features that can be transferred to various tasks. This approach has applications in robotic perception tasks, where robots can learn to identify object attributes and relationships without the need for manual labeling.

10. Capsule Networks are a novel architecture that addresses the limitations of traditional neural networks in handling spatial hierarchies and rotations. Unlike conventional neural networks that use individual neurons for feature detection, Capsule Networks group neurons into capsules, representing specific object parts or features. Capsules are designed to capture various attributes of objects, such as orientation and deformation. This allows Capsule Networks to handle spatial relationships more effectively, making them advantageous in tasks that require understanding complex spatial arrangements, such as pose estimation and object recognition in robotics.

11. Reinforcement Learning (RL) is a learning paradigm in neural networks where an agent learns to interact with an environment by taking actions to maximize a cumulative reward signal. RL differs from other learning paradigms by focusing on learning through interaction and feedback from the environment rather than from labeled data. This makes RL particularly suitable for training robots to perform tasks in dynamic and uncertain environments. RL has found applications in robotics for tasks like autonomous navigation, robotic control, and task optimization, where robots learn optimal policies through trial and error.

12. Continual Learning is an approach that enables robots to learn new tasks without forgetting previously acquired knowledge. In the context of neural networks, continual learning addresses the challenge of catastrophic forgetting, where learning new tasks erases knowledge of previously learned tasks. This is crucial for robots that need to adapt and learn new skills while retaining their existing capabilities. Continual learning methods involve strategies such as regularization, knowledge distillation, and generative replay to ensure that robots can accumulate knowledge over time without significant degradation in performance.

13. Ethics and regulations play a critical role in the deployment of AI-integrated robotics systems. Ethical considerations involve addressing questions related to the safety, transparency, and accountability of robotic actions. Regulations are necessary to ensure that AI-powered robots adhere to ethical guidelines and legal frameworks. For example, in the context of autonomous vehicles, ethical considerations involve decisions made by self-driving cars in life-threatening situations. Regulations must strike a balance between promoting innovation while safeguarding human safety and privacy.

14. AI-powered robotics is revolutionizing the medical field through emerging applications such as surgical robotics, diagnostics, and rehabilitation. Surgical robots enhance precision and minimize invasiveness during surgeries, leading to faster patient recovery. AI algorithms analyze medical images to aid in disease detection and diagnosis. Rehabilitation robots assist patients in regaining motor functions through personalized therapy. These applications improve patient outcomes, reduce medical errors, and enable new medical procedures that were previously unattainable.

15. Robotics has transformed space exploration by enabling the exploration of distant planets, asteroids, and celestial bodies. Robotic rovers and landers collect data, perform experiments, and transmit information back to Earth. Robots assist in building and maintaining space stations and telescopes. Drones explore planetary

atmospheres and conduct scientific research. The use of robotics in space missions reduces human risk, increases efficiency, and expands our understanding of the universe.

16. Robotics is revolutionizing industries such as manufacturing, logistics, and agriculture through automation and process optimization. In manufacturing, robots perform repetitive tasks with precision, leading to increased production efficiency and product consistency. In logistics, robots automate warehousing and distribution, speeding up order fulfillment and reducing operational costs. In agriculture, robots assist in planting, harvesting, and monitoring crops, increasing yields and reducing the need for manual labor.

17. The field of robotics research is marked by both challenges and opportunities. Challenges include issues related to safety, ethics, and the development of AI systems that can understand human intentions and preferences. Ensuring the safety of AI-integrated robots, establishing ethical guidelines for robot behavior, and fostering transparent decision-making are among the significant challenges. However, opportunities abound with the advancement of AI, machine learning, and sensor technologies. Opportunities include enhancing human-robot collaboration, developing robots for hazardous environments, and creating AI-powered assistive technologies for people with disabilities.

18. The interdisciplinary nature of robotics, encompassing AI, computer science, and machine learning, is a driving force behind its rapid advancement. The collaboration between these fields enables the development of sophisticated robotic systems that can perceive, reason, and act in complex environments. For example, the integration of AI algorithms with robotics enhances perception, decision-making, and adaptability. Computer science contributes to algorithm design and optimization, while machine learning enables robots to learn from data and improve over time.

19. The widespread deployment of AI-powered robotics systems raises several societal impacts and ethical considerations. Concerns include job displacement due to automation, bias in AI algorithms, and privacy breaches from data collection. As robots become more integrated into daily life, society must address these issues through policy frameworks, education, and responsible development. Ethical considerations involve determining robot behavior in ambiguous situations, ensuring transparency in decision-making, and establishing guidelines for human-robot interaction to prevent unintended harm.

20. The integration of advanced neural networks and AI technologies into robotics has the potential to reshape industries and job roles in the future. Industries such as healthcare, manufacturing, transportation, and entertainment will experience transformations as robots take on more tasks previously performed by humans. New job roles will emerge, focusing on robot design, maintenance, and supervision. However, this transition requires a collaborative effort between academia, industry, and policymakers to ensure a smooth transition, address societal challenges, and maximize the benefits of AI-integrated robotics.

Online Resources

Online Exam Papers

Link:https://forms.gle/kc49cRFEjD6YpQx89

QR Code:

YouTube Video Playlist Link

Link:https://www.youtube.com/playlist?list=PLzlh_31VWEFrgxK0YBD1DanI0VGZx5Ir3

QR Code:

Chapter 12: Robotics in Agriculture and Environmental Conservation

In recent years, the intersection of robotics and agriculture has emerged as a transformative force, reshaping traditional farming practices and contributing to environmental conservation efforts. The integration of robotics in agriculture, often referred to as agricultural robotics or agri-robotics, has the potential to revolutionize how food is grown, harvested, and managed while also promoting sustainability and environmental preservation. This innovative approach leverages the power of robotics, automation, and data-driven technologies to address the challenges facing the agriculture sector and contribute to global environmental conservation goals.

Precision farming is a cornerstone of modern agriculture, aiming to optimize resource use while minimizing waste. Robotics plays a pivotal role in achieving precision farming objectives by enabling precise and targeted interventions. Agricultural robots equipped with sensors, cameras, and GPS technology can monitor crop health, soil moisture levels, and environmental conditions in real time. This data is then processed using sophisticated algorithms to make informed decisions, such as precisely applying fertilizers, pesticides, and water only where needed. By minimizing excessive use of resources, precision farming reduces environmental pollution, conserves water, and promotes sustainable agricultural practices.

Traditional agricultural practices often rely on broad applications of chemicals and water to ensure crop health. This approach can lead to overuse of resources, soil degradation, and contamination of surrounding ecosystems. Agricultural robots are changing this paradigm by enabling site-specific management. For instance, robotic weeding systems use computer vision and AI to differentiate between crops and weeds, allowing targeted weed removal without the need for herbicides. Similarly, drones equipped with multispectral sensors can identify areas of water stress in fields, allowing farmers to address irrigation needs more efficiently.

The adoption of agricultural robots is contributing to practices that promote biodiversity and soil health. Precision planting robots ensure optimal spacing between seeds, leading to healthier and more vigorous plant growth. This, in turn, minimizes competition among plants and reduces the need for excessive chemical interventions. In orchards, robotic harvesters handle fruits delicately, preventing damage to trees and plants, and avoiding unnecessary disturbances to ecosystems. By reducing soil compaction through targeted robotic interventions, soil health is maintained, and the overall ecosystem is better preserved.

One of the challenges faced by the agricultural sector is the shortage of labor, particularly in rural areas. Agricultural robots address this issue by providing a reliable and tireless workforce. As a result, farming operations can continue smoothly, ensuring timely planting, harvesting, and maintenance activities. This not only improves efficiency but also reduces the pressure to clear new land for cultivation, thus preserving natural habitats and contributing to environmental conservation efforts.

While the benefits of robotics in agriculture are evident, the adoption of these technologies also brings economic advantages to farmers. While initial investment costs may be a concern, the long-term benefits, including increased yield, reduced resource consumption, and labor savings, can outweigh the initial expenses. Governments and organizations are also incentivizing the adoption of sustainable agricultural practices, including robotics, through grants, subsidies, and educational programs. As technology continues to evolve, the cost-effectiveness of agricultural robots is expected to improve, making them accessible to a wider range of farmers.

The integration of robotics in agriculture presents a transformative opportunity to enhance productivity, efficiency, and sustainability while contributing to environmental conservation efforts. By enabling precision farming, reducing resource waste, promoting soil health, and mitigating labor shortages, agricultural robots are redefining the way food is produced. As the agriculture sector faces the challenge of feeding a growing global population while preserving natural resources, robotics emerges as a key enabler of sustainable and responsible farming practices. As technology continues to advance, the seamless integration of robotics into farming ecosystems promises to shape a more resilient and environmentally conscious future for agriculture and beyond.

Agricultural Robots: The agriculture industry has undergone significant transformations over the centuries, from manual labor to mechanization. In recent years, the integration of robotics and automation technologies has brought about a new wave of innovation in farming practices. Agricultural robots, also known as agribots, have emerged as a promising solution to address various challenges faced by the modern agricultural sector. These robots are designed to perform a wide range of tasks, from planting and harvesting to monitoring crops and managing pests. By leveraging advanced technologies such as sensors, machine learning, and artificial intelligence (AI), agricultural robots are revolutionizing the way crops are cultivated, leading to increased efficiency, productivity, and sustainability.

The Need for Agricultural Innovation: The global population continues to rise, placing increased demands on the agricultural sector to produce more food to feed the world. Additionally, factors such as climate change, labor shortages, and the need to reduce the environmental impact of farming have further heightened the urgency to adopt innovative farming practices. Traditional farming methods, reliant on manual labor and heavy machinery, are often inefficient, resource-intensive, and labor-demanding. Agricultural robots offer a promising solution to address these challenges by automating various tasks and optimizing resource utilization.

Capabilities of Agricultural Robots: Agricultural robots encompass a diverse range of machines, each designed to perform specific tasks within the farming process. Some common types of agricultural robots include:

Harvesting Robots: These robots are equipped with sensors and computer vision technology to identify ripe fruits or vegetables and perform precise harvesting. They can work tirelessly and consistently, reducing the dependency on manual labor during the harvest season.

Weeding Robots: Weeding is a crucial but labor-intensive task in agriculture. Weeding robots use AI and computer vision to differentiate between crops and weeds, enabling targeted and precise weed removal without the need for herbicides.

Planting Robots: Planting seeds at consistent depths and spacing is essential for optimal crop growth. Planting robots ensure uniform planting, reducing seed wastage and enhancing crop yield.

Monitoring Robots: Equipped with various sensors, monitoring robots collect data on crop health, soil moisture, and other environmental factors. This data helps farmers make informed decisions and apply resources efficiently.

Spraying Robots: Precision in pesticide and fertilizer application is critical to avoid overuse and minimize environmental impact. Spraying robots use sensors and AI to apply chemicals only where needed, reducing waste and protecting the ecosystem.

Advantages of Agricultural Robots: The adoption of agricultural robots offers numerous advantages to farmers and the industry as a whole:

Increased Efficiency: Robots work tirelessly and consistently, eliminating the limitations of human labor such as fatigue and time constraints. This results in increased efficiency and productivity in farming operations.

Resource Optimization: Agricultural robots are equipped with sensors that enable them to collect real-time data on soil conditions, crop health, and weather patterns. This data allows farmers to optimize resource use, reducing water, fertilizer, and pesticide usage.

Labor Savings: Labor shortages in rural areas are a significant challenge for the agriculture industry. Agricultural robots alleviate the reliance on manual labor, ensuring that farming operations can continue even in the face of workforce limitations.

Precision Farming: Robots offer precise and targeted interventions, such as planting, spraying, and weeding. This precision reduces the use of chemicals, minimizes waste, and contributes to sustainable farming practices.

Reduced Environmental Impact: The accurate application of fertilizers and pesticides by agricultural robots reduces the environmental footprint of farming. This approach minimizes runoff and contamination of soil and water resources.

Challenges and Considerations: While the adoption of agricultural robots holds immense promise, there are several challenges that need to be addressed for their widespread implementation:

Cost: The initial investment in agricultural robots can be substantial, making it a barrier for small-scale farmers. The cost of purchasing, maintaining, and repairing robots needs to be carefully evaluated against the potential benefits.

Technical Expertise: Operating and maintaining agricultural robots require technical expertise. Farmers need training to operate and troubleshoot these machines effectively.

Adaptability: Agricultural robots need to be adaptable to different crop types, terrains, and weather conditions. Developing robots that can handle various scenarios without compromising efficiency is a significant engineering challenge.

Regulations: As with any technological innovation, regulations and standards for agricultural robots need to be developed to ensure safe and responsible use. This includes considerations for worker safety, data privacy, and environmental impact.

Integration: Integrating agricultural robots into existing farming practices and workflows requires careful planning. Farmers need to identify the tasks that can be automated and determine how robots will work alongside human labor.

Case Studies and Examples: Agricultural robots have already demonstrated their potential to revolutionize farming practices. In the wine industry, robots equipped with cameras and sensors navigate vineyards, inspecting grapes and determining their ripeness for optimal harvest times. In orchards, robotic harvesters delicately pick fruits, reducing labor costs and ensuring consistent quality. Additionally, drone technology is used to monitor large fields, identifying areas that require additional irrigation or pest control.

Future Outlook: The integration of robotics and AI into agriculture is expected to continue growing in the coming years. As technology advances, agricultural robots will become more sophisticated, capable of handling a broader range of tasks and adapting to diverse environments. The development of AI algorithms will enable robots to make real-time decisions based on complex data inputs, enhancing their autonomy and effectiveness. Additionally, collaborative efforts between researchers, engineers, and farmers will drive the development of affordable and adaptable robotic solutions tailored to the needs of different agricultural practices.

Conclusion: Agricultural robots are transforming farming practices by addressing the challenges of labor shortages, resource optimization, and sustainability. These robots offer increased efficiency, precision, and data-driven decision-making, contributing to higher crop yields and reduced environmental impact. While challenges such as cost, technical expertise, and regulations exist, the potential benefits of agricultural robots for the agriculture industry and food security are undeniable. As technology continues to evolve, the seamless integration of robots into existing farming ecosystems will shape the future of agriculture, ensuring that the world's growing population is fed sustainably and efficiently.

Environmental Monitoring Robots: Guardians of Nature: In an era characterized by rapid industrialization, urbanization, and climate change, the preservation of the environment has become a paramount concern. To effectively monitor and safeguard natural ecosystems, the integration of robotics has emerged as a powerful tool. Environmental monitoring robots, often referred to as "guardians of nature," are technological innovations designed to collect data, analyze environmental conditions, and contribute to the preservation of delicate ecosystems. These robots are deployed across various environments, from oceans and forests to urban areas, with the aim of gathering critical data to inform conservation efforts, policy-making, and sustainable resource management.

The Need for Environmental Monitoring: Human activities have significantly altered natural ecosystems, leading to habitat destruction, pollution, and biodiversity loss. Effective conservation and restoration efforts necessitate accurate and comprehensive data about the environment. Traditional monitoring methods are often limited by human capacity, access to remote areas, and the vastness of ecosystems. Environmental monitoring robots address these challenges by offering consistent, real-time data collection capabilities that span diverse terrains and challenging conditions.

Robotic Adaptability and Diversity: Environmental monitoring robots come in various forms, each tailored to specific environments and data collection requirements. Underwater autonomous vehicles (AUVs) explore marine ecosystems, gathering information about ocean temperature, salinity, and marine life distribution. Unmanned aerial vehicles (UAVs), or drones, capture aerial imagery of landscapes and detect forest fires or illegal deforestation activities. Ground-based robots navigate through rugged terrains, collecting data about soil quality, water pH levels, and pollution concentrations. This adaptability allows environmental monitoring robots to cover a wide spectrum of ecosystems and address diverse conservation challenges.

Technological Components and Capabilities: Environmental monitoring robots are equipped with advanced technologies that enable them to collect precise and meaningful data. Sensors, such as temperature and humidity sensors, spectrometers, and cameras, capture a range of environmental parameters. These sensors are integrated into the robot's hardware, allowing it to record data continuously and transmit it in real time. In some cases, robotic swarms are deployed, wherein multiple robots collaborate to cover larger areas or collect data from multiple points simultaneously.

Data Processing and Analysis: The data collected by environmental monitoring robots holds immense value for ecological research and conservation efforts. However, the sheer volume of data generated presents its own challenge. AI-driven algorithms play a crucial role in processing and analyzing this data. Machine learning techniques identify patterns, anomalies, and trends in the collected data, providing insights that inform conservation strategies and policy decisions. For example, analysis of satellite imagery captured by drones can identify deforestation areas, enabling authorities to take prompt action against illegal activities.

Applications in Biodiversity Conservation: Environmental monitoring robots play a pivotal role in biodiversity conservation by enabling researchers to study and protect various species and ecosystems. In marine environments, underwater robots gather data about coral reefs, marine life distribution, and oceanographic conditions. This data informs efforts to protect delicate marine ecosystems from pollution, overfishing, and climate change impacts. In terrestrial environments, robots help monitor endangered species, study migration patterns, and assess the impact of human activities on wildlife habitats.

Disaster Response and Mitigation: Environmental monitoring robots are also essential tools for disaster response and mitigation. During natural disasters such as wildfires, earthquakes, or tsunamis, robots can be deployed to assess damage, locate survivors, and provide critical information to emergency responders. These robots navigate hazardous environments that may be unsafe for humans, contributing to more efficient and effective disaster management strategies.

Challenges and Future Directions: While environmental monitoring robots offer immense potential, several challenges must be addressed for their widespread and effective use. Technological advancements are necessary to enhance battery life, navigation accuracy, and communication capabilities, especially in remote and extreme environments. Regulatory frameworks and ethical considerations surrounding data collection, privacy, and ecosystem disruption also need careful attention.

Conclusion: Environmental monitoring robots represent a revolutionary approach to preserving and safeguarding our planet's delicate ecosystems. With their adaptability, technological capabilities, and data-driven insights, these robots play a vital role in biodiversity conservation, disaster response, and sustainable resource management. As technology continues to evolve, environmental monitoring robots will contribute to a more comprehensive understanding of our environment, empower informed decision-making, and inspire collaborative efforts to ensure the well-being of our planet for generations to come.

Early Warning Systems for Natural Disasters: Natural disasters, ranging from earthquakes and tsunamis to hurricanes and wildfires, have the potential to cause immense destruction and loss of life. In the face of these unpredictable and catastrophic events, the development and implementation of early warning systems have emerged as crucial tools to mitigate the impact and save lives. Early warning systems are sophisticated technological frameworks that integrate real-time data collection, monitoring, communication, and prediction algorithms to provide timely alerts and critical information to at-risk populations and authorities. These systems play an indispensable role in enhancing preparedness, response, and recovery efforts, transforming the way societies manage natural disasters and safeguard their communities.

Understanding Early Warning Systems: Early warning systems are designed to detect the initiation of natural hazards, such as earthquakes, tsunamis, floods, and severe weather events, and rapidly disseminate alerts to affected populations and relevant authorities. These systems leverage a combination of data sources, including seismic sensors, weather satellites, ocean buoys, and environmental monitoring networks, to monitor potential triggers of natural disasters. The data collected is then processed and analyzed using sophisticated algorithms to generate accurate predictions and assessments of potential impacts.

Timely Alerts and Risk Communication: The core objective of early warning systems is to provide timely alerts to communities at risk. When a potential hazard is detected, the system triggers an alert mechanism that broadcasts warnings through various communication channels, such as text messages, sirens, social media, and mobile apps. These alerts offer valuable lead time for individuals to take protective measures, evacuate if necessary, and make informed decisions that can significantly reduce the impact of a disaster. Effective risk communication is a cornerstone of early warning systems, as clear and concise information empowers individuals to respond appropriately and seek safety.

Interdisciplinary Collaboration: Early warning systems require interdisciplinary collaboration among experts in meteorology, seismology, geology, hydrology, engineering, data science, and communication. These experts work together to develop models and algorithms that accurately predict the onset and impact of natural hazards. The integration of various data streams, including satellite imagery, seismic activity, oceanic conditions, and weather patterns, contributes to the holistic understanding of potential threats. Such collaboration ensures that the warning systems are reliable, robust, and capable of handling dynamic and complex situations.

Localized and Contextualized Alerts: One of the strengths of modern early warning systems is their ability to provide localized and context-specific alerts. These alerts take into consideration the geographic location, vulnerability, and infrastructure of the at-risk area. For example, an early warning system for flooding can account for factors such as elevation, topography, river flow rates, and soil saturation to accurately predict the extent of flooding and areas likely to be affected. This granularity ensures that alerts are relevant and actionable, minimizing false alarms and enhancing the credibility of the system.

Challenges and Innovations: Despite their effectiveness, early warning systems face challenges that demand ongoing innovation and improvement. Ensuring the reliability of data sources, addressing false positives and false negatives, and developing robust communication infrastructure in remote or vulnerable areas are key challenges. Additionally, the rapid evolution of technology and the need to adapt to changing environmental conditions necessitate continuous updates and enhancements to the system's algorithms and hardware.

Global Efforts and Success Stories: Early warning systems have demonstrated their value on a global scale. For instance, Japan's advanced earthquake early warning system provides seconds to tens of seconds of warning before an earthquake's destructive seismic waves reach populated areas, allowing individuals to seek cover and automated systems to shut down critical infrastructure. In the case of tsunamis, the Indian Ocean Tsunami Warning System has been instrumental in minimizing casualties by issuing timely alerts to coastal communities after the devastating 2004 Indian Ocean tsunami.

Beyond Natural Disasters: Early warning systems are not limited to natural disasters. They also extend to other areas such as disease outbreaks, public health emergencies, and even climate change impacts. For instance, systems that monitor air quality and detect disease outbreaks in real time contribute to the prevention and control of health crises. Similarly, climate early warning systems provide critical information to communities vulnerable to extreme weather events and support adaptation strategies in the face of a changing climate.

Conclusion: Early warning systems represent a pivotal shift in disaster management paradigms. They empower societies to anticipate, respond, and recover from natural disasters with greater resilience and efficiency. By harnessing the power of real-time data, predictive modeling, and effective communication, these systems contribute to saving lives, reducing economic losses, and enhancing the overall well-being of communities. As technology continues to advance and interdisciplinary collaborations deepen, the future of early warning systems holds the promise of even more accurate predictions, broader coverage, and a stronger collective response to the challenges posed by natural disasters.

Exercise

Multiple-choice Questions

1. What is the primary objective of integrating robotics in agriculture?
a) Increasing manual labor
b) Reducing crop yield
c) Enhancing efficiency and productivity
d) Decreasing technology usage
2. How do agricultural robots contribute to precision farming?
a) By randomly distributing pesticides
b) By using a one-size-fits-all approach for crops
c) By applying fertilizers uniformly across fields
d) By relying solely on human judgment
3. What is the significance of using drones in agriculture?
a) To create unnecessary noise pollution
b) To obstruct sunlight from reaching crops
c) To improve crop monitoring and yield estimation
d) To reduce access to remote areas
4. How do robotics systems contribute to reducing pesticide usage in agriculture?
a) By increasing pesticide application
b) By targeting specific areas for pesticide use
c) By applying pesticides indiscriminately
d) By advocating for pesticide-free farming
5. What role do robotic sensors play in environmental monitoring?
a) They increase pollution levels
b) They detect natural disasters in advance
c) They provide real-time data on environmental conditions
d) They encourage deforestation

Review Questions:

1. How does the integration of robotics in agriculture contribute to enhanced efficiency and productivity in farming practices?
2. Describe the concept of precision farming and how agricultural robots play a crucial role in achieving it.
3. What are some key advantages of using drones in agriculture, and how do they impact various aspects of farming?
4. How do robotic systems contribute to reducing pesticide usage and promoting sustainable farming practices?
5. Explain the role of robotic sensors in environmental monitoring and how they provide valuable insights into ecological conditions.
6. How do underwater robots benefit marine conservation efforts by collecting data and monitoring marine ecosystems?
7. In what ways do robotic systems contribute to monitoring and managing the health of forests and trees?
8. Describe the potential impact of robotic systems in managing soil health and irrigation practices for sustainable agriculture.
9. How does automation in harvesting through agricultural robots improve efficiency and address labor challenges in the agriculture sector?
10. Discuss the significance of using autonomous electric vehicles in agriculture and their potential environmental benefits.
11. How do robotic systems contribute to environmental conservation and ecological monitoring in sensitive ecosystems?

12. Explain how robotic systems can be deployed as early warning systems for natural disasters, providing timely alerts to communities.
13. In what scenarios are robotic systems particularly beneficial for conducting environmental monitoring tasks in hazardous environments?
14. How can robotics assist in disaster relief and response efforts, particularly in challenging and remote areas?
15. Describe the role of robotic systems in monitoring air and water quality, and their contribution to environmental conservation.
16. How does the integration of AI and data analytics enhance the effectiveness of environmental monitoring robots?
17. Explain the potential challenges and ethical considerations associated with the deployment of robotics in environmental conservation.
18. Discuss the interdisciplinary collaboration between robotics, environmental science, and conservation efforts for sustainable practices.
19. How do robotic systems contribute to combating illegal wildlife poaching and protecting endangered species?
20. Describe the potential future developments in robotics for environmental conservation and how they can shape the field in the coming years.

Answers Multiple-choice Questions

Answers:
1. c)
2. c)
3. c)
4. b)
5. c)

Review Questions:

1. The integration of robotics in agriculture revolutionizes farming practices by significantly enhancing efficiency and productivity. Through the use of robotic technologies, tasks such as planting, monitoring, irrigation, and harvesting can be performed with precision and accuracy. Agricultural robots equipped with advanced sensors and AI algorithms can analyze soil conditions, monitor crop health, and optimize resource usage. This enables farmers to make data-driven decisions, resulting in improved yields and resource conservation. For instance, autonomous robots can precisely apply fertilizers and pesticides only where needed, minimizing wastage and environmental impact. Additionally, robots can work tirelessly and consistently, overcoming the limitations of human labor and reducing operational delays during critical phases of cultivation.
2. Precision farming involves the application of technology to optimize farming practices based on real-time data and insights. Agricultural robots are pivotal in achieving precision farming as they enable targeted actions to be taken for specific crops and areas. These robots are equipped with various sensors, such as GPS, cameras, and soil monitors, which collect data on factors like soil moisture, nutrient levels, and crop health. This data is then analyzed using AI algorithms to provide actionable insights for farmers. Agricultural robots facilitate variable rate application of inputs like water and fertilizers, adjusting them according to the specific needs of different parts of a field. This concept not only boosts productivity but also reduces waste and environmental impact. By enabling precise interventions, such as targeted irrigation and efficient resource utilization, precision farming enhances crop quality, yield, and sustainability.
3. Drones offer numerous advantages in agriculture, transforming traditional farming practices. Firstly, drones equipped with high-resolution cameras and sensors can capture detailed aerial images of fields, allowing farmers to assess crop health and monitor growth patterns. This helps in identifying issues such as pest infestations and nutrient deficiencies early, enabling prompt action. Secondly, drones can rapidly cover large areas and collect data efficiently, significantly reducing the time required for traditional manual methods. Thirdly, drones facilitate the creation of detailed maps and 3D models of fields, aiding in precise planning and decision-making. Additionally, drones equipped with multispectral or thermal sensors can provide insights beyond what the human eye can perceive. These capabilities collectively lead to improved crop management, reduced costs, and optimized resource utilization in agriculture.
4. Robotic systems play a pivotal role in reducing pesticide usage and promoting sustainable farming practices. Traditional agricultural methods often involve blanket application of pesticides, leading to overuse and environmental harm. Agricultural robots equipped with advanced sensors and AI algorithms enable targeted and localized pesticide application. These robots can identify specific areas with pest infestations and deliver pesticides only to those areas, minimizing overall usage. This approach, known as site-specific or variable rate application, reduces chemical exposure to non-target areas and minimizes the environmental impact. Moreover, robotic systems

can work around the clock, ensuring timely intervention when pests are detected, preventing the need for continuous pesticide use. By contributing to precision pest management, these robots improve crop quality, reduce chemical residues, and support sustainable agricultural practices.

5. Robotic sensors play a critical role in environmental monitoring by providing valuable insights into ecological conditions. These sensors are designed to detect various environmental parameters such as air quality, water quality, temperature, humidity, and more. In ecosystems where manual data collection is challenging, robotic sensors offer the ability to gather data from remote or hazardous locations. For example, aquatic robots equipped with sensors can monitor water bodies for pollutants, temperature changes, and other vital parameters. These sensors collect data in real-time, enabling scientists and environmentalists to make informed decisions and respond promptly to changes in the environment. Furthermore, robotic sensors can be deployed for long durations, offering continuous monitoring capabilities that would be otherwise difficult to achieve through traditional methods. Through their ability to provide accurate, high-frequency data, robotic sensors contribute to a deeper understanding of ecosystems and support effective environmental conservation efforts.

6. Underwater robots, also known as underwater autonomous vehicles (UAVs) or autonomous underwater vehicles (AUVs), play a crucial role in marine conservation efforts by collecting data and monitoring marine ecosystems. These robots are equipped with a range of sensors, cameras, and scientific instruments that allow them to explore the underwater environment and gather valuable information. One of the primary contributions of underwater robots is their ability to access areas that are difficult or dangerous for human divers to reach. This includes deep-sea environments, underwater caves, and regions with extreme conditions. By collecting data on water temperature, salinity, currents, and marine life distribution, underwater robots help scientists gain insights into ocean dynamics, biodiversity, and the impact of climate change. Additionally, these robots support efforts to map and monitor marine protected areas, study underwater geological features, and assess the health of coral reefs. Overall, underwater robots contribute significantly to marine conservation by providing essential data for informed decision-making and policy formulation.

7. Robotic systems play a vital role in monitoring and managing the health of forests and trees, contributing to conservation efforts and sustainable forestry practices. These systems are equipped with various sensors, such as LiDAR (Light Detection and Ranging), cameras, and spectrographs that enable them to collect data about the vegetation, tree health, and overall ecosystem conditions. For instance, LiDAR-equipped drones can create detailed 3D models of forests, accurately measuring tree heights, canopy density, and terrain variations. This information is essential for forest management planning, including timber inventory, disease detection, and wildfire risk assessment. Moreover, robotic systems can be deployed in challenging terrain and remote areas, providing access to data that would be difficult to obtain using traditional methods. By offering insights into forest health, biodiversity, and ecosystem dynamics, robotic systems contribute to conservation efforts, enable informed decision-making, and support sustainable forest management practices.

8. Robotic systems have the potential to make a significant impact on managing soil health and irrigation practices for sustainable agriculture. Soil health is a critical factor in crop productivity and environmental sustainability. Robotic systems equipped with sensors, such as moisture probes and nutrient analyzers, can assess soil conditions in real-time. By continuously monitoring soil moisture levels, nutrient content, and pH levels, these robots provide valuable insights into the health of the soil. This information enables farmers to optimize irrigation and nutrient application, ensuring that crops receive the right amount of water and nutrients. Additionally, robots can assist in precision planting by determining optimal seed placement based on soil conditions. By reducing water wastage, preventing over-fertilization, and promoting healthy soil microbial activity, robotic systems contribute to sustainable agriculture practices and resource conservation.

9. Automation in harvesting through agricultural robots addresses efficiency challenges and labor shortages in the agriculture sector. Traditional harvesting methods often require a significant workforce during peak seasons, and labor availability can be unpredictable. Agricultural robots equipped with specialized grippers, sensors, and AI algorithms can perform selective and efficient harvesting. For instance, robots can detect the ripeness of fruits through computer vision and pick them with precision, reducing waste and ensuring higher-quality produce. Additionally, robotic harvesting can operate continuously, day and night, allowing farmers to complete harvesting tasks quickly and efficiently. By overcoming labor challenges and minimizing post-harvest losses, robotic harvesting systems contribute to increased farm productivity, reduced costs, and improved supply chain management.

10. The utilization of autonomous electric vehicles holds significant promise for revolutionizing agriculture and offering environmental benefits. Electric autonomous vehicles can be employed for various tasks such as planting, cultivating, spraying, and harvesting. These vehicles are powered by clean energy sources, reducing greenhouse gas emissions and minimizing the carbon footprint of agricultural operations. Moreover, their autonomous capabilities

enable them to operate with high precision, minimizing soil compaction and optimizing resource usage. Electric vehicles also have fewer moving parts compared to traditional combustion-engine vehicles, resulting in reduced maintenance needs and operational costs. Additionally, these vehicles can be integrated with renewable energy sources such as solar panels, making them even more sustainable. The adoption of autonomous electric vehicles in agriculture aligns with the global push for sustainability and reduced environmental impact, contributing to a greener and more efficient farming future.

11. Robotic systems play a crucial role in environmental conservation and ecological monitoring, particularly in sensitive ecosystems. These systems can navigate challenging terrains and remote areas, enabling researchers and conservationists to gather data without disrupting delicate ecosystems. For instance, robots equipped with specialized sensors can monitor fragile coral reefs without damaging the reefs themselves. These sensors collect data on factors such as water temperature, pH levels, and biodiversity, providing insights into the health of marine ecosystems. Furthermore, robotic systems contribute to habitat mapping and wildlife tracking, helping scientists understand migration patterns, behavior, and population dynamics. By offering a non-intrusive means of data collection, robotic systems support efforts to preserve and protect biodiversity in ecosystems that are vulnerable to human interference.

12.Robotic systems serve as valuable tools for deploying early warning systems for natural disasters, providing timely alerts to communities and potentially saving lives. These systems can monitor environmental conditions and detect anomalies that indicate the onset of disasters such as earthquakes, tsunamis, floods, and wildfires. For example, seismic sensors can be deployed to detect ground vibrations before an earthquake strikes, giving residents valuable seconds to take cover. Similarly, water level sensors can provide advance warning of potential floods, allowing residents to evacuate in a timely manner. Robotic systems can also be equipped with communication capabilities to send alerts to relevant authorities and communities. By enabling early detection and rapid response, robotic systems enhance disaster preparedness and response efforts, minimizing the impact of natural disasters on human lives and property.

13. Robotic systems are particularly beneficial for conducting environmental monitoring tasks in hazardous environments where human intervention may be unsafe. These environments include areas with extreme temperatures, toxic gases, radiation, and difficult terrain. For instance, robots equipped with sensors can be used to monitor active volcanoes, collecting data on gas emissions, temperature, and volcanic activity. In nuclear disaster sites, robots can assess radiation levels and structural integrity without exposing humans to danger. Furthermore, robotic systems can navigate areas with high pollution levels, tracking air quality and pollution sources. By performing data collection and analysis in hazardous environments, robots contribute to a deeper understanding of environmental conditions and help protect human workers from exposure to harm.

14. Robotic systems play a crucial role in disaster relief and response efforts, particularly in challenging and remote areas. During natural disasters such as earthquakes, hurricanes, or landslides, infrastructure can be severely damaged, and access to affected areas can be limited. Robotic systems equipped with sensors, cameras, and mobility capabilities can be deployed to assess damage, locate survivors, and provide critical data to first responders. For instance, drones can conduct aerial surveys of disaster-stricken regions, providing real-time images and mapping data. Ground-based robots can navigate debris and hazardous conditions to locate survivors and relay their positions to rescue teams. By functioning as eyes and ears in disaster-stricken areas, robotic systems enhance the efficiency and effectiveness of search and rescue operations, ultimately saving lives and reducing response times.

15. Robotic systems play a significant role in monitoring air and water quality, contributing to environmental conservation by assessing pollution levels and identifying sources of contamination. Air quality monitoring robots are equipped with sensors that measure parameters such as particulate matter, volatile organic compounds, and gas emissions. These robots can be deployed in urban areas, industrial zones, and near pollution sources to collect real-time data on air quality. Similarly, water quality monitoring robots equipped with water samplers and sensors can assess parameters such as pH levels, dissolved oxygen, and pollutant concentrations. These robots navigate water bodies such as rivers, lakes, and coastal areas, providing insights into the health of aquatic ecosystems. By continuously monitoring air and water quality, robotic systems enable proactive measures to address pollution and contribute to the preservation of natural resources.

16. The integration of AI and data analytics enhances the effectiveness of environmental monitoring robots by enabling real-time analysis of collected data. AI algorithms can process large volumes of data from sensors and cameras, identifying patterns, anomalies, and trends. For example, AI-powered image recognition can identify species in wildlife monitoring images, enabling researchers to track population dynamics. Additionally, AI algorithms can predict pollution levels based on historical data and current environmental conditions. Real-time data analysis allows for immediate responses to emerging issues, such as sudden increases in pollution or changing

weather patterns. By providing actionable insights, AI integration maximizes the value of data collected by robotic systems, supporting informed decision-making and efficient resource allocation.

17. The deployment of robotic systems in environmental conservation raises various challenges and ethical considerations. One key challenge is ensuring that the use of robots does not disrupt or harm ecosystems. Careful consideration must be given to the potential impacts of robotic presence on wildlife behavior and habitat. Moreover, the collection and transmission of data by robotic systems raise privacy concerns, as data on sensitive areas or endangered species could be misused. Ethical considerations also extend to the potential displacement of human workers in environmental monitoring and conservation roles. Balancing the benefits of robotic efficiency with the preservation of jobs and human expertise is a complex ethical issue. As robotic systems become more integrated into conservation efforts, a comprehensive ethical framework is essential to guide their deployment and use in a responsible and sustainable manner.

18. The interdisciplinary collaboration between robotics, environmental science, and conservation efforts is instrumental in achieving sustainable practices and effective environmental monitoring. Scientists, engineers, conservationists, and policymakers collaborate to develop robotic systems that address specific environmental challenges. The expertise of environmental scientists ensures that robotic systems collect relevant data and contribute to meaningful insights. Engineers design robotic platforms that can withstand challenging conditions while accurately collecting data. Conservationists provide insights into the ecological impact of robotic deployment and help define ethical guidelines. Policymakers play a role in shaping regulations and standards for the responsible use of robotic systems in conservation. The synergy between these disciplines leads to the development of innovative solutions that contribute to the preservation of natural resources and biodiversity.

19. Robotic systems contribute to combating illegal wildlife poaching and protecting endangered species through enhanced surveillance and deterrence measures. These systems are equipped with advanced sensors and cameras that can monitor wildlife habitats and detect unauthorized intrusions. For example, drones equipped with thermal cameras can track poachers in protected areas, enabling law enforcement agencies to respond swiftly. Ground-based robots can navigate dense forests and difficult terrains, aiding in tracking and deterring illegal activities. Additionally, robotic systems can collect data on wildlife behavior and population dynamics, assisting conservationists in making informed decisions about protection strategies. By acting as guardians of wildlife, robotic systems play a critical role in safeguarding endangered species and preserving biodiversity.

20. The potential future developments in robotics for environmental conservation are diverse and hold promise for addressing complex challenges. One area of advancement is the integration of advanced AI algorithms for predictive modeling and decision support. AI-powered models can simulate the impact of climate change, habitat loss, and pollution, aiding in the formulation of proactive conservation strategies. Furthermore, robotic systems may become more versatile, with the ability to adapt to different terrains and ecosystems. For example, modular robotic platforms could be designed to be deployed in forests, oceans, and urban areas for diverse monitoring purposes. Additionally, improvements in battery technology and energy efficiency will enable longer missions and enhanced capabilities for robotic systems. As technology advances, robotic systems will continue to evolve as powerful tools for environmental conservation, playing a crucial role in preserving the planet's natural resources for future generations.

Online Resources

Online Exam Papers

Link: https://forms.gle/bBxwa8gktdqPpBzM9

QR Code:

YouTube Video Playlist Link

Link: https://www.youtube.com/playlist?list=PLzlh_31VWEFoMrahvi_w3RBJb97Vw79IJ

QR Code:

Chapter 13: Robotic Artistry and Entertainment

Robotic artistry and entertainment represent a captivating intersection of technology and creativity, where cutting-edge robotics blend seamlessly with artistic expression to create unique and engaging experiences for audiences worldwide. This convergence has given rise to a new era in entertainment, where robots are not just tools, but performers and collaborators that captivate and inspire. The evolution of robotic artistry and entertainment is rooted in a rich history that spans decades, from the early experimental endeavors to the sophisticated and awe-inspiring productions of today.

The origins of robotic artistry can be traced back to the mid-20th century, when artists and engineers began to explore the possibilities of incorporating machines into their creative processes. Early pioneers, such as Nam June Paik and Nicolas Schöffer, experimented with kinetic sculptures and installations that merged mechanical movement with artistic expression. These groundbreaking works laid the foundation for what would eventually become a dynamic field where technology and aesthetics converged.

One of the earliest instances of robotic entertainment emerged with the advent of animatronics. Theme parks and entertainment venues began to introduce animatronic characters, creatures, and displays that brought a touch of magic and lifelike motion to their attractions. These early animatronics featured basic movements and predefined sequences, captivating audiences with their seemingly autonomous actions.

The 21st century marked a significant turning point in the evolution of robotic artistry and entertainment. As technology advanced, robotics became more sophisticated, allowing for greater levels of interactivity, expressiveness, and realism. Robotic artists and engineers collaborated to create performances that transcended the boundaries of traditional art forms, seamlessly blending music, dance, visual effects, and robotics.

Robotic entertainment experienced a surge in popularity through shows like Cirque du Soleil's "KÀ," which integrated robotics and acrobatics to create a mesmerizing spectacle that pushed the boundaries of what was thought possible on stage. Additionally, robot competitions like the FIRST Robotics Competition and BattleBots introduced audiences to the excitement of robotic competitions, further enhancing the public's interest in this emerging field.

Today, robotic artistry and entertainment encompass a diverse range of forms, from robotic dancers and musicians to interactive installations and immersive experiences. Artists, engineers, and creative technologists collaborate to design robotic systems that evoke emotion, tell stories, and engage audiences on a profound level. These productions often incorporate AI algorithms, advanced sensors, and intricate mechanical designs to create performances that are both technically impressive and artistically impactful.

As the field continues to evolve, the possibilities for robotic artistry and entertainment are boundless. The integration of virtual reality (VR) and augmented reality (AR) technologies opens up new avenues for immersive experiences, allowing audiences to interact with robotic performers and environments in unprecedented ways. The marriage of robotics and AI also holds the potential to create truly autonomous performances, where robots respond to the audience's reactions and adapt their actions in real-time.

In conclusion, the journey of robotic artistry and entertainment from its humble beginnings to its current state of technological marvels reflects the innovation and creativity of human minds. The combination of artistic vision and technological prowess has yielded a vibrant and dynamic field that continues to push the boundaries of what is possible in the realm of entertainment. As technology advances and artistic expression evolves, the realm of robotic artistry promises to remain a captivating arena where the imagination knows no bounds.

Robotic Art and Creativity: Merging Technology and Expression: The realm of art has always been a playground for human creativity, a canvas on which ideas, emotions, and imagination come to life. In recent years, however, a new and exciting chapter has unfolded in the world of art – one that involves the fusion of technology and artistic expression through robotic art. This innovative intersection has led to the creation of mesmerizing and thought-provoking artworks that challenge our perceptions of both art and technology. Robotic art represents a unique blend of creativity and engineering, where machines are not just tools, but active participants in the artistic process.

Historical Context and Early Experimentation: The concept of robots engaging in artistic endeavors is not entirely new. In the mid-20th century, artists like Nam June Paik and Nicolas Schöffer experimented with kinetic sculptures and installations that incorporated mechanical movement and interactivity. These early pioneers set the stage for the integration of technology into artistic expression. Paik's "Robot K-456," created in the 1960s, featured a modified television set on wheels that moved around the space, projecting light and sound onto the walls. Schöffer's "CYSP 1," exhibited at the World's Fair in 1958, combined light and movement to create a dynamic visual experience.

The Evolution of Robotic Art: As technology progressed, so did the possibilities for robotic art. The development of advanced sensors, actuators, and programming languages enabled artists and engineers to create artworks that responded to their environment, engaged with viewers, and even showcased a level of autonomy. Robotic sculptures began to incorporate complex movements, interactive elements, and AI algorithms, blurring the lines between the creator, the machine, and the audience.

Robotic Performers and Collaborators: Robotic art has also ventured into the realm of performance. The emergence of robotic performers has challenged traditional notions of live entertainment. Companies like Boston Dynamics have developed robots capable of dynamic and fluid movements, which have been incorporated into dance performances. These collaborations between humans and robots create a mesmerizing synergy where the precision of technology is combined with the emotional depth of human expression.

Interactive and Participatory Art: Interactive and participatory art has taken on new dimensions through robotics. Installations and exhibits invite viewers to engage with robotic artworks, often allowing them to influence the artwork's behavior and appearance. Robotic art installations equipped with sensors can react to the presence and movements of viewers, generating a unique and evolving experience for each individual. This interactivity blurs the line between observer and participant, granting viewers agency over the artistic process.

Challenges and Ethical Considerations: While robotic art opens up exciting possibilities, it also raises ethical and philosophical questions. As machines become more integrated into creative processes, debates arise about the role of human intent, the authenticity of art created by machines, and the potential loss of the human touch in artistic expression. Questions of authorship, creativity, and the potential for robotic art to replace human artists underscore the complex relationship between technology and art.

The Intersection of Art, Technology, and Society: Robotic art not only challenges artistic conventions but also invites us to reflect on the broader implications of technological advancement in society. It prompts discussions about the changing nature of work, the relationship between humans and machines, and the ways in which technology can amplify our creative capabilities. Robotic art provides a tangible glimpse into the future where technology is not just functional but serves as a vessel for human expression and emotion.

In conclusion, robotic art represents a captivating fusion of technology and artistic expression that transcends traditional boundaries. From early kinetic sculptures to interactive installations and robotic performers, the evolution of robotic art showcases the remarkable synergy between human creativity and technological innovation. As this field continues to evolve, it prompts us to rethink our understanding of art, creativity, and the potential for machines to contribute to the rich tapestry of human expression.

Robots in Entertainment and Media: The world of entertainment and media has been profoundly impacted by the rapid advancements in robotics and technology. What was once the realm of science fiction has now become a tangible reality, as robots are increasingly integrated into various aspects of the entertainment industry, From film and television to theme parks and live performances, robots have taken center stage, captivating audiences and transforming the way we experience and interact with entertainment content. This convergence of technology and storytelling has opened up new frontiers for creativity, engagement, and immersive experiences.

A Transformative Presence on Screen: The integration of robots into the entertainment industry has revolutionized the visual and storytelling elements of film and television. In the past, animatronics and practical effects were limited by their static nature, often requiring elaborate setups and extensive human involvement. However, the advent of advanced robotics has enabled filmmakers to bring lifelike and dynamic robotic characters to the screen. From the charming droids of the "Star Wars" franchise to the intricately designed robots in "I, Robot," robots have become compelling characters that resonate with audiences on emotional and narrative levels.

Enhancing Immersive Experiences: Beyond traditional storytelling, robots have ushered in a new era of immersive experiences in entertainment. Theme parks around the world are embracing robotic technology to create interactive attractions that transport visitors into fantastical worlds. These attractions often feature robotic characters that engage with guests, blurring the lines between fiction and reality. For example, Disney's animatronic figures have evolved to exhibit lifelike movements, expressions, and interactions, creating a truly immersive and magical experience for park visitors.

Live Performances and Theatrical Productions: Robots have also found their way into live performances and theatrical productions, pushing the boundaries of what is possible on stage. Cutting-edge technologies enable robots to seamlessly collaborate with human performers, enhancing the visual spectacle and narrative impact of performances. Dancing robots, expressive animatronics, and even robotic aerial performers have added a new layer of excitement and wonder to live shows, captivating audiences with their precision and creativity.

Virtual and Augmented Reality: The convergence of robots and virtual/augmented reality (VR/AR) technologies has paved the way for entirely new forms of entertainment experiences. In VR and AR environments, robots can be virtually brought to life, allowing users to interact with them in immersive digital spaces. These experiences bridge the gap between physical and digital realities, enabling users to engage with robots in novel and innovative ways. Additionally, robots can serve as avatars in VR environments, giving users the sensation of embodying a robotic character.

Challenges and Considerations: While robots in entertainment offer exciting possibilities, they also present challenges and considerations. The integration of robots into creative industries raises questions about the role of human performers, the authenticity of experiences, and potential disruptions to traditional forms of entertainment. Ethical concerns related to human-robot interactions, job displacement, and the impact on human creativity also come to the forefront.

The Future of Entertainment: As technology continues to evolve, the entertainment industry stands on the cusp of transformative changes. Robots are poised to play increasingly central roles in entertainment experiences, blurring the lines between fiction and reality, physical and digital realms. Whether it's through lifelike characters on screen, interactive attractions, or immersive VR/AR experiences, robots are redefining the way we engage with entertainment content. As creators and engineers collaborate, the potential for new narratives, experiences, and forms of engagement remains limitless.

In conclusion, the integration of robots into entertainment and media is reshaping the way we consume and interact with creative content. From enhancing storytelling on screen to providing immersive experiences in theme parks and live performances, robots are pushing the boundaries of what is possible in the realm of entertainment. As technology continues to advance, the future promises even more captivating and innovative ways for robots to captivate and engage audiences in new and unexpected ways.

Robotic Art Installations and Interactive Exhibits: The world of art has always been a canvas for innovation, challenging the boundaries of human imagination and expression. In recent years, the integration of robotics and technology into artistic endeavors has ushered in a new era of creativity, giving rise to captivating robotic art installations and interactive exhibits that push the envelope of what is possible in the realm of artistic expression. This fusion of technology and art has not only redefined how audiences engage with artworks but has also paved the way for thought-provoking dialogues about the intersections of humanity, technology, and culture.

The Emergence of Robotic Art: Robotic art installations and interactive exhibits have gained significant attention in the art world for their ability to create immersive and dynamic experiences that captivate and engage viewers on multiple levels. These creations often utilize cutting-edge technologies, including robotics, sensors, artificial intelligence, and augmented reality, to transform traditional art forms into living, evolving, and interactive pieces that respond to the presence and actions of the audience.

Breaking the Fourth Wall: One of the defining features of robotic art installations and interactive exhibits is their ability to break down the traditional boundaries between the artwork and the audience. In traditional art, viewers are passive observers, whereas robotic art installations invite active participation and engagement. Whether it's through touch, movement, voice commands, or even biometric data, these artworks encourage viewers to become co-creators, shaping the evolution of the artwork in real-time.

Exploring Themes and Concepts: Robotic art installations often delve into a wide range of themes and concepts, from the exploration of human emotions to societal issues and philosophical inquiries. These installations provide a unique platform for artists to convey complex ideas through the interplay of technology and artistic expression. For example, an interactive exhibit could explore the relationship between humans and machines, prompting viewers to reflect on their own interactions with technology and the potential implications for the future.

Dynamic and Evolving Creations: What sets robotic art installations apart is their dynamic and ever-changing nature. Unlike static artworks, these installations evolve over time, responding to the input and engagement of the audience. This dynamic quality not only challenges traditional notions of artistic permanence but also emphasizes the transient and ephemeral nature of technological interactions.

Challenges and Ethical Considerations: As with any integration of technology and art, there are challenges and ethical considerations to address. The reliance on technology introduces questions about the longevity and maintenance of these artworks, as well as concerns about the potential obsolescence of the technology powering them. Additionally, ethical dilemmas related to privacy, data collection, and the potential for algorithmic biases must be navigated in the creation of interactive exhibits that gather and process audience data.

Inspiring New Perspectives: Robotic art installations and interactive exhibits offer a fresh and dynamic way for audiences to engage with art, often inspiring new perspectives and conversations. By encouraging viewers to actively participate, these creations create a sense of agency and personal connection, forging deeper emotional ties between the audience and the artwork. This engagement also democratizes the artistic experience, making it accessible to a broader range of individuals who may not have engaged with traditional art forms.

Future Directions: The world of robotic art installations and interactive exhibits is continuously evolving, with artists, engineers, and creators pushing the boundaries of what is possible. As technology advances, the potential for even more immersive, responsive, and intricate installations grows. With the advent of virtual and augmented reality, artists can create digital environments that blur the lines between physical and virtual spaces, offering audiences unprecedented opportunities for exploration and interaction.

In conclusion, robotic art installations and interactive exhibits represent a groundbreaking fusion of technology and creativity, transforming how audiences experience and engage with art. These dynamic and immersive experiences challenge traditional notions of artistic expression, inviting viewers to become active participants in the creative process. As the boundaries of technology continue to expand, the future promises even more innovative and awe-inspiring installations that push the envelope of artistic possibility and redefine the relationship between humans, technology, and art.

Ethical Debates Surrounding Robot-Generated Art: The realm of art has long been a domain characterized by human ingenuity, creativity, and emotional expression. However, the advent of technology and the integration of robotics into artistic practices have given rise to a new wave of creativity—one that blurs the lines between human and machine-generated art. As robotic systems and artificial intelligence algorithms become more adept at producing artworks, ethical debates have emerged surrounding the authenticity, value, and implications of robot-generated art. These debates reflect broader discussions about the role of automation in creative endeavors and the complex relationship between technology, creativity, and human agency.

The Rise of Robot-Generated Art: Robot-generated art, often referred to as "Robo-art," encompasses a range of creative outputs produced by machines or algorithms. These creations can span various forms of art, including visual arts, music, literature, and more. Machine learning algorithms, neural networks, and generative models have enabled robots to analyze vast datasets of existing artworks and produce new pieces that mimic established artistic styles or generate entirely novel compositions. This transformative shift in the art landscape has ignited both excitement and apprehension within the artistic community and beyond.

Ethical Considerations in Robot-Generated Art: As robot-generated art gains prominence, ethical questions have arisen that challenge conventional understandings of creativity, authorship, and originality. Some of these considerations include:

Authenticity and Originality: Critics argue that artworks created by machines lack the human experience and emotional depth that traditionally define artistic expression. The question of whether robot-generated art can truly be considered authentic and original raises concerns about the role of human intention and emotion in the creative process.

Authorship and Attribution: Determining the rightful authorship of robot-generated art is a complex issue. While the algorithm's creators or operators play a role in shaping the AI's abilities, the algorithm itself generates the final output. This raises questions about how credit and recognition should be allocated in a collaborative human-machine endeavor.

Value and Significance: The value and significance of robot-generated art are subjects of ongoing debate. Some argue that these creations lack the depth and meaning that human artists infuse into their work. Others contend that robot-generated art contributes to the diversification of artistic expression and challenges traditional notions of value.

Aesthetic Judgment: Evaluating the artistic merit of robot-generated art requires a reevaluation of aesthetic judgment. Can algorithms truly possess aesthetic sensibilities, or do they merely replicate patterns present in existing art? This challenge raises concerns about the role of human aesthetic preferences in assessing art.

Impact on Human Creativity and Innovation: The integration of robots and algorithms into the artistic landscape has sparked discussions about how automation may impact human creativity and innovation. Critics fear that the proliferation of robot-generated art could lead to a devaluation of human artistic efforts and discourage originality. However, proponents argue that automation can serve as a tool to inspire new creative directions and push artists to explore uncharted territories.

Expanding the Boundaries of Artistic Expression: Despite the ethical debates, robot-generated art has undeniably expanded the boundaries of artistic expression. Artists and technologists are collaborating to explore the possibilities that emerge at the intersection of human and machine creativity. Some artists embrace the potential of AI as a collaborator, using algorithms to amplify their creative visions or challenge their artistic assumptions.

Ethical Frameworks and Guidelines: Addressing the ethical complexities of robot-generated art requires the establishment of clear frameworks and guidelines. Some propose the development of "artistic AI ethics," which would encompass considerations such as transparency in algorithmic processes, proper attribution of authorship, and the preservation of human agency and intention in the creative process.

The Future Landscape: As robot-generated art continues to evolve, it is essential to approach these ethical debates with a holistic perspective. Discussions should consider the implications for artists, audiences, and the broader cultural landscape. While ethical concerns are valid, they should not overshadow the potential of AI and robotics to amplify artistic creativity, foster new collaborations, and challenge preconceived notions of what constitutes art.

In conclusion, the rise of robot-generated art has ignited a complex and nuanced discussion about the intersection of technology and creativity. Ethical considerations surrounding authenticity, authorship, and value intersect with broader conversations about human creativity, innovation, and the role of automation in artistic expression. Navigating these debates requires a balanced approach that respects both the contributions of technology and the unique attributes of human creativity, ultimately shaping the future landscape of art in ways that reflect the complex interplay between human ingenuity and technological advancement.

Exercise

Multiple-choice Questions

1. What is the primary focus of "Robotic Artistry and Entertainment"?
a) Robotic Ethics
b) Robotic Engineering
c) Robotic Artistry and Entertainment
d) Robotic Programming
2. Which term refers to the use of robots in creative and entertainment-related tasks?
a) Robotic Automation
b) Robotic Artistry
c) Robotic Intelligence
d) Robotic Ethics
3. What is the potential benefit of using robots in the field of art and entertainment?
a) Decreased human involvement
b) Reduced creativity
c) Enhanced efficiency
d) Decreased audience engagement
4. Which of the following is an example of robotic artistry?
a) Automated manufacturing
b) Autonomous vehicles
c) Robot-assisted surgery
d) Robot-created paintings
5. In the context of entertainment, what is the role of social robots?
a) Enhancing industrial processes
b) Providing companionship and engagement
c) Performing complex mathematical calculations
d) Monitoring environmental conditions

Review Questions:

1. What is the central focus of the "Robotic Artistry and Entertainment"?
2. How does the integration of robotics impact the field of art and entertainment?
3. Explain the concept of "robotic artistry" and provide an example.
4. What potential benefits can robots bring to the realm of creative endeavors?
5. Differentiate between traditional forms of entertainment and robotic entertainment.
6. How can robots enhance audience engagement in live performances?
7. Describe the role of social robots in providing entertainment and engagement.
8. What challenges might arise when using robots for creative tasks in entertainment?
9. Provide an example of a social robot designed specifically for companionship and entertainment.
10. How do robots contribute to the creation of visual art forms?
11. Explain the technology behind robots mimicking human facial expressions and gestures.
12. In what ways can robotics influence the creativity of human artists?
13. Discuss the ethical implications of using robots to replace human performers in entertainment.
14. How does the development of lifelike humanoid robots impact the entertainment industry?
15. What potential economic changes might occur due to the integration of robotic artistry?
16. How can robots be employed to provide immersive experiences in theme parks and exhibitions?
17. What role can AI play in generating artistic content for entertainment purposes?
18. Describe the potential risks associated with relying heavily on robotic artistry in entertainment.
19. Discuss the balance between human creativity and robotic assistance in the artistic process.
20. How can robotic artistry contribute to the cultural and social aspects of entertainment?

Answers Multiple-choice Questions

Answers:

1. c)
2. b)
3. c)
4. d)
5. b)

Review Questions:

1. The central focus of the chapter "Robotic Artistry and Entertainment" is to delve into the convergence of robotics and the world of art and entertainment. This chapter explores the intricate ways in which technological advancements in robotics are reshaping creative processes and redefining audience experiences within the entertainment landscape.
2. The integration of robotics introduces a transformative impact on the field of art and entertainment by revolutionizing the methods of creation, presentation, and interaction. Robotics enables the infusion of automation and interactivity into artistic endeavors, allowing artists to explore innovative mediums, generate novel narratives, and create immersive experiences that captivate audiences in entirely new ways.
3. "Robotic artistry" embodies the incorporation of robotic systems and technologies in the creative processes of artistic expression. An illustrative example of this concept is a robotic arm equipped with various brushes and color dispensers, programmed to autonomously create intricate paintings by dynamically applying colors and brush strokes based on predefined algorithms.
4. Robots bring forth an array of potential benefits to the realm of creative endeavors. Their precision and repeatability allow for intricate detail work and the execution of repetitive tasks, enabling human artists to focus on more conceptual and imaginative aspects. Moreover, robots can introduce new techniques and mediums, expanding the horizons of artistic exploration.
5. Traditional forms of entertainment predominantly feature human performers, while robotic entertainment involves the incorporation of robots as performers, creators, or interactive elements. Traditional entertainment engages audiences through human emotions, while robotic entertainment creates engagement through a blend of human emotions and technological innovation.
6. Robots enhance audience engagement in live performances by seamlessly executing choreographed movements, synchronized with visual and auditory elements. This synchronicity creates a captivating visual spectacle that immerses the audience and augments their connection with the performance.
7. Social robots play a pivotal role in providing entertainment and engagement by offering companionship, interactive experiences, and emotional connections to users. Through conversations, storytelling, expressive gestures, and personalized interactions, social robots create meaningful and dynamic engagements for individuals.
8. Challenges in using robots for creative tasks in entertainment encompass preserving authenticity and emotional resonance in the artistic experience, ensuring that technology enhances rather than supersedes the artistic intent, and addressing potential backlash arising from excessive reliance on automation.
9. A notable example of a social robot designed for companionship and entertainment is a lifelike humanoid robot programmed to hold conversations, share jokes, play games, and respond to human emotions through facial expressions and gestures, thereby providing a sense of companionship and entertainment to users.
10. Robots contribute to the creation of visual art forms through programmed algorithms that control their brushwork, color mixing, and spatial positioning. This precision allows them to generate intricate patterns and designs that may be challenging for human artists to achieve consistently.
11. The technology behind robots mimicking human facial expressions and gestures relies on artificial neural networks. These networks are trained on extensive datasets of human expressions, enabling robots to learn and replicate the subtleties of various emotions and actions.
12. Robotics can influence the creativity of human artists by providing innovative tools, techniques, and perspectives that can inspire unconventional artistic approaches. Collaborative endeavors between humans and robots result in novel artworks that merge technological sophistication with artistic ingenuity.
13. The ethical implications of using robots to replace human performers in entertainment touch upon authenticity, emotional resonance, and societal values. The substitution of human performers with robots may challenge the perception of what constitutes genuine artistic expression and potentially lead to concerns over job displacement.
14. The development of lifelike humanoid robots exerts a transformative impact on the entertainment industry by enabling interactive, realistic, and lifelike experiences. Such robots blur the line between reality and technology, offering audiences a new dimension of engagement.
15. The integration of robotic artistry could bring about potential economic changes, including shifts in employment dynamics within the creative and technological sectors. Demand for skilled workers in both robotics and creative fields may increase, while traditional roles might undergo transformation.

16. Robots can enhance immersive experiences in theme parks and exhibitions through interactive installations, augmented reality scenarios, and multisensory engagements that transport visitors into dynamic and captivating environments.
17. AI can contribute to generating artistic content by analyzing existing artistic works, identifying patterns and styles, and then creating new and original pieces that draw inspiration from these patterns, thus adding a fresh dimension to artistic expression.
18. Relying heavily on robotic artistry presents potential risks, including the potential loss of human authenticity, emotional depth, and unique perspectives that human artists bring to their creations. Over-reliance on automation might lead to the homogenization of artistic output.
19. Striking a balance between human creativity and robotic assistance involves harnessing technology to amplify human artistic capabilities rather than overshadowing them. This balance ensures that technology remains a tool that enhances and augments human expression.
20. Robotic artistry contributes to the cultural and social aspects of entertainment by fostering discussions on the relationship between technology and creativity, challenging traditional notions of art, and forging novel interactive experiences that cater to diverse audiences and cultural contexts.

Online Resources

Online Exam Papers

Link: https://forms.gle/MP9wmuUqF7EXhHGR9

QR Code:

YouTube Video Playlist Link

Link: https://www.youtube.com/playlist?list=PLzlh_31VWEFrEisZSYFfg5v0i0vQzsoiv

QR Code:

Chapter 14: Future Horizons of Robotics Research and Innovation

The field of robotics has witnessed remarkable growth and innovation over the past few decades, transforming various industries and aspects of human life. As we look to the future, the horizons of robotics research and innovation are expanding rapidly, promising to redefine the way we live, work, and interact with technology. From advancements in artificial intelligence (AI) and machine learning to the development of highly adaptable and autonomous robots, the future of robotics holds immense potential across diverse domains.

The convergence of robotics and AI is shaping the future of this field. Advanced AI algorithms are enabling robots to understand and respond to complex environments, enhancing their decision-making capabilities. This fusion allows robots to autonomously adapt to dynamic scenarios, making them suitable for tasks that require cognitive skills, such as medical diagnostics, disaster response, and logistics optimization.

Future robotics research emphasizes the collaboration between humans and robots in shared workspaces. Innovations in collaborative robotics, or cobots, are making it possible for humans and robots to work side by side, complementing each other's strengths. This trend will revolutionize manufacturing, healthcare, and even everyday life, where robots and humans collaborate to achieve tasks more efficiently and safely.

Traditional rigid robots are being supplemented by soft robotics, which involve flexible and pliable materials. These robots mimic the properties of natural organisms, enabling them to navigate complex and unpredictable environments more effectively. Soft robotics has significant potential in healthcare, where they can interact gently with the human body for surgical assistance and rehabilitation.

The future of transportation is closely tied to robotics. Autonomous vehicles and drones are being developed to enhance transportation efficiency, safety, and sustainability. Self-driving cars and aerial drones promise to revolutionize the way goods and people are transported, reducing congestion and lowering carbon emissions.

The evolution of service and social robots is poised to redefine assistance in various domains. From healthcare and elderly care to customer service and education, robots equipped with natural language processing and understanding capabilities will be capable of providing personalized support and engagement.

As robots become more integrated into society, the ethical and regulatory aspects of their use are gaining prominence. Ensuring that robots are designed and programmed with ethical considerations in mind is crucial to prevent unintended consequences. Research in this area will shape the guidelines and regulations governing the deployment of robots in different contexts.

The study of biomechanics is driving innovations in Biorobotics, which involve creating robots inspired by biological systems. These robots emulate the structures and mechanisms found in nature, leading to advancements in areas like prosthetics, exoskeletons, and biomechanical simulations for scientific research.

As humanity ventures deeper into space exploration, robotics will play a vital role in enabling missions on distant planets, moons, and asteroids. Research is focused on developing robots that can navigate challenging extraterrestrial environments, conduct scientific experiments, and build infrastructure for future space colonies.

The application of robotics in environmental monitoring and agriculture is gaining traction. Autonomous drones and ground-based robots can monitor ecosystems, detect pollution, and aid in precision agriculture by optimizing crop management and resource utilization.

The interface between robotics and neuroscience is an emerging field known as neuro-robotics. This area explores the interaction between robots and the human nervous system, leading to breakthroughs in neuroprosthetics, brain-computer interfaces, and the understanding of human cognition.

The future of robotics research and innovation is boundless, offering solutions to some of the most complex challenges humanity faces. From AI-driven decision-making and human-robot collaboration to the development of versatile and adaptable robots, the advancements in robotics are set to revolutionize industries and reshape our daily lives. The synergy between technology and human creativity is propelling robotics into uncharted territories, opening new horizons that hold the potential to enhance efficiency, safety, and the quality of life on a global scale. As researchers, engineers, and innovators continue to push boundaries, the transformational impact of robotics is poised to transcend our current understanding and lead us into a future of unprecedented possibilities.

AI-Powered Cognitive Systems for Complex Tasks: systems have emerged as a transformative force. These systems, often powered by advanced machine learning techniques, are designed to emulate human-like cognitive abilities, enabling them to tackle complex tasks that were once considered beyond the scope of automation. This article explores the concept of AI-powered cognitive systems, their applications in handling intricate tasks, and the challenges and future prospects associated with their implementation.

Understanding Cognitive Systems: Cognitive systems in AI refer to technology that can mimic human cognitive functions such as learning, reasoning, problem-solving, and decision-making. Unlike traditional rule-based systems, cognitive systems possess the ability to adapt and improve their performance based on experience and data analysis.

AI and Complex Tasks: Complex tasks typically involve intricate decision-making processes that require context awareness, nuanced understanding, and the ability to process vast amounts of information. AI-powered cognitive systems excel in addressing these tasks by leveraging advanced algorithms and data-driven insights.

Applications in Healthcare: AI-powered cognitive systems are making significant contributions to the healthcare sector. These systems can analyze medical records, images, and patient data to assist in diagnosis, treatment planning, and drug discovery. Cognitive systems can also enhance patient care by providing real-time insights to medical professionals.

Financial Analysis and Decision-Making: The financial industry benefits from AI-powered cognitive systems by enabling rapid analysis of market trends, risk assessment, and investment strategies. These systems can sift through vast amounts of financial data to provide informed recommendations for traders and investors.

Natural Language Processing (NLP): NLP is a crucial component of cognitive systems, allowing machines to understand and generate human language. Applications range from sentiment analysis in social media to language translation and ChatBot interactions, where cognitive systems can process and respond to natural language input.

Supply Chain Optimization: The complexities of supply chain management often involve dynamic variables and data from various sources. Cognitive systems can optimize supply chain operations by analyzing demand patterns, inventory levels, and logistics data to ensure efficient and cost-effective processes.

Autonomous Vehicles: Cognitive systems are pivotal in the development of autonomous vehicles. These systems process real-time sensor data, interpret road conditions, and make split-second decisions to navigate safely and efficiently through complex environments.

Manufacturing and Quality Control: Cognitive systems are revolutionizing manufacturing by enabling predictive maintenance and quality control. By analyzing sensor data from equipment and production lines, these systems can identify anomalies and predict maintenance needs, minimizing downtime.

Challenges and Considerations:

Data Quality and Bias: Cognitive systems heavily rely on data. Poor data quality or biased datasets can lead to inaccurate insights and decisions.

Interpretability: The complexity of cognitive systems can make it challenging to understand how they arrive at particular conclusions, which is a concern in fields like healthcare and law.

Ethical Concerns: As cognitive systems handle sensitive tasks, ethical considerations surrounding data privacy and potential consequences of automated decisions arise.

Human-Cognitive System Collaboration: The future lies in human-cognitive system collaboration. These systems can complement human expertise by analyzing vast amounts of data, providing insights, and assisting in decision-making, while humans offer context, creativity, and ethical judgment.

Emerging Technologies: Advances in AI, particularly deep learning and neural networks, continue to enhance the capabilities of cognitive systems. These technologies enable more accurate and context-aware reasoning, contributing to the performance of complex tasks.

The Role of Big Data: The synergy between cognitive systems and big data is pivotal. The abundance of data generated by various sources enhances the learning and decision-making capabilities of cognitive systems, making them more adept at addressing complex tasks.

Regulatory Landscape: As cognitive systems become more integrated into critical sectors, regulatory frameworks must be established to ensure accountability, transparency, and ethical use of AI-powered technologies.

Future Horizons:

Healthcare Diagnostics and Treatment: Cognitive systems will continue to revolutionize healthcare, aiding in personalized treatment plans and drug discovery.

Environmental Monitoring: Cognitive systems can analyze vast environmental data to predict natural disasters, manage resources, and mitigate climate change impacts.

Education and Learning: AI-powered cognitive systems can offer personalized and adaptive learning experiences for students, tailoring education to individual needs.

AI-powered cognitive systems represent a groundbreaking frontier in technology, enabling machines to replicate human-like cognitive abilities and perform complex tasks with unparalleled accuracy and efficiency. These systems are revolutionizing industries such as healthcare, finance, manufacturing, and transportation, pushing the boundaries of what automation can achieve. While challenges like data quality, interpretability, and ethical concerns remain, the collaboration between humans and cognitive systems offers a promising future where technology complements human ingenuity. As advancements in AI and big data continue to evolve, the horizons of AI-powered cognitive systems are set to reshape industries, drive innovation, and elevate the potential of human-machine collaboration.

Learning from Experience and Adaptation: The evolution of artificial intelligence (AI) has given rise to systems that can learn from experience and adapt to dynamic environments. This capability represents a significant shift in the AI landscape, where machines are no longer limited to programmed rules but can acquire knowledge, refine their performance, and respond intelligently to changes. This article delves into the concept of learning from experience and adaptation in AI, exploring its implications, applications, challenges, and the future prospects of this transformative approach.

Learning from Experience: Learning in AI involves the process of extracting patterns, insights, and knowledge from data. Traditional rule-based systems lack the ability to learn from new information, making them rigid in dynamic scenarios. Learning from experience empowers AI systems to continuously update their knowledge and behavior based on incoming data.

Machine Learning Paradigms: The paradigm of machine learning encompasses various techniques, including supervised, unsupervised, and reinforcement learning. These techniques enable AI systems to process large datasets, identify patterns, and make predictions or decisions based on learned models.

Adaptation to Dynamic Environments: The ability to adapt to changing circumstances is crucial for AI systems to remain effective over time. Adaptable AI systems can adjust their strategies, behaviors, and decisions in response to new challenges, unexpected events, or shifts in the environment.

Reinforcement Learning: Reinforcement learning is a cornerstone of learning from experience and adaptation. This approach involves training AI agents to interact with an environment and learn optimal actions to maximize rewards. Reinforcement learning has found applications in robotics, gaming, and autonomous systems.

Natural Language Processing (NLP): NLP is a domain where learning from experience is especially relevant. AI models can learn linguistic patterns and nuances from vast amounts of text data, enabling them to understand context, sentiment, and even generate human-like language.

Predictive Analytics: In business and finance, learning from historical data enables AI to predict future trends, customer behavior, and market dynamics. Adaptive models can refine predictions as new data becomes available, leading to more accurate insights.

Autonomous Systems: Learning and adaptation are integral to the development of autonomous systems, such as self-driving cars and drones. These systems learn from real-world interactions and adapt their behaviors to navigate complex environments safely and efficiently.

Healthcare Diagnostics: AI-driven diagnostic tools learn from medical images, patient data, and historical cases to enhance accuracy in identifying diseases and conditions. Adaptation ensures that these models stay up-to-date with evolving medical knowledge.

Challenges and Considerations:

Data Quality and Bias: Learning from experience relies on data quality. Biased or incomplete datasets can lead to inaccurate learning and biased decision-making.

Interpretable AI: As AI models become more complex, understanding their decision-making processes becomes challenging. Interpretability is essential for trust and accountability.

Ethical Concerns: Adaptable AI systems may inadvertently learn and perpetuate biases present in data, leading to discriminatory outcomes. Ensuring fairness and ethical use is imperative.

Transfer Learning: Transfer learning is a technique where AI models trained on one task can adapt their knowledge to perform well on related tasks. This approach reduces the need for extensive data collection and training for each new scenario.

Continual Learning: Continual learning focuses on adapting AI systems over time without forgetting previous knowledge. This approach is essential to ensure that systems remain competent as they encounter new information.

The Role of Big Data: Big data plays a pivotal role in learning from experience. The abundance of data allows AI systems to capture diverse patterns and insights, leading to robust learning and adaptation.

Human-Machine Collaboration: While AI systems can learn from experience, human expertise remains indispensable. Collaboration between humans and AI facilitates ongoing learning, fine-tuning, and addressing scenarios that AI might struggle with.

Future Horizons:

Personalized Medicine: AI models will learn from patient data to provide personalized treatment recommendations, accounting for genetic factors, medical history, and evolving medical research.

Smart Cities: AI-driven urban systems will adapt to real-time data on traffic, energy consumption, and public services to optimize city operations and enhance quality of life.

Education: Adaptive learning platforms will personalize educational content based on students' strengths, weaknesses, and learning styles, revolutionizing the education sector.

The evolution of AI from rule-based systems to adaptable, learning entities is reshaping industries and driving innovation. Learning from experience and adaptation empower AI systems to handle dynamic and complex scenarios that were once considered beyond the realm of automation. As AI models process massive amounts of data, they gain the ability to identify patterns, make predictions, and adapt strategies in response to changing circumstances. Challenges related to data quality, interpretability, and ethics require careful consideration, but ongoing research and collaboration between humans and machines hold the promise of overcoming these hurdles.

The future horizons of learning from experience and adaptation are vast. From personalized medicine to smart cities and education, AI's ability to learn and adapt is poised to create more efficient, responsive, and intelligent systems. As AI continues to learn and evolve, it will complement human expertise, drive efficiencies, and unlock new possibilities across diverse domains. With the synergy between AI and human intelligence, we are witnessing a transformative era where technology's ability to learn and adapt mirrors the essence of human cognition.

Implications for Human-Robot Collaboration: As robotics and artificial intelligence (AI) advance, the landscape of human-robot collaboration is undergoing a profound transformation. The integration of robots into various sectors, ranging from manufacturing to healthcare, is reshaping how humans and machines interact and work together. This article explores the implications of human-robot collaboration, examining its potential benefits, challenges, ethical considerations, and the future prospects of this evolving partnership.

Enhanced Efficiency and Productivity: Human-robot collaboration holds the promise of significantly boosting efficiency and productivity across industries. Robots excel in repetitive, precise tasks, while humans bring creativity, critical thinking, and decision-making skills. When combined, this partnership optimizes processes and reduces human error.

Manufacturing and Industry: One of the most significant areas of human-robot collaboration is manufacturing. Collaborative robots, or cobots, work alongside human workers to perform tasks that require strength and precision. This synergy enhances production speed, quality control, and workplace safety.

Healthcare and Medical Assistance: Robots are becoming increasingly important in healthcare. Surgical robots assist surgeons in complex procedures, improving precision and minimizing invasiveness. Robotic exoskeletons aid in rehabilitation, enabling patients to regain mobility and strength.

Autonomous Vehicles: Human-robot collaboration is at the forefront of autonomous vehicle development. While AI systems handle navigation and control, human oversight is critical for handling exceptional situations and making complex decisions.

Agriculture: In agriculture, robots are used for tasks such as planting, harvesting, and monitoring crops. Human-robot collaboration in this sector leads to increased crop yield, reduced labor costs, and optimized resource utilization.

Space Exploration: Human-robot collaboration is vital for space exploration. Robots can perform tasks in environments too hazardous for humans, while astronauts oversee and manage their activities from a distance.

Challenges of Collaboration:

Interoperability: Integrating robots into existing workflows requires ensuring compatibility with existing systems and processes.

User Interface and Experience: Creating user-friendly interfaces that allow humans to interact intuitively with robots is a challenge.

Ethical Considerations: The potential displacement of human jobs by robots raises ethical questions about the social implications of widespread collaboration.

Ethical Considerations:

Job Displacement: The introduction of robots can lead to concerns about job loss. Balancing automation with job creation and retraining programs is crucial.

Human Oversight: Ensuring that humans maintain control over robots, especially in critical tasks, is an ethical imperative.

Safety and Security: Collaborative robots should be designed to prioritize human safety and prevent potential security breaches.

Trust and Acceptance:

User Acceptance: For human-robot collaboration to succeed, humans must feel comfortable and confident working alongside robots.

Transparency: Clearly communicating the capabilities and limitations of robots fosters trust and prevents misunderstandings.

Human-Centered Design:

Ergonomics: Designing robots that complement human physical abilities and movements enhances collaboration and reduces physical strain.

Natural Interaction: Intuitive interfaces and natural language processing enable seamless communication and interaction between humans and robots.

Collaborative Learning: Robots can learn from human expertise and adapt their behavior based on human feedback. This symbiotic learning process enhances the capabilities of both humans and machines.

Future Prospects:

Personalized Healthcare: Collaborative robots will assist medical professionals in delivering personalized care, from diagnostics to treatment plans.

Elderly Care: Robots will support elderly individuals in daily tasks, enhancing their quality of life and relieving the burden on caregivers.

Humanitarian Aid: Collaborative robots will play a crucial role in disaster response and humanitarian missions, where they can navigate hazardous environments and assist rescue efforts.

Human-Robot Teams in Research: Collaborative robots are revolutionizing research, working alongside scientists in data analysis, laboratory tasks, and even creativity-driven projects.

Education and Training: Robots can serve as educational tools, providing hands-on experience for students in various fields, from programming to engineering.

Innovation and Creativity: Human-robot collaboration has the potential to foster innovative thinking by combining human creativity with robotic precision and efficiency.

Societal Impacts:

Redefining Work: Collaborative robots have the potential to shift the nature of work, requiring societies to adapt to new employment paradigms.

Education and Skills: Preparing the workforce for human-robot collaboration necessitates investing in relevant education and skills development.

The evolution of human-robot collaboration is redefining industries, changing the nature of work, and pushing the boundaries of innovation. The synergy between human creativity, critical thinking, and decision-making, combined with robotic precision and efficiency, holds immense potential to revolutionize various sectors. While challenges such as ethical considerations, trust-building, and human-centered design must be addressed, the trajectory of human-robot collaboration is undeniably transformative. As technology advances and robots become increasingly integrated into our daily lives, the ethical and societal implications of collaboration must be carefully navigated. By fostering an ecosystem of collaboration that prioritizes human well-being, safety, and creativity, we can harness the full potential of this partnership to shape a future where humans and robots work together harmoniously, driving progress, innovation, and positive societal change.

Manipulating Matter at the Nanoscale: The ability to manipulate matter at the nanoscale has opened up new frontiers in science, engineering, and technology. Nanotechnology, the science of controlling and designing materials and devices at nanometer dimensions, holds promise across a wide range of applications, from medicine and electronics to energy and materials science. This article explores the significance, methods, applications, challenges, and future prospects of manipulating matter at the nanoscale.

Nanoscale and Its Significance: The nanoscale refers to dimensions on the order of nanometers (one billionth of a meter). At this scale, materials exhibit unique properties and behaviors due to quantum effects and surface interactions. Manipulating matter at the nanoscale allows scientists and engineers to harness these properties for innovative applications.

Top-Down and Bottom-Up Approaches: Nano manipulation is achieved through two main approaches: top-down and bottom-up. Top-down methods involve carving or patterning larger materials to create nanoscale structures, while bottom-up methods assemble nanoparticles and molecules to build nanoscale structures.

Methods of Nano manipulation:

Photolithography: In the semiconductor industry, photolithography is a top-down technique used to create nanoscale patterns on silicon wafers.

Atomic Force Microscopy (AFM): AFM enables the manipulation of individual atoms and molecules, offering precision at the atomic level.

Chemical Vapor Deposition (CVD): CVD is a bottom-up technique used to deposit thin films of materials onto substrates, allowing precise control over material growth.

Self-Assembly: Self-assembly relies on the inherent properties of nanoparticles and molecules to arrange themselves into desired structures.

DNA Nanotechnology: DNA molecules are used as building blocks to create nanoscale structures with high precision and programmability.

Applications in Electronics:

Nano electronics: Nano manipulation enables the fabrication of transistors, sensors, and memory devices at the nanoscale, enhancing performance and energy efficiency.

Quantum Dots: Nanoscale semiconductor materials, known as quantum dots, emit light of specific wavelengths, making them valuable for displays, imaging, and solar cells.

Medical Applications:

Drug Delivery: Nanoparticles can be engineered to carry drugs and target specific cells or tissues, improving the efficacy and reducing side effects of treatments.

Nano medicine: Nanoscale devices can be used for diagnostics, imaging, and even the treatment of diseases at the molecular level.

Materials Science and Engineering:

Nanocomposites: Manipulating matter at the nanoscale allows the incorporation of nanoparticles into materials, enhancing their mechanical, electrical, and thermal properties.

Coatings and Thin Films: Nanoscale coatings provide enhanced wear resistance, corrosion protection, and improved optical properties.

Energy Applications:

Nanomaterials for Energy Storage: Nanoscale materials, such as graphene and nanowires, are being explored for advanced energy storage solutions, including batteries and super capacitors.

Solar Cells: Nano manipulation enables the design of solar cells that capture light more efficiently, enhancing their conversion efficiency.

Challenges and Considerations:

Precise Control: Manipulating matter at the nanoscale requires precise control over individual atoms and molecules, which can be challenging.

Scalability: Methods that work well in the lab may face challenges when scaled up for mass production.

Ethical and Environmental Concerns: Nanoparticles and nanomaterials may have unknown effects on human health and the environment, necessitating careful study and regulation.

Emerging Trends:

Nanorobotics: The integration of Nano manipulation with robotics enables the creation of nanoscale machines for medical procedures, environmental monitoring, and more.

DNA Origami: DNA molecules are engineered to fold into specific shapes, creating a platform for designing nanoscale structures with high precision.

Nanoscale 3D Printing: 3D printing at the nanoscale allows the fabrication of intricate structures for applications in electronics, photonics, and biomedical devices.

Future Prospects:

Advanced Electronics: Nano manipulation will continue to play a pivotal role in pushing the boundaries of miniaturization and enhancing electronic device performance.

Revolutionizing Medicine: The precision of Nano manipulation holds the potential to revolutionize personalized medicine, diagnostics, and treatments.

Sustainable Energy Solutions: Nanomaterials and Nano manipulation will contribute to the development of more efficient energy storage and conversion technologies.

Ethical and Societal Implications:

Safety and Regulation: The introduction of nanoscale materials and devices requires thorough safety assessments and regulatory frameworks.

Equity and Accessibility: Ensuring equitable access to nanotechnology advancements is essential to prevent technological disparities.

Manipulating matter at the nanoscale represents a frontier of scientific and technological exploration with far-reaching implications. From electronics and medicine to energy and materials science, Nano manipulation offers unprecedented opportunities for innovation. The ability to engineer materials and devices at the nanoscale not only revolutionizes industries but also holds the potential to address pressing global challenges, from sustainable energy solutions to personalized healthcare. As researchers and engineers continue to advance nanotechnology, ethical considerations, safety protocols, and collaboration across disciplines will play a crucial role in harnessing the full potential of this groundbreaking field.

Applications in Medicine, Electronics, and Materials Science: The world of science and technology is witnessing an era of remarkable advancement, with breakthroughs in nanotechnology revolutionizing multiple domains. This article delves into the myriad applications of nanotechnology in medicine, electronics, and materials science. From targeted drug delivery in medicine to the miniaturization of electronic devices and the enhancement of material properties, nanotechnology is shaping the future in unprecedented ways.

Nanotechnology in Medicine: Nanotechnology has unleashed a paradigm shift in the field of medicine, enabling novel approaches for diagnostics, treatment, and drug delivery.

Targeted Drug Delivery: Nanoparticles can be engineered to deliver drugs specifically to diseased cells, minimizing side effects and maximizing therapeutic efficiency.

Nano medicine: Nanoparticles and nanoscale devices are used for imaging, diagnostics, and disease monitoring at the molecular level.

Cancer Therapies: Nanoparticles can deliver chemotherapeutic agents directly to tumor sites, reducing damage to healthy tissues and enhancing treatment effectiveness.

Regenerative Medicine: Nanomaterials are used to create scaffolds for tissue engineering, aiding in the regeneration of damaged tissues and organs.

Personalized Medicine: Nanotechnology allows for the customization of treatments based on an individual's genetic makeup, leading to more effective and tailored medical interventions.

Nanotechnology in Electronics: The electronics industry has embraced nanotechnology to create smaller, more powerful devices with enhanced performance characteristics.

Miniaturization: Nanotechnology enables the fabrication of nanoscale transistors and components, paving the way for more compact and energy-efficient electronics.

Quantum Computing: Nanoscale structures are crucial for the development of quantum computing, which holds the potential to solve complex problems beyond the capabilities of classical computers.

Flexible Electronics: Nanomaterials and nanofabrication techniques are utilized to create flexible and wearable electronics that can conform to various shapes and surfaces.

Optoelectronics: Nanotechnology is integral to the development of optoelectronic devices such as LEDs, solar cells, and photodetectors, improving efficiency and performance.

Nanotechnology in Materials Science: Nanotechnology has unlocked the potential to engineer materials with enhanced properties and functionalities.

Enhanced Mechanical Properties: Nanomaterials exhibit unique mechanical properties due to their high surface area and quantum effects. They find applications in lightweight composites, durable coatings, and advanced textiles.

Self-Healing Materials: Nanoparticles can be incorporated into materials to enable self-healing capabilities, prolonging the lifespan of structures and reducing maintenance costs.

Nanocomposites: By integrating nanoparticles into matrices, nanocomposites with improved electrical, thermal, and mechanical properties are created.

Energy Storage and Conversion: Nanotechnology contributes to the development of high-performance batteries, super capacitors, and fuel cells with improved energy density and efficiency.

Challenges and Considerations: While the applications of nanotechnology are promising, several challenges must be addressed:

Safety and Toxicity: The potential health and environmental impacts of nanoparticles require rigorous investigation to ensure their safe use.

Regulation: Developing regulatory frameworks that balance innovation with safety is crucial to prevent unintended consequences.

Manufacturing Scalability: Transitioning from laboratory-scale production to mass manufacturing of nanomaterials and devices poses scalability challenges.

Synergy and Convergence: Nanotechnology's impact is not limited to individual sectors; rather, it fosters convergence and collaboration across fields.

Lab-on-a-Chip Devices: Combining nanotechnology with microfluidics enables the creation of portable diagnostic devices for point-of-care testing.

Bionics and Neuro technology: Nanotechnology interfaces with neuroscience, leading to developments in brain-computer interfaces and prosthetics.

Smart Materials: Materials engineered at the nanoscale exhibit responsiveness to external stimuli, allowing for applications in sensors and actuators.

Future Prospects:

Nanorobotics in Medicine: Nano robots designed to perform specific tasks within the human body hold potential for targeted drug delivery, surgery, and diagnostics.

Nanoscale Computing: Advancements in quantum computing and neuromorphic computing promise to revolutionize computation and data processing.

Sustainable Materials: Nanotechnology will contribute to the development of environmentally friendly materials and energy solutions.

Ethical and Societal Implications:As nanotechnology advances, ethical considerations surrounding privacy, security, and equitable access to its benefits must be carefully addressed.

The applications of nanotechnology in medicine, electronics, and materials science are reshaping industries and enhancing human capabilities. From precise drug delivery to the miniaturization of electronics and the creation of advanced materials, nanotechnology's influence is far-reaching. Collaborations between researchers, engineers, and professionals from various disciplines are driving innovation and convergence, unlocking novel possibilities at the nanoscale. As society navigates the ethical, safety, and regulatory challenges associated with nanotechnology, the potential to revolutionize medicine, electronics, and materials science remains both exciting and transformative.

Challenges and Potential Ethical Concerns: The rapid advancement of nanotechnology has opened up exciting possibilities across various industries. However, along with its promising potential, nanotechnology also brings forth a series of challenges and ethical concerns that must be carefully examined. This article delves into the key challenges and potential ethical considerations associated with nanotechnology, shedding light on the importance of responsible development and adoption.

Safety and Toxicity: As nanomaterials become more prevalent, understanding their safety profile is paramount. Nanoparticles can behave differently from bulk materials due to their unique size and properties, potentially leading to unforeseen health and environmental risks.

Health Concerns: Nanoparticles may enter the body through inhalation, ingestion, or skin contact. Their small size can allow them to reach sensitive areas and potentially cause adverse health effects.

Environmental Impact: The release of nanoparticles into the environment raises concerns about their impact on ecosystems and potential bioaccumulation in organisms.

Regulatory Challenges: The rapidly evolving nature of nanotechnology poses challenges for regulatory bodies striving to ensure safety and effectiveness.

Lack of Standardization: Due to the diverse nature of nanomaterials, standardizing testing and evaluation methods becomes challenging.

Risk Assessment: Developing appropriate risk assessment frameworks for nanomaterials requires adapting existing regulations to accommodate their unique characteristics.

Ethical Considerations: Nanotechnology's potential impact on society raises important ethical questions that warrant thorough exploration.

Privacy and Surveillance: The development of nanoscale sensors and devices raises concerns about privacy invasion and surveillance capabilities.

Equity and Access: Ensuring equitable access to nanotechnology advancements is essential to prevent technological disparities between different communities.

Dual-Use Concerns: Nanotechnology could potentially be used for both beneficial and harmful purposes, raising concerns about its dual-use nature.

Environmental Implications: The widespread use of nanomaterials can lead to unintended consequences for the environment.

Eco-Toxicity: Nanoparticles may have adverse effects on aquatic life and soil ecosystems, impacting food chains and biodiversity.

Nanoparticle Mobility: Nanoparticles' small size and mobility can lead to contamination of water bodies and soil, potentially affecting human health.

Healthcare Ethics: In the realm of Nano medicine, unique ethical considerations arise that require careful examination.

Informed Consent: The complexity of Nano medicine treatments may pose challenges to obtaining truly informed consent from patients.

Human Enhancement: The potential for nanotechnology to enhance human capabilities raises ethical questions about the boundaries between therapy and enhancement.

Nano-Ethics: The interdisciplinary nature of nanotechnology demands a specialized focus on ethical considerations.

Transparency: Ensuring transparent communication about the risks and benefits of nanotechnology is crucial for building public trust.

Unintended Consequences: The unpredictability of nanoscale interactions calls for anticipating unintended consequences and ethical implications.

Intellectual Property and Ownership: As nanotechnology innovations accelerate, issues related to intellectual property rights and ownership of nanomaterials and devices emerge.

Patent System Challenges: Defining patent rights for nanotechnology innovations that involve interdisciplinary approaches can be complex.

Open Access vs. Profit: Balancing open access to knowledge with commercial interests is a challenge in the development of nanotechnologies.

Education and Awareness: Addressing the ethical concerns of nanotechnology requires raising awareness among professionals, policymakers, and the general public.

Ethics Education: Integrating ethics education into nanotechnology curricula is essential to cultivate responsible research and innovation.

Public Engagement: Engaging the public in discussions about the ethical implications of nanotechnology encourages broader perspectives and informed decision-making.

Global Governance and Collaboration: Nanotechnology's global nature necessitates international collaboration to address its ethical and regulatory challenges effectively.

Harmonized Regulations: Developing harmonized international regulations can prevent disparities in safety standards and facilitate responsible development.

Ethical Frameworks: Collaborative efforts are needed to establish ethical frameworks that guide the responsible use of nanotechnology on a global scale.

Responsible Innovation: Emphasizing responsible innovation ensures that ethical considerations are an integral part of the research and development process.

Ethical Impact Assessments: Incorporating ethical impact assessments into the development of nanotechnology products helps identify and mitigate potential risks.

Multi-Stakeholder Dialogue: Involving various stakeholders, including scientists, policymakers, industry representatives, and ethicists, fosters comprehensive ethical discussions.

Nanotechnology's transformative potential comes with a set of challenges and ethical considerations that must be addressed proactively. From ensuring the safety of nanomaterials to navigating complex regulatory landscapes and ethical dilemmas, responsible development and adoption are imperative. Collaboration among scientists, policymakers, ethicists, and the public is essential to strike a balance between innovation and ethical integrity. By prioritizing safety, equity, transparency, and responsible innovation, the full potential of nanotechnology can be harnessed while upholding ethical standards that benefit humanity and the environment.

Quantum Robotics: The Quantum Computing Revolution: The convergence of quantum computing and robotics is shaping a new era of technological innovation – Quantum Robotics. This groundbreaking field explores the marriage of quantum mechanics and robotics, offering unparalleled capabilities that extend beyond the limits of classical computing. This article delves into the concepts, implications, challenges, and future prospects of Quantum Robotics, highlighting its potential to revolutionize industries and reshape the landscape of automation.

Understanding Quantum Computing: Quantum computing leverages the principles of quantum mechanics to perform computations that are exponentially faster than classical computers. Unlike classical bits, quantum bits or qubits can exist in multiple states simultaneously, enabling quantum computers to process vast amounts of information in parallel.

Superposition: Qubits exist in a superposition of states, allowing them to represent both 0 and 1 at the same time.

Entanglement: Entanglement is a phenomenon where qubits become correlated, even when separated by significant distances, leading to enhanced computational power.

The Quantum Robotics Paradigm:

Quantum Robotics combines the principles of quantum computing with the capabilities of robotics, enabling systems to perform complex tasks with unprecedented efficiency and accuracy.

Quantum Sensors: Quantum-enhanced sensors can provide ultra-precise measurements, enabling robots to interact with the environment at an unprecedented level of detail.

Optimization: Quantum algorithms can optimize complex tasks, such as path planning for autonomous vehicles and logistics optimization.

Machine Learning: Quantum machine learning allows robots to learn patterns and behaviors from data more efficiently, improving their decision-making abilities.

Applications in Automation and AI: Quantum Robotics has transformative potential across multiple industries.

Drug Discovery: Quantum computing can simulate molecular interactions with high accuracy, accelerating the drug discovery process.

Financial Modeling: Quantum algorithms can solve complex optimization problems in financial markets, enhancing risk analysis and portfolio management.

Materials Design: Quantum Robotics can accelerate the development of advanced materials with specific properties, revolutionizing industries from aerospace to electronics.

Challenges in Quantum Robotics:

Despite its promises, Quantum Robotics faces several challenges that need to be addressed for widespread adoption.

Quantum Error Correction: Quantum systems are prone to errors due to decoherence and noise. Developing effective error correction mechanisms is crucial.

Hardware Limitations: Quantum computers are still in their infancy, and creating scalable, fault-tolerant quantum hardware is a significant challenge.
Algorithm Development: Creating quantum algorithms that exploit the unique capabilities of quantum computers requires interdisciplinary expertise.
Ethical and Security Implications: As Quantum Robotics develops, ethical considerations and security implications must be carefully examined.
Cryptographic Impact: Quantum computers can potentially break classical cryptographic systems, necessitating the development of quantum-safe encryption methods.
Bias and Fairness: Ensuring that quantum algorithms used in robotics are unbiased and fair is essential to prevent discriminatory outcomes.
Future Prospects: Quantum Robotics holds immense promise for reshaping industries and advancing technological frontiers.
Autonomous Systems: Quantum-enhanced sensors and optimization algorithms will drive the development of more capable and efficient autonomous robots.
Healthcare: Quantum-enhanced simulations can lead to personalized medical treatments and faster drug discovery.
Environmental Solutions: Quantum Robotics can contribute to solving complex environmental challenges by optimizing resource allocation and simulations.
Collaboration and Interdisciplinary:
Quantum Robotics requires collaboration between experts in quantum physics, robotics, computer science, and various application domains.
Quantum Skills: Developing a skilled workforce proficient in both quantum computing and robotics is essential for advancing Quantum Robotics.
Public-Private Partnerships: Collaborations between academia, industry, and governments will drive research, development, and innovation in this field.
Educational Imperative:
Preparing the next generation of scientists and engineers to work at the intersection of quantum computing and robotics is vital.
Interdisciplinary Curricula: Educational programs that bridge quantum mechanics and robotics will foster innovation and breakthroughs.
Establishing dedicated centers for Quantum Robotics research and training will accelerate advancements.
Quantum Robotics is a transformative field that promises to revolutionize industries, enable previously unthinkable capabilities, and reshape the landscape of automation. The fusion of quantum computing's computational power with the versatility of robotics opens doors to unparalleled applications in fields ranging from medicine and finance to materials science and environmental solutions. However, realizing the full potential of Quantum Robotics requires addressing challenges in quantum hardware, algorithm development, ethics, and security. With interdisciplinary collaboration, investment in education, and a commitment to responsible development, Quantum Robotics has the potential to create a future where robots equipped with quantum capabilities operate at the cutting edge of technological innovation.

Exercise

Multiple-choice Questions

1. What is the main focus of "Future Horizons of Robotics Research and Innovation"?
a) Historical Evolution of Robotics
b) Current Trends in Robotics Manufacturing
c) Ethical Dilemmas in Robotics
d) Future Prospects of Robotics Research and Innovation
2. What does the term "quantum robotics" refer to?
a) Robotics with superhuman strength
b) Robotics utilizing quantum physics principles
c) Robotics focused on nanoscale manipulation
d) Robotics powered by renewable energy sources
3. What distinguishes quantum computing from classical computing?
a) Quantum computing uses binary code.
b) Quantum computing is slower than classical computing.
c) Quantum computing processes information sequentially.
d) Quantum computing uses qubits that can exist in multiple states.
4. How does quantum computing enhance computational power?

a) It uses larger transistors.
b) It relies on high-speed processors.
c) It allows for parallel processing with qubits.
d) It increases the clock speed of the computer.
5. What is "superposition" in the context of quantum computing?
a) The ability to store vast amounts of data
b) The principle of entanglement between qubits
c) The capability to perform calculations faster
d) The simultaneous existence of multiple states in a qubit
6. What is a potential application of quantum sensors in robotics?
a) Enhancing robotic physical strength
b) Enabling robots to predict the future
c) Providing robots with emotions
d) Improving a robot's interaction with its environment
7. How do quantum algorithms differ from classical algorithms?
a) Quantum algorithms cannot solve optimization problems.
b) Quantum algorithms are slower than classical algorithms.
c) Quantum algorithms can solve complex problems more efficiently.
d) Quantum algorithms only work with binary data.
8. Which industry could benefit from quantum computing's ability to simulate molecular interactions?
a) Agriculture
b) Fashion
c) Medicine
d) Automotive
What is a major challenge in the development of quantum computers?
a) The lack of interest from researchers
b) The inability to create qubits
c) Ensuring compatibility with classical computers
d) Overcoming quantum errors and noise
10. What is a potential impact of quantum computing on cryptography?
a) It will make all encryption methods obsolete.
b) It will have no impact on cryptography.
c) It will make encryption methods more secure.
d) It will only affect paper-based cryptography.
11. In Quantum Robotics, what does the concept of "entanglement" enable?
a) Robots to communicate with each other using light signals
b) Robots to perform tasks without human intervention
c) Robots to predict future events
d) Robots to process information in parallel
12. Which industry could potentially benefit from Quantum Robotics in terms of optimizing supply chain logistics?
a) Fashion
b) Construction
c) Finance
d) Agriculture
13. What is a crucial aspect of addressing ethical concerns in Quantum Robotics?
a) Ignoring ethical considerations to prioritize technological advancements
b) Developing a skilled workforce without ethical training
c) Fostering public engagement and transparent communication
d) Focusing solely on technological development and ignoring ethics
14. What type of sensors can provide ultra-precise measurements in Quantum Robotics?
a) Classical sensors
b) Analog sensors
c) Quantum-enhanced sensors
d) Thermal sensors
15.Which field could benefit from Quantum Robotics in terms of improving risk analysis and portfolio management?

a) Agriculture
b) Finance
c) Healthcare
d) Construction

Review Questions:

1. What are the key driving forces behind the convergence of quantum computing and robotics?
2. How does quantum computing differ from classical computing, and how does this difference impact the field of robotics?
3. What is the concept of "superposition" in quantum computing, and how does it contribute to the capabilities of quantum robots?
4. Explain the term "quantum entanglement" and its potential implications for quantum robotics applications.
5. How does Quantum Robotics leverage quantum-enhanced sensors to improve the interaction of robots with their environment?
6. In what ways can quantum algorithms optimize tasks performed by robots in various industries?
7. Describe the potential applications of quantum computing in the fields of drug discovery and healthcare.
8. What challenges does the development of quantum computers pose in the context of Quantum Robotics?
9. How can Quantum Robotics impact industries such as finance and supply chain logistics through optimization and risk analysis?
10. Discuss the ethical and security implications associated with the convergence of quantum computing and robotics.
11. What is the significance of addressing ethical considerations in Quantum Robotics, and how can it shape responsible innovation?
12. How might Quantum Robotics play a role in the advancement of personalized medicine and medical treatments?
13. Explain the potential impact of quantum computers on cryptography and the need for quantum-safe encryption methods.
14. What is the role of interdisciplinary collaboration in the successful development of Quantum Robotics?
15. How can quantum-enhanced simulations contribute to the design and development of advanced materials in different industries?
16. Discuss the importance of public engagement and transparent communication in navigating the ethical and societal aspects of Quantum Robotics.
17. How might Quantum Robotics enable robots to perform complex tasks with a level of efficiency that was previously unattainable?
18. What are some of the challenges related to quantum error correction and noise reduction in Quantum Robotics?
19. How could Quantum Robotics contribute to addressing environmental challenges and optimizing resource allocation in different sectors?
20. What strategies can educational institutions and research centers adopt to prepare the workforce for the interdisciplinary demands of Quantum Robotics?

Answers Multiple-choice Questions

Answers:
1. d)
2. b)
3. d)
4. c)
5. d)
6. d)
7. c)
8. c)
9. d)
10. c)
11. d)
12. c)
13. c)
14. c)
15. b)

Review Questions:

1. The convergence of quantum computing and robotics is driven by a synergy between the extraordinary computational power of quantum computers and the complex decision-making and physical capabilities of robots. Quantum computing's ability to process information in parallel, harnessing the principles of superposition and entanglement, presents a revolutionary approach to solving complex problems that have been challenging for classical computers. On the other hand, robotics leverages mechanical, electrical, and software systems to interact with and manipulate the physical world. Combining these two domains opens up new avenues for efficient problem-solving, optimization, and innovation across various industries.

2. Quantum computing differs significantly from classical computing in terms of its fundamental principles and processing capabilities. Unlike classical bits that represent either 0 or 1, quantum bits or qubits can exist in multiple states simultaneously due to superposition. This enables quantum computers to perform vast calculations in parallel, offering unprecedented computational speed for certain problems. This distinction profoundly impacts robotics, as tasks that require complex calculations, optimization, and simulations can be executed exponentially faster with quantum algorithms. This means that robots can make quicker decisions, optimize routes, and solve intricate problems more efficiently, revolutionizing their capabilities in various industries such as manufacturing, logistics, and healthcare.

3. "Superposition" in quantum computing refers to the unique property of qubits that allows them to exist in multiple states at once. This property enables quantum computers to perform calculations in parallel, which is fundamentally different from classical computers that process information sequentially. In the context of quantum robots, superposition empowers robots to consider multiple solutions simultaneously, accelerating tasks that involve evaluating numerous possibilities. For example, a quantum robot performing path planning can simultaneously explore multiple routes and identify the most efficient path through a complex environment, enhancing its decision-making speed and efficiency.

4. Quantum entanglement is a phenomenon in quantum mechanics where two or more qubits become correlated in such a way that the state of one qubit is instantaneously connected to the state of another, even if they are physically separated by a considerable distance. In quantum robotics, entanglement holds immense potential for enabling secure and instantaneous communication between robots. This could have applications in remote control scenarios, where a robot's actions can be influenced by a remote operator in real-time, even over vast distances. Moreover, entanglement may enable robots to collaborate and synchronize actions with remarkable precision, enhancing their collective capabilities for tasks such as swarm robotics and collaborative assembly.

5. Quantum Robotics harnesses quantum-enhanced sensors to elevate robots' interactions with their environment to unprecedented levels of precision. Quantum sensors utilize the principles of quantum mechanics to detect and measure physical properties with exceptional sensitivity. This capability enables robots to gather detailed and accurate data from their surroundings, which is particularly valuable in fields such as environmental monitoring, medical imaging, and quality control in manufacturing. By leveraging quantum-enhanced sensors, robots can make more informed decisions, adapt to dynamic environments, and contribute to higher levels of precision and accuracy in various applications.

6. Quantum algorithms have the potential to optimize tasks performed by robots across multiple industries due to their inherent ability to solve complex problems more efficiently than classical algorithms. In industries like logistics and supply chain management, quantum algorithms can efficiently optimize routes for delivery trucks, reducing fuel consumption and delivery times. In drug discovery, quantum algorithms can simulate molecular interactions with high accuracy, accelerating the identification of potential drug candidates. These optimization capabilities translate into enhanced efficiency, reduced costs, and improved decision-making for robots operating in diverse scenarios.

7. Quantum computing holds immense promise for revolutionizing drug discovery and healthcare applications. In drug discovery, quantum computers can simulate the behavior of molecules and chemicals at a level of detail that is unattainable by classical computers. This capability enables rapid and accurate screening of potential drug candidates, expediting the drug development process and potentially leading to the discovery of novel treatments for diseases. Additionally, quantum computing's immense computational power can aid in analyzing complex biological systems, contributing to advancements in personalized medicine by tailoring treatments to individual patients' genetic profiles and medical histories.

8. The development of quantum computers for Quantum Robotics poses several challenges that need to be addressed for effective implementation. One of the major challenges is quantum error correction. Quantum systems are highly susceptible to errors due to DE coherence and noise, which can cause qubits to lose their quantum states and result in inaccurate calculations. Developing robust error correction mechanisms is essential to ensure the reliability and stability of quantum algorithms running on quantum robots. Additionally, creating scalable and fault-tolerant quantum hardware is a significant hurdle, as quantum computers are still in their early stages of development and

require technological advancements to reach a state where they can handle complex tasks required by robotics applications.
9. Quantum Robotics has the potential to transform industries such as finance and supply chain logistics through optimization and risk analysis. In finance, quantum algorithms can solve complex optimization problems that arise in portfolio management, risk assessment, and algorithmic trading. Quantum computers can efficiently analyze large datasets and assess risk factors, enabling financial institutions to make informed decisions in real-time. In supply chain logistics, quantum algorithms can optimize the allocation of resources, route planning for delivery vehicles, and inventory management, resulting in reduced operational costs, enhanced resource utilization, and improved overall efficiency.
10. The convergence of quantum computing and robotics raises important ethical and security implications that must be carefully addressed. On the ethical front, questions arise about the responsible use of quantum robotics, particularly in scenarios where robots make autonomous decisions based on quantum algorithms. Ensuring that these decisions align with human values and ethical standards becomes crucial to prevent unintended consequences. In terms of security, quantum computers have the potential to break classical cryptographic systems, which could compromise data privacy and cybersecurity. The development of quantum-safe encryption methods becomes imperative to ensure that sensitive information remains secure in a quantum-powered world.
11. Addressing ethical considerations in Quantum Robotics is of paramount importance as this field advances. Ethical considerations guide the development, deployment, and use of quantum robots in ways that align with societal values and avoid potential harm. By incorporating ethical frameworks into the design and operation of quantum robots, responsible innovation can be achieved. Ethical considerations include transparency in algorithmic decision-making, ensuring fairness and unbiased outcomes, and addressing potential biases in data used for training quantum algorithms. Ethical guidelines also play a critical role in safeguarding against unintended consequences and potential risks associated with the convergence of quantum computing and robotics.
12 Quantum Robotics holds the potential to significantly contribute to the advancement of personalized medicine and medical treatments. Quantum computers can simulate the interactions between drugs and biological systems at a level of detail that classical computers cannot achieve. This capability accelerates drug discovery by identifying potential drug candidates that are tailored to an individual patient's genetic profile and medical history. Additionally, quantum computing can optimize treatment plans by considering a patient's unique physiological responses, leading to more effective and personalized medical interventions. Quantum-enhanced medical imaging techniques can also improve diagnostics and provide deeper insights into medical conditions.
13. The potential impact of quantum computers on cryptography has significant implications for data security. Quantum computers possess the ability to efficiently factor large numbers, a task that forms the foundation of many classical encryption methods. This capacity jeopardizes the security of sensitive data protected by classical cryptographic systems. To address this challenge, the development of quantum-safe encryption methods is essential. These methods use cryptographic algorithms that remain secure even in the presence of quantum computers. Implementing quantum-safe encryption safeguards against the potential threats posed by the cryptographic vulnerabilities introduced by the quantum computing revolution.
14. Interdisciplinary collaboration plays a pivotal role in the successful development of Quantum Robotics. Quantum Robotics is a convergence of quantum physics, computer science, robotics, and various application domains. Collaborations between experts in these diverse fields enable the exchange of knowledge, expertise, and insights, fostering innovation and breakthroughs. Academic institutions, research organizations, and industries need to collaborate to address the complex challenges posed by Quantum Robotics, such as quantum error correction, algorithm development, and ethical considerations. Interdisciplinary collaboration accelerates the development of practical quantum algorithms, quantum-enhanced sensors, and robust quantum hardware that collectively drive the advancement of Quantum Robotics.
15. Quantum-enhanced simulations offer a unique opportunity to accelerate the design and development of advanced materials in various industries. In fields like materials science and engineering, quantum computers can simulate the behavior of molecules and materials at the quantum level, providing insights into their properties and interactions. This quantum-level simulation enables researchers to design materials with specific desired characteristics, such as enhanced strength, conductivity, or thermal properties. By guiding the creation of novel materials, Quantum Robotics can revolutionize industries ranging from aerospace to electronics, unlocking new possibilities for lightweight and high-performance materials with applications in diverse sectors.
16. Public engagement and transparent communication are integral components of navigating the ethical and societal aspects of Quantum Robotics. As Quantum Robotics advances, it is essential to involve the public in discussions surrounding its implications, benefits, risks, and potential challenges. Transparent communication fosters understanding and awareness among stakeholders, including policymakers, researchers, industry leaders, and the

general public. Addressing concerns related to job displacement, ethical dilemmas, and the potential impact on society's fabric requires open dialogue and collaboration. Public input and feedback contribute to the development of ethical guidelines, regulatory frameworks, and policies that ensure responsible development and deployment of Quantum Robotics technologies.

17. Quantum Robotics enables robots to perform complex tasks with unprecedented levels of efficiency and precision. The combination of quantum computing's computational power and robotics' physical capabilities empowers robots to tackle intricate problems that were previously considered infeasible due to computational complexity. For instance, in manufacturing, quantum robots can optimize production processes, leading to reduced waste and increased product quality. In healthcare, quantum-enhanced medical imaging can provide clearer and more detailed insights into the human body, enhancing diagnostic accuracy. By leveraging quantum algorithms and sensors, robots can achieve levels of performance that were previously unattainable using classical computing and conventional robotics.

18. Quantum error correction and noise reduction are critical challenges that Quantum Robotics faces in harnessing the potential of quantum computers. Quantum systems are highly sensitive to environmental disturbances, leading to errors in quantum operations. Quantum error correction techniques involve encoding quantum information in such a way that errors can be detected and corrected. The development of robust error correction mechanisms is crucial to ensure the accuracy and reliability of quantum algorithms used in Quantum Robotics applications. Overcoming noise and DE coherence is essential to maintain the integrity of quantum states and computations, especially in scenarios where robots rely on quantum-enhanced sensors and algorithms for decision-making.

19. Quantum Robotics has the potential to address environmental challenges and optimize resource allocation in various sectors. By harnessing quantum algorithms, robots can optimize energy consumption, waste management, and resource distribution, contributing to sustainable practices. In agriculture, quantum-enhanced simulations can model ecosystems and predict optimal planting and harvesting times, promoting efficient land utilization. Additionally, quantum robotics can play a role in optimizing energy grids by efficiently managing the distribution of renewable energy sources. By addressing environmental challenges through quantum-enhanced solutions, Quantum Robotics contributes to a greener and more sustainable future.

20. Educational institutions and research centers must adopt multifaceted strategies to prepare the workforce for the interdisciplinary demands of Quantum Robotics. These strategies encompass curriculum development, research collaborations, and industry partnerships. Interdisciplinary curricula should be designed to bridge the gap between quantum computing, robotics, and other relevant fields. Students should receive training in both quantum mechanics and robotics, enabling them to apply quantum algorithms to robotic systems effectively. Research centers should foster collaboration between quantum physicists, roboticists, computer scientists, and domain experts to address real-world challenges. Industry partnerships offer students practical exposure to Quantum Robotics applications, facilitating the development of hands-on skills and a deep understanding of the field's potential.

Online Resources

Online Exam Papers

Link:https://forms.gle/7awYv73CgEZS1sD89

QR Code:

YouTube Video Playlist Link

Link:https://www.youtube.com/playlist?list=PLzlh_31VWEFrHTaIRTXs0FBIASAnq_nDJ

QR Code:

Chapter 15: Robotics Education and Outreach

The rapid evolution of robotics technology has ushered in a new era of innovation and transformation across various industries. As the field continues to expand its boundaries, it becomes increasingly essential to cultivate a workforce equipped with the knowledge, skills, and passion to drive its progress forward. This pivotal role of education in shaping the future of robotics has led to the emergence of an overarching theme: Robotics Education and Outreach. This chapter delves into the significance, approaches, challenges, and outcomes of robotics education, shedding light on how educational initiatives and outreach programs are not only shaping the next generation of roboticists but also fostering broader societal understanding and engagement with this transformative technology.

Robotics has transitioned from being a niche field to an integral part of our everyday lives, with applications ranging from manufacturing and healthcare to agriculture and space exploration. In light of this evolution, education systems across the globe are recognizing the need to incorporate robotics into their curricula at various levels. Robotics education encompasses a diverse array of activities, including classroom learning, hands-on workshops, competitions, and experiential projects that provide students with practical insights into the world of robotics. These initiatives not only cultivate technical skills but also nurture critical thinking, problem-solving abilities, and teamwork – attributes that are crucial in preparing students for the challenges of the modern world.

Beyond formal education, robotics outreach initiatives play a pivotal role in democratizing access to robotics knowledge and fostering inclusivity. Outreach programs aim to engage individuals of all ages, backgrounds, and skill levels, providing them with opportunities to interact with robots, learn about their capabilities, and appreciate their potential impact on society. These programs often bridge the gap between academia, industry, and the general public, facilitating the dissemination of cutting-edge research and real-world applications in an accessible manner. By demystifying robotics and inspiring curiosity, outreach initiatives contribute to a more informed and enthusiastic society that is better equipped to embrace the opportunities and challenges posed by advanced robotics technologies.

As the chapter unfolds, it explores the multifaceted dimensions of robotics education and outreach. It delves into the strategies employed to integrate robotics into formal education, from primary schools to universities, emphasizing the significance of tailored curricula and hands-on experiences. The chapter also delves into the world of robotics competitions and hackathons, which offer students a platform to apply theoretical knowledge in practical scenarios while fostering healthy competition and innovation. Furthermore, it examines the role of mentorship, partnerships with industry, and collaboration with research institutions in enriching robotics education with real-world insights and experiences.

The chapter also addresses the broader implications of robotics education and outreach, considering their influence on diversity and inclusion within the robotics community. It underscores the importance of breaking down barriers and promoting equity in opportunities, ensuring that individuals from all walks of life can engage with robotics and contribute to its advancement. Additionally, the chapter explores the reciprocal relationship between robotics research and education, highlighting how educators and researchers collaborate to develop effective pedagogical methods and curricula that align with the rapid advancements in robotics technology.

Ultimately, Robotics Education and Outreach serve as a cornerstone for the continued growth and sustainability of the robotics field. By nurturing a diverse and knowledgeable workforce, instilling a sense of wonder and exploration, and promoting active engagement with robotics technology, education and outreach initiatives are shaping the trajectory of innovation while fostering a society that is well-prepared to navigate the exciting challenges and opportunities presented by the ever-evolving world of robotics.

Educational Robotics: Fostering STEM Learning: Education forms the cornerstone of technological advancement and societal progress. In an era marked by rapid technological transformations, fostering a deep understanding of Science, Technology, Engineering, and Mathematics (STEM) subjects has become paramount. Educational robotics, a dynamic and engaging approach, has emerged as a powerful tool for cultivating STEM learning from early education to higher levels. This section, "Educational Robotics: Fostering STEM Learning," delves into the pivotal role of robotics in shaping STEM education, highlighting its significance, approaches, benefits, and implications.

The Significance of STEM Education: STEM education is widely recognized as a catalyst for equipping students with the skills needed to thrive in a technology-driven world. It nurtures critical thinking, problem-solving, creativity, and collaboration – essential attributes for addressing complex global challenges. As industries increasingly rely on technology, a robust STEM foundation has become vital for fostering innovation and preparing a workforce that can contribute meaningfully to various sectors.

Educational Robotics as a Pedagogical Tool: Educational robotics capitalizes on the innate curiosity and affinity for hands-on learning that many students possess. It leverages robotic platforms and kits to create interactive and experiential learning environments. By immersing students in real-world applications, educational robotics bridges the gap between theoretical knowledge and practical implementation. This approach not only enriches students' understanding of STEM concepts but also ignites a passion for learning by fostering a tangible connection to technology.

Hands-On Learning and Experiential Education: One of the defining features of educational robotics is its emphasis on hands-on learning. Traditional pedagogical approaches often focus on theoretical concepts, which can sometimes disengage students. Educational robotics, on the other hand, enables students to apply theoretical knowledge immediately, creating a direct link between concepts and their real-world relevance. This experiential education method promotes deeper comprehension, as students actively engage with robots, troubleshoot challenges, and witness cause-and-effect relationships firsthand.

Interdisciplinary Learning: Educational robotics encourages interdisciplinary learning by integrating various STEM subjects into a cohesive narrative. While building and programming robots, students encounter principles from mathematics, physics, engineering, and computer science. For instance, designing a robot that follows a precise path involves geometry, while programming it requires understanding algorithms and logic. This holistic approach not only enriches students' learning experience but also mirrors the collaborative nature of technological innovation in the real world.

Benefits of Educational Robotics: The adoption of educational robotics offers a multitude of benefits that extend beyond traditional classroom settings. Among these advantages are:

Active Engagement: Educational robotics captivates students' interest and promotes active engagement, making learning enjoyable and meaningful.

Problem-Solving Skills: Students learn to identify challenges, develop strategies, and iterate solutions, enhancing their problem-solving capabilities.

Collaboration: Robotics projects often require teamwork, fostering collaboration, communication, and the ability to work harmoniously in groups.

Critical Thinking: Designing, building, and programming robots prompt students to think critically, analyze situations, and make informed decisions.

Creative Expression: Students have the freedom to design and customize robots, allowing for creative expression and innovation.

Future-Ready Skills: Exposure to robotics technologies equips students with skills that align with the demands of the modern workforce, especially in fields such as automation, artificial intelligence, and robotics.

Inclusivity: Educational robotics can be adapted to accommodate diverse learning styles and abilities, promoting inclusivity within STEM education.

Approaches to Educational Robotics: The implementation of educational robotics varies across educational levels and institutions. Different approaches cater to distinct learning objectives and resources:

Curricular Integration: Some educational institutions integrate robotics into existing curricula, weaving robotics activities into subjects like mathematics or science. This approach contextualizes STEM concepts, making them more relatable and engaging.

Extracurricular Activities: Robotics clubs, competitions, and workshops provide students with opportunities to explore robotics beyond the classroom. These extracurricular activities promote deeper involvement, allow for creative exploration, and encourage friendly competition.

Dedicated Courses: Some institutions offer standalone robotics courses that provide in-depth exploration of robotics principles, programming languages, and engineering concepts. These courses cater to students with a keen interest in robotics as a focal area of study.

Project-Based Learning: Project-based learning involves assigning students complex challenges that they must solve through robotics. This approach nurtures critical thinking and problem-solving skills while fostering autonomy and creativity.

Implications and Future Directions: As educational robotics gains prominence, educators, policymakers, and stakeholders must consider several key implications:

Teacher Training: Effective integration of robotics into education requires well-trained educators who are familiar with both the technology and pedagogical techniques. Professional development for teachers ensures that they can maximize the benefits of educational robotics.

Equity and Access: Ensuring equitable access to educational robotics is crucial for preventing technological disparities. Efforts should be made to provide resources and opportunities to underserved communities, promoting inclusivity in STEM learning.

Ethical Considerations: Introducing robotics in education necessitates discussions about ethical dimensions, including the responsible use of technology, privacy concerns, and potential biases in algorithms.

Research and Evaluation: Ongoing research is essential to assess the effectiveness of educational robotics in enhancing STEM learning outcomes. Data-driven insights guide the refinement of approaches and curricula.

Educational robotics stands at the intersection of technology and education, offering a transformative approach to STEM learning. By immersing students in interactive and experiential experiences, it cultivates critical skills required for success in the modern world. Through hands-on learning, interdisciplinary exploration, and a strong emphasis on problem-solving, educational robotics not only shapes the next generation of innovators but also contributes to a society equipped to address complex challenges through technological advancement. As educators, researchers, and policymakers continue to harness the potential of educational robotics, they pave the way for a future where STEM education is dynamic, inclusive, and deeply impactful.

Designing Robotics Curricula for Different Age Groups: Education is not a one-size-fits-all endeavor; it must be tailored to the developmental stages, interests, and cognitive abilities of learners. In the realm of robotics education, designing curricula that cater to different age groups is a nuanced and essential task. This section, "Designing Robotics Curricula for Different Age Groups," explores the intricacies of crafting educational frameworks that align with the needs and capacities of learners at various stages of their educational journey.

Understanding Developmental Stages: Children undergo distinct developmental stages as they grow, each characterized by specific cognitive, emotional, and social milestones. Educators must consider these stages when developing robotics curricula to ensure that the content is both meaningful and effective. These stages include:

Early Childhood (Ages 3-6): During this phase, children are curious explorers. Curricula for this age group focus on introducing basic concepts such as shapes, colors, and simple cause-and-effect relationships through engaging and hands-on activities.

Elementary School (Ages 6-11): Children in elementary school are developing logical reasoning and critical thinking skills. Curricula should integrate robotics to teach foundational programming concepts, teamwork, and problem-solving through interactive projects.

Middle School (Ages 11-14): Middle school students are ready for more complex challenges. Curricula for this age group can delve deeper into programming languages, sensors, and mechanics, allowing students to build more advanced robotic systems.

High School (Ages 14-18): High school students can engage in sophisticated robotics projects that encompass topics like artificial intelligence, machine learning, and advanced programming. Curricula should also emphasize real-world applications and career pathways in robotics.

Guiding Principles for Curriculum Design: Designing robotics curricula for different age groups requires adherence to several guiding principles that optimize learning outcomes and engagement:

Progressive Complexity: Curricula should follow a logical progression, starting with simple concepts and gradually introducing more complex topics. This scaffolding approach ensures that students build upon prior knowledge as they advance.

Hands-On Exploration: Regardless of age, hands-on exploration is paramount. Interactive projects, experiments, and challenges allow students to apply theoretical concepts practically, enhancing their understanding and retention.

Inquiry-Based Learning: Encouraging students to ask questions, explore possibilities, and discover solutions fosters an intrinsic motivation to learn. Robotics projects that prompt students to identify problems and design their solutions cultivate critical thinking skills.

Real-World Relevance: Content should connect to real-world applications, demonstrating the tangible impact of robotics in various industries. Relating classroom learning to practical scenarios enhances students' engagement and appreciation for the subject.

Collaboration and Communication: Curricula should incorporate collaborative projects that encourage teamwork and communication. Group work not only mirrors real-world collaborative environments but also enhances social skills.

Creativity and Customization: Allowing students to customize and personalize their projects encourages creativity and ownership. This approach empowers learners to explore their interests while solving problems uniquely.

Curricular Approaches for Different Age Groups: Designing robotics curricula for various age groups requires tailoring content, tools, and pedagogical methods to suit learners' capabilities:

Early Childhood: Exploratory Play and Simple Robotics: For the youngest learners, curricula should emphasize hands-on play with age-appropriate robotics kits. These kits introduce basic building and coding concepts through colorful and engaging activities. Activities might include creating robots that move in response to sound or light, fostering sensory exploration and simple problem-solving.

Elementary School: Introduction to Coding and Logic: In elementary school, curricula can introduce block-based coding platforms that enable students to program robots using visual blocks. Projects might involve designing robots to navigate mazes or complete challenges, teaching programming logic and sequential thinking.

Middle School: More Complex Robotics Systems: Middle school curricula can delve into more complex robotics systems, incorporating elements like sensors, motors, and more advanced programming languages. Projects might focus on creating robots that perform specific tasks, fostering a deeper understanding of robotics components and capabilities.

High School: Advanced Robotics and Specialization: High school curricula should challenge students with advanced topics such as machine learning, computer vision, and AI integration. Projects might involve designing robots that recognize objects or navigate autonomously. Additionally, curricula can offer specialization options, allowing students to focus on specific areas of interest like robotics engineering, automation, or robotics in healthcare.

Adapting to Learning Environments: The design of robotics curricula must consider the context in which learning takes place, whether in traditional classrooms, online environments, or hybrid models:

Traditional Classrooms: In traditional classrooms, educators can incorporate hands-on robotics activities, demonstrations, and group projects. Hands-on experience with physical robots fosters a tangible connection to technology and encourages peer learning.

Online Learning: Online robotics curricula can leverage virtual robotics platforms and simulators. These platforms enable students to design, program, and test robots in virtual environments, facilitating remote learning and accessibility.

Hybrid Models: Hybrid models combine in-person and online elements. Educators can assign virtual robotics projects for remote learning and use classroom time for discussions, collaborative activities, and physical robot interactions.

Challenges and Considerations: Designing robotics curricula for different age groups presents various challenges and considerations:

Resource Constraints: Limited access to robotics kits, computers, or other technology resources can hinder effective implementation. Educators must consider how to maximize learning despite resource limitations.

Teacher Training: Educators need training to effectively teach robotics concepts and technologies. Professional development programs ensure that teachers are well-prepared to facilitate robotics learning experiences.

Inclusivity: Curricula should be inclusive, addressing diverse learning styles, abilities, and cultural backgrounds. Accommodations for different learning needs ensure equitable access to robotics education.

Assessment Strategies: Developing appropriate assessment methods that measure both technical skills and conceptual understanding is crucial for evaluating student progress accurately.

Designing robotics curricula for different age groups is a complex endeavor that requires a deep understanding of developmental stages, educational principles, and pedagogical methods. A well-designed curriculum not only imparts robotics knowledge but also cultivates critical thinking, problem-solving, creativity, and collaboration skills. By aligning content with learners' capabilities and creating meaningful learning experiences, educators pave the way for a generation that is well-equipped to navigate the technological landscape and contribute meaningfully to the world through robotics and STEM education.

Cross-Disciplinary Integration and Real-World Applications: The dynamic landscape of robotics education extends far beyond the confines of traditional subject boundaries. To prepare students for the complex challenges of the modern world, educators are embracing cross-disciplinary integration and real-world applications within robotics curricula. This section, "Cross-Disciplinary Integration and Real-World Applications," delves into the transformative potential of merging diverse fields and connecting classroom learning to practical contexts, creating a holistic educational experience that transcends traditional silos.

The Power of Cross-Disciplinary Integration: Cross-disciplinary integration involves combining knowledge, methods, and perspectives from multiple academic domains to create a more comprehensive and interconnected learning experience. This approach recognizes that real-world challenges often require solutions that draw from diverse fields, breaking down the barriers between subjects and fostering a deeper understanding of complex issues.

Robotics as an Interdisciplinary Nexus: Robotics serves as an ideal nexus for cross-disciplinary integration due to its inherently multifaceted nature. A robotics project can involve mathematics, physics, computer science, engineering, biology, art, and more. By designing curricula that highlight these connections, educators enable students to see how concepts from various disciplines come together to create functional solutions.

Mathematics and Physics Integration: Mathematics and physics play a crucial role in robotics, underpinning concepts like kinematics, mechanics, and sensors. For instance, understanding the relationship between velocity and acceleration is vital when programming robots to move smoothly. By integrating mathematical and physical principles into robotics curricula, educators provide students with a tangible context for applying abstract concepts.

Computer Science and Programming: Computer science is at the heart of robotics, as programming is essential to control robotic systems. Integrating programming languages into robotics curricula empowers students to develop algorithmic thinking, logic, and problem-solving skills. Concepts like loops, conditionals, and variables take on real-world significance as students use them to program robots to perform specific tasks.

Engineering and Design: Engineering principles guide the design and construction of robots, incorporating aspects like structural integrity, material selection, and mechanical systems. Integrating engineering concepts into robotics curricula challenges students to consider both form and function. They learn to balance aesthetics with functionality, mirroring the thought processes of real-world engineers.

Biology and Biomimicry: Biomimicry involves drawing inspiration from nature's designs to create innovative solutions. Robotics curricula can incorporate biological concepts to inspire students to design robots that emulate natural forms and behaviors. For instance, students might explore how the anatomy of animals like insects or birds can inform the design of robots capable of unique locomotion or navigation.

Art and Aesthetics: Incorporating art into robotics education nurtures creativity and encourages students to view robots as vehicles for creative expression. By combining technical skills with artistic sensibilities, students can design robots that are not only functional but also aesthetically pleasing. Integrating art also encourages students to explore the social and ethical implications of robotics in artistic contexts.

Real-World Applications: The value of robotics education extends beyond theoretical concepts, manifesting in real-world applications that resonate with students' interests and aspirations. By showcasing practical applications of robotics, educators make learning relevant and inspire students to envision their roles in shaping the future.

Industry Applications: Introducing students to robotics' role in various industries – from manufacturing and healthcare to agriculture and space exploration – opens their eyes to potential career paths. By demonstrating how robots are used to enhance efficiency, accuracy, and safety, educators provide students with a tangible connection between classroom learning and professional opportunities.

Humanitarian and Social Impact: Highlighting robotics' potential to address humanitarian and social challenges engages students in meaningful ways. For example, robots can be designed to assist people with disabilities or aid in disaster relief efforts. By showcasing the positive impact robots can have on society, educator's foster empathy and a sense of responsibility among students.

Environmental Stewardship: Robotics has a role to play in addressing environmental challenges, such as monitoring ecosystems, managing waste, and promoting sustainable practices. Educators can illustrate how robots are used to collect data on air quality, clean up ocean debris, or plant trees in deforested areas. By linking robotics to environmental stewardship, educators inspire students to consider the ecological implications of technology.

Innovation and Entrepreneurship: Robotic technology has paved the way for innovative startups and entrepreneurial ventures. Educators can introduce students to robotics entrepreneurship by exploring case studies of individuals who have developed novel robotic solutions to real-world problems. This approach not only exposes students to the entrepreneurial spirit but also empowers them to see themselves as future innovators.

Ethical and Societal Considerations: Real-world applications of robotics also provide opportunities to discuss ethical and societal considerations. Educators can guide students through thought-provoking discussions about privacy, job displacement, bias in algorithms, and the responsible use of technology. Encouraging these conversations equips students with a holistic perspective on the impact of robotics on society.

Challenges and Considerations: Integrating cross-disciplinary concepts and real-world applications into robotics education presents educators with a set of challenges and considerations:

Curricular Coordination: Cross-disciplinary integration requires collaboration between educators from different fields, ensuring that the curriculum seamlessly merges diverse perspectives.

Resource Availability: Real-world applications often require access to specialized equipment, materials, or facilities. Educators must consider how to provide students with these resources, especially in resource-constrained settings.

Pedagogical Expertise: Educators may need to acquire expertise in disciplines beyond their primary field to effectively integrate cross-disciplinary concepts. Professional development and training programs can bridge these knowledge gaps.

Time Constraints: Balancing core content with cross-disciplinary integration can be challenging within the constraints of a fixed curriculum. Educators must strategize how to allocate time effectively.

Assessment Strategies: Assessing cross-disciplinary projects requires innovative assessment methods that evaluate both disciplinary knowledge and interdisciplinary collaboration.

Cross-disciplinary integration and real-world applications are the cornerstones of a transformative robotics education. By dismantling traditional academic silos, educators empower students to view robotics as a convergence of diverse fields, fostering a holistic understanding of complex challenges. Linking classroom learning to practical contexts through real-world applications not only makes education engaging but also equips students with the knowledge, skills, and ethical awareness needed to navigate the intricate intersections of technology, society, and innovation. As robotics continues to shape the future, educators play a pivotal role in preparing the next generation for a world where boundaries are blurred, and opportunities are limitless.

Challenges and Success Stories in Robotics Education: The landscape of robotics education is marked by both challenges that educators must navigate and inspiring success stories that demonstrate the transformative potential of this approach. This section, "Challenges and Success Stories in Robotics Education," explores the obstacles that educators encounter in their pursuit of effective robotics instruction, as well as notable instances where robotics education has yielded remarkable outcomes that highlight its significance.

Challenges in Robotics Education: While robotics education offers immense benefits, it is not without its challenges. Educators must grapple with various obstacles to ensure that the learning experience is impactful and meaningful:

Resource Constraints: Robotics requires equipment, software, and facilities that may be financially demanding. Limited resources can hinder educators from providing students with hands-on experiences and access to cutting-edge technology.

Teacher Training: Educators must possess a deep understanding of robotics concepts, programming languages, and engineering principles to effectively teach robotics. Inadequate training can impede the quality of instruction and student engagement.

Curricular Integration: Integrating robotics into existing curricula requires careful planning to align with learning objectives and schedules. Navigating this integration while maintaining a coherent curriculum can be challenging.

Equity and Access: Ensuring equitable access to robotics education for all students, regardless of socioeconomic status, gender, or location, is essential. Overcoming barriers to participation requires proactive efforts and innovative solutions.

Interdisciplinary Collaboration: Collaborating with educators from different disciplines to integrate robotics can be complex. Effective communication, coordination, and alignment of teaching methods are crucial.

Assessment Complexity: Traditional assessment methods may not adequately capture the multifaceted skills developed in robotics education, such as problem-solving, collaboration, and creative thinking. Developing assessment strategies that encompass these skills is a challenge.

Success Stories in Robotics Education: Despite these challenges, numerous success stories underscore the value of robotics education in inspiring students, fostering innovation, and preparing them for the future:

FIRST Robotics Competition: The FIRST Robotics Competition is an international competition that challenges high school students to design, build, and program robots to compete in various tasks. This competition has not only nurtured technical skills but also instilled teamwork, leadership, and a passion for STEM fields among participants.

VEX Robotics: VEX Robotics offers a range of educational platforms that empower students to design and build robots. The program emphasizes problem-solving and collaboration, enabling students to apply classroom learning to real-world challenges.

LEGO Education: LEGO Education offers robotics kits that allow students of all ages to build and program robots. These kits provide an accessible entry point into robotics education and introduce young learners to STEM concepts in an engaging manner.

Girls Who Code: Girls Who Code is a nonprofit organization that aims to close the gender gap in technology by providing girls with opportunities to learn coding and robotics. The program empowers young women to become future leaders in technology and innovation.

MIT App Inventor: MIT App Inventor is a visual programming platform that allows students to create mobile apps and control robots using a simple interface. This tool democratizes coding and robotics education, making it accessible to a wide range of learners.

NASA's Robotics Alliance Project: NASA's Robotics Alliance Project offers educational resources, competitions, and challenges that inspire students to explore robotics and engineering. By engaging students in authentic space-related tasks, the program ignites curiosity and encourages pursuit of STEM careers.

Impact and Significance: The success stories in robotics education showcase its impact on students, educators, and the broader community:

Student Engagement: Robotics education captivates students' interest and provides them with tangible, hands-on experiences that deepen their understanding of STEM concepts. These experiences spark curiosity and enthusiasm for learning.

Skill Development: Robotics education cultivates a range of skills, including problem-solving, critical thinking, creativity, collaboration, and communication. These skills are transferable to various contexts, contributing to students' overall academic and personal growth.

Diversity and Inclusion: Successful robotics education initiatives strive to create diverse and inclusive learning environments. Programs that engage underrepresented groups in STEM fields contribute to a more equitable and representative workforce.

Innovation: Robotics education encourages innovative thinking and empowers students to devise solutions to real-world problems. Many success stories demonstrate students' ability to apply robotics concepts creatively and develop solutions that have practical applications.
Career Pathways: By exposing students to robotics concepts and technology, robotics education paves the way for potential careers in fields such as engineering, computer science, robotics research, and technological innovation.
Overcoming Challenges and Scaling Success: Addressing challenges and scaling success in robotics education require a multi-faceted approach:
Teacher Training: Providing educators with comprehensive training in robotics concepts, pedagogy, and technology equips them to deliver high-quality instruction.
Equitable Access: Initiatives that provide robotics kits, resources, and mentorship to underserved communities enhance access and promote diversity in robotics education.
Curricular Integration: Collaborative efforts among educators, administrators, and curriculum developers ensure that robotics education aligns with learning objectives and complements existing curricula.
Assessment Innovation: Developing assessment methods that capture the full spectrum of skills nurtured in robotics education promotes accurate evaluation of student progress.
Partnerships: Collaborations with industry, academia, and nonprofit organizations can provide funding, expertise, and resources to support robotics education initiatives.
Challenges in robotics education are opportunities for growth, innovation, and transformation. Success stories demonstrate that effective robotics education can ignite students' passion for learning, equip them with invaluable skills, and inspire them to pursue meaningful careers. By addressing challenges, fostering collaboration, and scaling successful initiatives, educators and stakeholders can ensure that robotics education continues to shape the next generation of thinkers, innovators, and problem solvers who will shape the future.
Robotics Outreach Programs: Inspiring the Next Generation: The journey of robotics education extends far beyond classroom walls, reaching out to inspire the next generation through innovative outreach programs. This section, "Robotics Outreach Programs: Inspiring the Next Generation," explores the profound impact of outreach initiatives that engage students, ignite curiosity, and create pathways for young minds to explore the world of robotics, fostering a culture of innovation, exploration, and discovery.
The Essence of Robotics Outreach: Robotics outreach programs bridge the gap between education and inspiration, reaching students in diverse settings, including schools, community centers, museums, and public events. These initiatives go beyond traditional curricula, using hands-on activities, interactive demonstrations, and engaging challenges to spark interest in robotics and STEM fields. The essence of robotics outreach lies in its ability to make technology accessible, relatable, and exciting for learners of all backgrounds.
Hands-On Engagement and Exploration: At the heart of robotics outreach lies the principle of hands-on engagement. Outreach programs offer students the opportunity to interact with robots, build simple machines, and solve challenges through active participation. By engaging in tangible activities, students experience the thrill of discovery and develop a deeper understanding of concepts that might otherwise seem abstract.
Interactive Demonstrations and Role Models: Robotics outreach often features interactive demonstrations where participants can observe robots in action and even control them. These experiences allow students to witness the practical applications of robotics technology and envision themselves as creators of innovative solutions. Additionally, exposure to role models, such as robotics experts and professionals, offers young learners tangible examples of the diverse career paths available in robotics-related fields.
Diverse Settings and Inclusivity: Robotics outreach programs are intentionally designed to reach a wide audience, transcending geographic, socioeconomic, and cultural boundaries. These initiatives take place in schools, libraries, community centers, science museums, and even online platforms. The inclusivity of robotics outreach ensures that students from all walks of life have the opportunity to engage with technology and explore their interests.
STEM Advocacy and Inspiration: Robotics outreach programs serve as powerful advocates for STEM education. By showcasing the excitement and potential of robotics, these initiatives inspire students to pursue STEM subjects in their formal education and beyond. Exposure to robotics sparks curiosity, cultivates a growth mindset, and encourages learners to embrace challenges as opportunities for learning and growth.
Notable Robotics Outreach Initiatives: Numerous robotics outreach programs have made a lasting impact on students' lives and contributed to the cultivation of a generation of innovators:
For Inspiration and Recognition of Science and Technology (FIRST): FIRST Robotics encompasses a range of programs, including the FIRST LEGO League and the FIRST Tech Challenge. These programs engage students in designing, building, and programming robots to complete challenges. FIRST fosters teamwork, creativity, and problem-solving skills while promoting the values of gracious professionalism.

RoboCup: RoboCup is an international robotics competition that aims to advance robotics research and education. It offers multiple leagues, including soccer, rescue, and home assistance, where robots compete in complex tasks. RoboCup not only challenges participants to develop advanced robotics systems but also stimulates research in AI, computer vision, and autonomous navigation.

Robotics Roadshow: The Robotics Roadshow brings robotics and STEM education to rural and underserved communities. It involves traveling to schools and community centers, providing interactive robotics demonstrations and hands-on activities. This initiative breaks down barriers to access and brings the wonders of robotics to students who may have limited exposure to technology.

Girls Who Code: While Girls Who Code primarily focuses on coding, it also introduces participants to robotics concepts. This initiative aims to close the gender gap in technology fields by offering girls a supportive environment to explore coding, robotics, and computer science.

National Robotics Week: National Robotics Week is an annual event that celebrates robotics and its impact on society. It features a diverse range of activities, including competitions, workshops, demonstrations, and exhibitions, aimed at engaging students and the general public in robotics-related experiences.

Impact and Significance: Robotics outreach programs have a profound impact on participants, educators, and society as a whole:

Early Exposure: Outreach initiatives provide students with early exposure to robotics, influencing their career choices and sparking an interest in STEM subjects.

Confidence Building: Hands-on experiences and successes in outreach programs build participants' confidence in their problem-solving and technical skills.

Inclusive Participation: Outreach programs promote diversity and inclusivity by reaching underrepresented groups in STEM fields, including girls, minorities, and underserved communities.

Community Engagement: Robotics outreach strengthens ties between educational institutions, communities, and organizations, fostering a sense of unity and collaboration.

STEM Pipeline: By inspiring young learners, robotics outreach programs contribute to a healthy pipeline of future STEM professionals and innovators.

Overcoming Challenges and Scaling Impact: To maximize the impact of robotics outreach programs, educators and stakeholders must address challenges and leverage strategies for expansion:

Resource Allocation: Securing funding, materials, and personnel is essential for sustaining and expanding outreach efforts.

Inclusivity: Efforts should be made to ensure that outreach programs are accessible to diverse populations and promote equal participation.

Sustainability: To achieve lasting impact, outreach programs need a sustainable model that considers long-term funding, partnerships, and scalability.

Evaluation and Improvement: Continuous assessment of program effectiveness allows for refinement and improvement based on participant feedback and outcomes.

Collaboration: Collaboration between educational institutions, community organizations, industry partners, and government agencies enhances the reach and scope of outreach programs.

Robotics outreach programs serve as beacons of inspiration, sparking curiosity and igniting the flames of innovation in the hearts and minds of young learners. By bringing technology to the fingertips of students, these initiatives open doors to new possibilities, encourage exploration, and create pathways to STEM fields. The significance of robotics outreach lies not only in the skills and knowledge it imparts but also in the cultural shift it generates – a shift towards a future where young minds are empowered to shape the world through technology, creativity, and collaboration. As robotics outreach continues to flourish, it contributes to a brighter and more inclusive future, where the next generation of thinkers, creators, and problem solvers emerge with the tools they need to make a lasting impact.

Exercise

Multiple-choice Questions

1. What is the primary goal of robotics outreach programs?

a) To develop advanced robots for industrial applications

b) To introduce students to coding and programming concepts

c) To inspire interest and engagement in robotics and STEM fields

d) To promote competitions among robotic enthusiasts

2. How do robotics outreach programs typically engage participants?

a) Through theoretical lectures and presentations

b) By providing passive reading materials

c) With hands-on activities and interactive demonstrations
d) By conducting online quizzes and assessments
3. What is the essence of hands-on engagement in robotics outreach?
a) Observing robots remotely through virtual reality
b) Learning about robots from textbooks and articles
c) Interacting with robots and solving challenges actively
d) Watching videos of robots in action on the internet
4. Why are interactive demonstrations important in robotics outreach?
a) To replace hands-on activities with virtual simulations
b) To discourage participants from engaging with robots
c) To showcase real-world applications of robotics technology
d) To solely focus on theoretical concepts of robotics
5. Which factor contributes to the inclusivity of robotics outreach programs?
a) Limiting access to certain socioeconomic groups
b) Focusing only on participants with prior robotics experience
c) Requiring participants to pay a high enrollment fee
d) Reaching diverse learners from different backgrounds
6. What does STEM advocacy refer to in the context of robotics outreach?
a) Advocating for stricter regulations in the robotics industry
b) Advocating for increased funding for robotics research
c) Promoting the importance of science, technology, engineering, and mathematics education
d) Encouraging participants to focus solely on robotics careers
7. Which organization offers a range of programs like FIRST LEGO League and FIRST Tech Challenge to engage students in robotics challenges?
a) NASA
b) RoboCup
c) VEX Robotics
d) FIRST Robotics
8. What is the primary focus of the RoboCup international robotics competition?
a) Creating robots for home assistance tasks
b) Developing robots for space exploration
c) Advancing robotics research and education
d) Designing robots to assist in disaster relief efforts
9. How does the Robotics Roadshow initiative bring robotics education to underserved communities?
a) By exclusively hosting events in well-funded schools
b) By providing online tutorials and workshops
c) By traveling to schools and community centers for interactive demonstrations
d) By offering high-cost robotics kits to interested participants
10. What is the purpose of the Girls Who Code initiative?
a) To create coding platforms exclusively for boys
b) To encourage girls to pursue careers in robotics manufacturing
c) To close the gender gap in technology and introduce girls to coding and robotics
d) To focus solely on advanced coding skills for girls
11. How does National Robotics Week contribute to robotics education?
a) By hosting international robot-building competitions
b) By providing scholarships to robotics enthusiasts
c) By celebrating robotics through a variety of engaging activities and events
d) By exclusively targeting professional roboticists
12. What is the primary impact of robotics outreach programs on participants?
a) Developing expertise in theoretical robotics concepts
b) Exclusively preparing participants for robotics careers
c) Igniting interest in robotics and inspiring pursuit of STEM subjects
d) Focusing solely on coding and programming skills
13. What role do interactive demonstrations play in robotics outreach?
a) They solely provide theoretical explanations of robotics concepts.
b) They engage participants in hands-on activities and challenges.

c) They discourage participants from exploring robotics.
d) They focus on discussing the history of robotics.
14. How do successful robotics outreach programs contribute to a diverse STEM workforce?
a) By excluding underrepresented groups from participation
b) By providing expensive robotic kits to a select few
c) By promoting inclusivity and reaching underrepresented communities
d) By focusing solely on experienced roboticists
15. What is the primary impact of robotics outreach on participants' confidence?
a) Decreasing their interest in STEM fields
b) Exclusively enhancing their technical skills
c) Building confidence in problem-solving and technical abilities
d) Focusing solely on passive learning experiences

Review Questions:

1. What is the primary goal of robotics outreach programs?
2. How do interactive demonstrations contribute to robotics outreach?
3. What is the significance of hands-on engagement in robotics outreach?
4. How do robotics outreach Programs Bridge the gap between education and inspiration?
5. Why is inclusivity important in robotics outreach initiatives?
6. What is the impact of robotics outreach programs on participants' confidence?
7. How do robotics outreach programs advocate for STEM education?
8. What is the role of role models in inspiring participants in robotics outreach?
9. Name a notable robotics competition that engages students in building and programming robots.
10. How does the Robotics Roadshow initiative extend robotics education to underserved communities?
11. What is the purpose of the Girls Who Code initiative in the context of robotics?
12. How does National Robotics Week celebrate the impact of robotics on society?
13. What impact do robotics outreach programs have on participants' early exposure to technology?
14. How does robotics outreach contribute to building a diverse STEM workforce?
15. Explain the importance of scaling impact in robotics outreach programs.
16. How do robotics outreach programs go beyond traditional curricula?
17. What is the significance of interactive demonstrations in robotics outreach?
18. How do robotics outreach programs encourage participants to explore their interests in technology?
19. How does robotics outreach foster a culture of innovation and exploration?
20. What strategies can be employed to address challenges in scaling robotics outreach programs?

Answers Multiple-choice Questions

Answers:
1. c)
2. c)
3.c)
4.c)
5.d)
6.c)
7.d)
8.c)
9.c)
10.c)
11.c)
12.c)
13.b)
14.c)
15.c)

Review Questions:

1. The primary goal of robotics outreach programs is to inspire interest and engagement in robotics and STEM fields among participants. These programs are designed to ignite curiosity, foster creativity, and encourage individuals to explore the exciting world of technology, engineering, and problem-solving. By providing hands-on experiences, interactive activities, and exposure to real-world applications, robotics outreach aims to spark a lifelong passion for learning and innovation.

2. Interactive demonstrations play a pivotal role in robotics outreach by offering participants a tangible and captivating experience with robotics technology. These demonstrations allow participants to witness robots in action, understand their capabilities, and even interact with them directly. By engaging multiple senses and offering a dynamic learning environment, interactive demonstrations make complex concepts more accessible and relatable, enhancing participants' understanding and enthusiasm for robotics.

3. The significance of hands-on engagement in robotics outreach cannot be overstated. Hands-on activities provide participants with the opportunity to actively engage with robots, solve challenges, and create projects. This immersive form of learning goes beyond theoretical concepts and empowers participants to apply their knowledge in practical scenarios. Hands-on engagement nurtures critical skills such as problem-solving, teamwork, and creativity, fostering a deeper connection to the subject matter.

4. Robotics outreach programs bridge the gap between education and inspiration by combining theoretical learning with hands-on experiences. While traditional education often focuses on theoretical concepts, robotics outreach provides participants with real-world applications, allowing them to interact with robots, create solutions, and witness the tangible impact of their efforts. This approach not only educates participants but also inspires them to explore STEM fields further and consider pursuing careers in related domains.

5. Inclusivity is of paramount importance in robotics outreach initiatives. These programs aim to provide equal access and opportunities for individuals from diverse backgrounds, regardless of gender, socioeconomic status, or ethnicity. By fostering inclusivity, robotics outreach breaks down barriers and reaches underrepresented groups, creating a diverse pool of future innovators and problem solvers who bring a wide range of perspectives and experiences to the field.

6. Robotics outreach programs have a significant impact on participants' confidence. Through hands-on activities, interactive challenges, and successful project completion, participants gain a sense of accomplishment that boosts their self-esteem and self-efficacy. The ability to tackle complex tasks, collaborate with peers, and witness their creations come to life cultivates a sense of empowerment that extends beyond robotics to various aspects of their academic and personal lives.

7. Robotics outreach programs advocate for STEM education by actively promoting the importance of science, technology, engineering, and mathematics in various contexts. These programs emphasize the practical applications of STEM subjects, encouraging participants to explore and excel in these fields. By showcasing the diverse and exciting possibilities within STEM, robotics outreach advocates for the cultivation of skills essential for the future workforce.

8. Role models play a crucial role in inspiring participants in robotics outreach. These role models, often accomplished individuals in the field of robotics or related domains, serve as living examples of what can be achieved through dedication, perseverance, and a passion for innovation. Their personal stories and experiences resonate with participants, motivating them to overcome challenges, pursue excellence, and envision themselves as future contributors to the field.

9. A notable robotics competition that engages students in building and programming robots is the FIRST Robotics Competition. FIRST (For Inspiration and Recognition of Science and Technology) aims to inspire young people to become science and technology leaders through hands-on challenges that encourage teamwork, innovation, and problem-solving. Participants design, build, and program robots to complete specific tasks in a competitive and collaborative environment.

10. The Robotics Roadshow initiative extends robotics education to underserved communities by bringing robotics directly to schools, community centers, and other venues. This initiative acknowledges that not all students have equal access to robotics programs due to geographical, financial, or other limitations. The Robotics Roadshow ensures that students in underserved areas have the opportunity to engage with robotics technology and experience the excitement of STEM learning.

11. The Girls Who Code initiative aims to bridge the gender gap in technology and robotics. This initiative recognizes the historical underrepresentation of women in STEM fields and seeks to empower girls by providing them with opportunities to learn coding, robotics, and computer science. By creating a supportive and inclusive environment, Girls Who Code fosters confidence, skill development, and a sense of belonging for young girls interested in technology.

12. National Robotics Week celebrates the impact of robotics on society by organizing a wide range of activities and events. These events showcase the latest advancements in robotics technology, provide hands-on experiences with robots, and highlight their contributions to various industries. National Robotics Week aims to inspire participants of all ages to engage with robotics and explore its applications in real-world contexts.

13. Robotics outreach programs have a significant impact on participants' early exposure to technology. By offering hands-on experiences, interactive demonstrations, and engagement with robots, these programs introduce

participants to the exciting world of technology from a young age. This exposure ignites curiosity, fosters familiarity with technology, and encourages participants to consider pursuing STEM subjects in their academic and professional journeys.

14. Robotics outreach contributes to building a diverse STEM workforce by promoting inclusivity and reaching out to underrepresented communities. These programs provide access to robotics education for individuals from various socioeconomic backgrounds, ethnicities, genders, and cultures. By fostering a diverse learning environment, robotics outreach helps cultivate a workforce that reflects the richness of perspectives found in society.

15. Scaling impact in robotics outreach programs is essential to extend the reach and benefits of these initiatives to a larger audience. Scaling can involve partnering with educational institutions, community organizations, and corporations to expand the geographical reach of programs. Online platforms, resources, and virtual engagement methods can also be employed to reach a broader demographic and ensure sustainability.

16. Robotics outreach programs go beyond traditional curricula by offering experiential and hands-on learning experiences. Unlike traditional classroom settings, robotics outreach emphasizes active participation, problem-solving, and collaboration. Participants engage in projects, challenges, and interactive demonstrations that allow them to apply theoretical concepts to real-world scenarios, fostering a deeper understanding and appreciation for robotics and STEM.

17. Interactive demonstrations in robotics outreach hold immense significance as they offer participants a firsthand experience of robotics technology. These demonstrations allow participants to witness robots in action, interact with them, and understand their capabilities. The dynamic and engaging nature of interactive demonstrations leaves a lasting impression, inspiring participants to explore robotics further and consider pursuing related subjects.

18. Robotics outreach programs encourage participants to explore their interests in technology by providing a platform for hands-on engagement, challenges, and creative problem-solving. By presenting participants with opportunities to interact with robots, build projects, and solve real-world challenges, these programs spark curiosity and empower participants to delve deeper into their technological interests.

19. Robotics outreach fosters a culture of innovation and exploration by nurturing participants' creative thinking, problem-solving skills, and willingness to experiment. Through hands-on activities, participants learn to tackle challenges, adapt to new situations, and develop a mindset that values innovation and continuous learning. This culture of innovation extends beyond robotics and influences participants' approach to various aspects of their lives.

20. To address challenges in scaling robotics outreach programs, several strategies can be employed. Partnerships with educational institutions, corporations, and community organizations can help expand the reach of programs. Utilizing online platforms and resources allows for virtual engagement and remote learning opportunities. Customizing programs to cater to different age groups, regions, and demographics ensures relevance and accessibility for a diverse audience. Regular assessment and feedback mechanisms help refine programs for maximum impact and sustainability.

Online Resources

Online Exam Papers

Link: https://forms.gle/7awYv73CgEZS1sD89

QR Code:

YouTube Video Playlist Link

Link: https://www.youtube.com/playlist?list=PLzlh_31VWEFqgX0Wms2BMn4pmS-lJP4LF

QR Code:

Chapter 16: Roboethics and Societal Implications

The convergence of robotics and artificial intelligence (AI) has ushered in an era of technological innovation that promises to reshape industries, enhance human capabilities, and revolutionize the way we interact with machines. As robots and AI systems become increasingly integrated into our lives, from factories and hospitals to homes and public spaces, profound ethical considerations and societal implications have emerged. These considerations go beyond the technical aspects of engineering and programming; they delve into the heart of how we interact with, coexist alongside, and entrust decision-making to machines that are becoming more autonomous and intelligent.

This intersection of technology and ethics has given rise to the field of Roboethics—a discipline that seeks to explore, understand, and address the ethical, social, and cultural dimensions of robotics and AI. Roboethics recognizes that as these technologies become ubiquitous and influential, we must grapple with questions that extend beyond their technical capabilities. These questions touch on issues of responsibility, accountability, human values, and the broader impact on society.

In this exploration of Roboethics and its societal implications, we embark on a journey that takes us beyond the realm of code and algorithms. We delve into the ethical dilemmas posed by autonomous machines, the intricate web of relationships that humans are forming with robots, and the implications of embedding AI systems into critical decision-making processes. Furthermore, we examine the roles of legislation, education, and public discourse in shaping the future of robotics and AI in a manner that aligns with our moral compass and societal values.

As we navigate the intricacies of Roboethics, it becomes evident that the challenges are multifaceted. From questions about the allocation of responsibility when an autonomous car makes a split-second ethical decision on the road, to concerns about the impact of AI on employment and human identity, Roboethics invites us to reflect on the values and principles that guide our technological progress. Moreover, the societal implications of robotics extend to issues of equity, bias, privacy, and the potential erosion of human agency in a world increasingly governed by machines.

This exploration is not just theoretical; it is grounded in the practical realities of a world where robots are assisting surgeons in delicate procedures, AI algorithms are influencing credit decisions, and social robots are offering companionship to the elderly. The decisions we make today about the design, deployment, and governance of these technologies will shape the ethical contours of our future. Through informed conversations, responsible innovation, and collective action, we have the opportunity to steer the course of robotics and AI towards a future that enhances human well-being, upholds our shared values, and respects the dignity of both humans and machines.

In the pages that follow, we will delve into the core principles of Roboethics, examining the ethical considerations that underpin human-robot interactions, the challenges of ensuring transparency and accountability in AI decision-making, and the ways in which societies are grappling with the transformative potential of these technologies. Through this exploration, we hope to shed light on the complex landscape of Roboethics and encourage readers to engage in a thoughtful dialogue about the choices we make as we forge ahead into an increasingly automated and intelligent world.

Roboethics: Ethical Considerations in Robotics: The integration of robotics and artificial intelligence (AI) technologies into various aspects of human life has led to a fundamental transformation of how we interact with machines, make decisions, and shape the world around us. As robots become more autonomous and capable of performing complex tasks, a pressing concern arises: how should we navigate the ethical dimensions of these advancements? The field of Roboethics emerges as a critical discipline that addresses the intricate ethical considerations inherent in robotics and AI.

Understanding Roboethics:Roboethics, a portmanteau of "robotics" and "ethics," is an interdisciplinary field that explores the moral and ethical implications of designing, developing, and deploying robotic systems and AI technologies. It examines questions related to responsibility, accountability, decision-making, human-robot relationships, and the broader societal impact of these technologies. As robots become more integrated into our lives, from medical procedures and autonomous vehicles to customer service and entertainment, the ethical dimensions of their actions and interactions come to the forefront.

The Challenge of Autonomy: One of the central ethical considerations in robotics revolves around the increasing autonomy of machines. As robots and AI systems become more capable of making decisions without direct human intervention, questions arise about the criteria these systems use to make choices. This poses challenges not only in terms of ensuring that these decisions align with human values but also in determining who should be held accountable when things go wrong. The idea of robots making ethical decisions evokes philosophical discussions about the nature of morality, the role of consciousness, and the responsibilities of their creators.

Responsibility and Accountability: When a robotic system makes an autonomous decision that has ethical implications, the question of responsibility becomes complex. Traditional notions of accountability, which often apply to human actions, need to be adapted to accommodate the involvement of machines. Should the responsibility lie with the programmer who designed the algorithm, the manufacturer of the robot, the owner of the machine, or even the robot itself? Determining who is liable for the consequences of a robot's actions has implications for legal frameworks, insurance policies, and the establishment of clear guidelines for the responsible use of technology.

Human-Robot Relationships: The integration of robots into various roles, including companionship, caregiving, and emotional support, raises questions about the nature of human-robot relationships. As robots become more sophisticated in mimicking human emotions and behaviors, users may develop emotional bonds with these machines. This prompts discussions about the authenticity of these relationships, the potential for exploitation by designers, and the ethical implications of forming attachments to non-human entities. Exploring the boundaries of human-robot relationships leads to inquiries about consent, manipulation, and the moral obligations of robot designers.

Bias and Fairness: The data used to train AI algorithms can sometimes contain biases present in the real world. This introduces ethical concerns when robots and AI systems perpetuate and amplify biases in decision-making processes. For example, AI algorithms used in hiring processes or loan approvals may inadvertently discriminate against certain groups. Ensuring fairness and equity in the outcomes of these systems requires a concerted effort to address bias in data collection, algorithm design, and model evaluation.

Impact on Employment: The automation of jobs through robotics and AI technologies has the potential to reshape labor markets and lead to job displacement. This raises ethical concerns about the societal impact of technological advancements. While automation can increase efficiency and productivity, it can also result in unemployment and economic inequalities. Addressing these concerns involves striking a balance between technological progress, human well-being, and the ethical responsibility to ensure that the benefits of automation are distributed equitably.

Transparency and Accountability: Ensuring that robots and AI systems operate transparently and can be held accountable for their actions is a critical ethical consideration. When machines make decisions that impact individuals or society at large, it is essential to understand the reasoning behind those decisions. This is particularly important in contexts such as healthcare, where AI systems make diagnostic and treatment recommendations. Developing transparent AI systems that can provide explanations for their decisions helps build trust and fosters responsible use.

Privacy and Data Security: As robots and AI systems collect and process vast amounts of data, concerns about privacy and data security emerge. Robots equipped with cameras, sensors, and microphones raise questions about surveillance, data ownership, and the potential for misuse of personal information. Ethical considerations encompass issues of consent, data anonymization, and protecting individuals from unauthorized access to their private information.

Roboethics stands at the crossroads of technological advancement and ethical responsibility. As robots and AI technologies continue to evolve, the need to address the ethical considerations they raise becomes increasingly urgent. Striking a balance between innovation and ethical safeguards is essential to ensure that the integration of robotics and AI enhances human well-being, upholds fundamental values, and promotes a just and equitable society. By engaging in robust discussions, fostering interdisciplinary collaboration, and making informed decisions, we can navigate the intricate ethical landscape of robotics and AI and shape a future that reflects our shared ethical principles.

Social and Cultural Impact of Robotics: The integration of robotics into various spheres of human life has ushered in a transformative technological paradigm shift, reshaping not only industries and economies but also the very fabric of societies and cultures worldwide. The profound impact of robotics reaches beyond economic considerations and technical capabilities, touching upon social structures, cultural norms, ethical values, and the fundamental ways in which humans interact with the world around them. This evolution invites a comprehensive exploration of the social and cultural implications that arise as robots increasingly become integral members of our communities and households.

Redefined Human-Robot Interactions: The advancement of robotics technology has introduced a new dimension to human interactions—interactions with machines. As robots assume roles in caregiving, companionship, and service, they participate in social exchanges that extend beyond mere utility. Humans project emotions, intentions, and expectations onto these machines, forming emotional bonds and relationships that transcend traditional notions of human-to-machine interaction. The emergence of social robots prompts questions about the boundaries between humans and machines, the authenticity of these interactions, and the potential for robots to fulfill emotional and psychological needs.

Cultural Perceptions and Attitudes: Different cultures hold distinct perceptions and attitudes toward robotics, influenced by historical, religious, and philosophical contexts. For instance, cultures that embrace animism or anthropomorphism may find it easier to form emotional connections with robots, whereas those with reservations about human-like technology may view these interactions with skepticism. Cultural values shape perceptions of ethical boundaries, appropriate use of robots, and the role of technology in society. Understanding these cultural nuances is pivotal in designing robots that align with diverse cultural norms and values.

Impact on Labor and Employment: The introduction of robots into the workforce challenges established notions of employment and labor dynamics. Automation and robotics have the potential to replace or augment human labor in various industries, leading to shifts in employment patterns and the types of skills in demand. This has social and cultural ramifications, as communities that rely on specific industries may face economic upheaval and displacement. Reskilling and upskilling become essential components of navigating these changes, preserving cultural identities while adapting to new economic realities.

Ethics and Value Systems: As robots take on roles that involve decision-making, particularly in fields like healthcare and finance, ethical dilemmas arise. The values embedded in the algorithms that guide these decisions reflect the priorities and ethical considerations of the programmers and designers. This opens a conversation about the cultural biases present in AI systems and the importance of inclusive and diverse development teams to ensure fairness and equity in decision-making processes. It also invites discussions about the preservation of cultural values in a world increasingly influenced by AI.

Evolving Social Norms: The proliferation of robots and AI technologies challenges traditional social norms and conventions. For instance, the use of social robots in caregiving for the elderly or individuals with disabilities may challenge prevailing attitudes toward the roles of family members and caregivers. Accepting robots as participants in caregiving forces societies to reevaluate cultural expectations, intergenerational dynamics, and notions of care and empathy. This disruption of norms can lead to both resistance and innovation as societies adapt to new paradigms.

Digital Divides and Inequities: The integration of robotics has the potential to exacerbate existing digital divides and socio-economic inequalities. Access to advanced technologies is not evenly distributed across societies, and marginalized communities may have limited exposure to and interaction with robots. This raises concerns about equitable access to the benefits of technological advancements, the potential reinforcement of social hierarchies, and the importance of addressing these inequities through inclusive design, education, and policy initiatives.

Shaping Identity and Belonging: The proliferation of robotics technology shapes personal and collective identities. The attachment formed with robots challenges traditional understandings of what it means to be human and connected. This transformation raises questions about the implications of these relationships on human self-perception, the redefinition of companionship, and the boundaries between human and machine experiences. Additionally, robots designed to engage with culturally specific rituals and practices contribute to a sense of belonging and cultural continuity.

Cultural Expression and Creativity: Robotics technology introduces new possibilities for artistic expression, storytelling, and cultural preservation. Robots can be programmed to create music, produce visual art, and even participate in cultural performances. This integration of technology and culture prompts reflections on the evolving nature of creativity, the fusion of traditional artistic forms with modern technology, and the preservation of cultural heritage in innovative ways.

Public Perception and Acceptance: The acceptance of robots in society is influenced by public perceptions, which in turn are shaped by cultural attitudes, media portrayals, and personal experiences. The level of comfort people have with robots varies, and societal norms surrounding their use may shift as familiarity grows. Educating the public about the capabilities and limitations of robots, as well as addressing concerns related to job displacement, privacy, and ethical use, plays a crucial role in fostering positive perceptions and widespread acceptance.

The integration of robotics into the social and cultural fabric of societies is a dynamic and multifaceted process. As robots increasingly inhabit spaces once exclusively occupied by humans, their impact extends beyond economic considerations to encompass the very essence of what it means to be human. Understanding and addressing the social and cultural implications of robotics is imperative to ensure that technological progress aligns with human values, respects cultural diversity, and contributes to a harmonious coexistence between humans and machines. By engaging in interdisciplinary dialogues, cross-cultural collaborations, and thoughtful design, we can navigate this transformative landscape and shape a future where robotics enhances human well-being while preserving the rich tapestry of cultural identities that define us.

Economic Impact and Job Transformation: The integration of robotics and automation technologies into various industries has ushered in a transformative wave of change, reshaping economic landscapes and redefining traditional notions of work and employment. This shift, commonly referred to as the "robotic revolution," brings both promises and challenges to economies across the globe. The economic impact of robotics extends beyond mere productivity gains, touching upon labor markets, job transformation, and the distribution of wealth. As industries harness the potential of automation, it becomes crucial to understand and navigate the intricate economic implications that arise from this technological evolution.

The Promise of Increased Productivity: One of the primary drivers behind the adoption of robotics and automation is the potential for increased productivity. Robots excel in repetitive and precise tasks, allowing businesses to streamline operations, reduce errors, and enhance efficiency. Automation also enables 24/7 production cycles, reducing downtime and contributing to higher output. As companies integrate robots into their workflows, they can scale their operations, meet growing demand, and improve their competitive edge in the global marketplace.

Labor Displacement and Job Transformation: While increased productivity is a significant benefit of automation, it is accompanied by concerns about job displacement. As robots take over routine and repetitive tasks, the nature of work is transformed. Some jobs become obsolete, leading to workforce challenges in industries heavily reliant on manual labor. This displacement necessitates a strategic approach to retraining and reskilling the workforce for roles that require creativity, problem-solving, and human interaction—skills that are less susceptible to automation.

Shift in Skill Demand: The robotic revolution reshapes the skillset demanded in the labor market. Traditional jobs that focus on routine manual tasks may dwindle, while new opportunities emerge in fields such as robotics maintenance, programming, data analysis, and human-robot collaboration. This shift underscores the importance of education and lifelong learning to equip individuals with the skills needed to thrive in a technologically advanced economy. Governments, educational institutions, and industries must work collaboratively to bridge the skills gap and ensure that workers can adapt to changing demands.

Economic Resilience and Competitiveness: Countries that embrace robotics and automation stand to gain economic resilience and competitiveness on the global stage. Automation enhances production efficiency, reduces operational costs, and accelerates innovation. This fosters an environment where businesses can rapidly adapt to changing market dynamics and respond to consumer demands. The ability to produce goods and services efficiently not only strengthens domestic economies but also positions nations as leaders in innovation and technology.

Small and Medium Enterprises (SMEs) and Innovation: While large corporations often have the resources to invest in robotics, the impact on SMEs is equally significant. Automation technologies can level the playing field, allowing smaller businesses to access advanced capabilities, streamline processes, and enter new markets. This democratization of technology fosters innovation, as SMEs can experiment with robotics to develop novel products and services. It also enables them to compete with larger entities, spurring economic growth and diversification.

Redefining Value Chains and Global Trade: The integration of robotics disrupts traditional value chains and global trade patterns. Automation enables on shoring or nearshoring of production, reducing the reliance on labor cost differentials across countries. This shift has implications for international trade dynamics, as the cost advantage of outsourcing labor-intensive tasks decreases. Countries that embrace robotics can establish competitive advantages based on technology and innovation, reshaping the global economic landscape.

Challenges of Job Transition: While the benefits of automation are evident, the challenges of transitioning from traditional jobs to roles that require advanced skills cannot be overlooked. Workers displaced by automation may experience difficulties in finding suitable employment opportunities. The economic burden of unemployment, coupled with the need for retraining, presents challenges that require coordinated efforts from governments, industries, and educational institutions to provide comprehensive support.

Wealth Distribution and Inequality: The economic impact of robotics is not uniform across all segments of society. The concentration of automation benefits in certain industries or regions can exacerbate existing inequalities. Additionally, the ownership of automation technologies can lead to the accumulation of wealth in the hands of a few, further widening the wealth gap. Addressing these challenges requires policies that promote equitable access to technology, ensure fair wages, and provide social safety nets for those affected by job displacement.

Innovation Ecosystems and Entrepreneurship: The robotic revolution fosters innovation ecosystems and entrepreneurial endeavors. Startups and entrepreneurs can leverage automation technologies to develop novel products and services that cater to changing market demands. These innovations can lead to the creation of new industries, the expansion of existing ones, and the emergence of disruptive business models. Encouraging a culture of innovation and providing support for startups can contribute to economic diversification and growth.

The integration of robotics and automation into economies marks a significant inflection point in the trajectory of global industries. The economic impact extends beyond mere productivity gains, reshaping labor markets, job roles, skill requirements, and global trade dynamics. As we navigate the challenges and opportunities of the robotic revolution, a proactive and strategic approach is essential. Governments, industries, educational institutions, and individuals must collaborate to ensure that the benefits of automation are widely distributed, workers are equipped with the skills of the future, and societies are resilient in the face of technological change. By embracing automation while addressing its social and economic implications, we can navigate this transformative phase and shape a future where technology enhances human prosperity and societal well-being.

Exercise

Multiple-choice Questions

1. What is the primary focus of"Roboethics and Societal Implications"?
a) Historical evolution of robotics technology
b) Ethical considerations in artificial intelligence
c) Economic impact of automation in industries
d) The role of robots in healthcare
2. What term describes the transformative shift caused by the integration of robotics and automation?
a) Digital revolution
b) Industrial revolution
c) Robotic uprising
d) Technological evolution
3. What is a primary driver behind the adoption of robotics and automation in various industries?
a) Cost reduction
b) Elimination of human labor
c) Aesthetic appeal
d) Resource depletion
4. How do robots contribute to increased productivity in industries?
a) By creating more job opportunities
b) By conducting repetitive tasks with precision
c) By replacing skilled workers
d) By solely focusing on creative tasks
5. What challenge arises from the displacement of jobs due to automation?
a) Overcrowding in job markets
b) A surplus of skilled workers
c) The need for retraining and reskilling
d) An oversupply of manual labor
6. What skills are likely to be in demand as automation transforms job roles?
a) Routine manual tasks
b) Creativity and problem-solving
c) Repetitive data entry
d) Unskilled labor
7. What impact does automation have on the traditional skillset demanded in the labor market?
a) No impact on skill requirements
b) Focus on routine manual skills
c) Shift towards creative and analytical skills
d) Disregard for technical expertise
8. How does the integration of robotics enhance economic resilience and competitiveness?
a) By increasing labor costs
b) By slowing down innovation
c) By enhancing production efficiency and innovation
d) By decreasing operational efficiency
9. What group benefits from the democratization of technology through automation?
a) Large corporations only
b) Small and Medium Enterprises (SMEs)
c) Government agencies
d) Educational institutions
10. How does automation impact value chains and global trade patterns?
a) It encourages offshoring of production.
b) It reduces the cost of labor-intensive tasks.
c) It promotes labor-intensive industries.
d) It decreases the importance of innovation.
11. What challenge is associated with the displacement of workers due to automation?
a) A surplus of highly-skilled workers

b) Difficulty finding suitable employment opportunities
c) A decrease in the need for creative skills
d) A reduced need for retraining programs
12. What is a potential consequence of automation's impact on value chains?
a) Increased dependency on outsourcing
b) Unchanged global trade dynamics
c) No effect on labor cost differentials
d) A decrease in innovation-driven economies
13. How do robotics outreach programs contribute to economic resilience?
a) By automating all aspects of business operations
b) By focusing solely on large corporations
c) By fostering innovation and adaptability
d) By decreasing production efficiency
14. What challenge arises from the concentration of automation benefits?
a) Decreased wealth inequality
b) A reduction in technology ownership
c) Widening wealth inequalities
d) Equitable distribution of automation benefits
15. How does the integration of robotics encourage innovation and entrepreneurship?
a) By limiting access to automation technologies
b) By narrowing the scope of business opportunities
c) By providing job security for traditional roles
d) By fostering the development of novel products and services

Review Question:

1. What is the central focus of "Roboethics and Societal Implications"?
2. How does the integration of robotics and automation impact economies on a global scale?
3. What is the promise of increased productivity associated with the adoption of robotics?
4. Explain the concept of labor displacement and job transformation in the context of robotics.
5. How does the demand for different skill sets shift due to the integration of robotics?
6. Discuss the role of automation in fostering economic resilience and competitiveness.
7. What benefits do Small and Medium Enterprises (SMEs) gain from the integration of automation?
8. How does the integration of robotics impact traditional value chains and global trade patterns?
9. What challenges arise from the displacement of workers by automation, and how can they be addressed?
10. Explain how wealth distribution and inequality can be affected by the integration of robotics.
11. How does automation contribute to the innovation ecosystem and entrepreneurship?
12. In what ways can governments and industries support workers facing job displacement due to automation?
13. Describe the significance of bridging the skills gap and providing education in a changing job landscape.
14. How does the integration of robotics and automation impact global economic dynamics?
15. Discuss the concept of equitable distribution of benefits and its role in addressing economic challenges.
16. How can automation technologies democratize opportunities for Small and Medium Enterprises (SMEs)?
17. What is the importance of innovation and adaptability in the face of automation?
18. Explain how the integration of robotics can impact on shoring and nearshoring of production.
19. What is the significance of economic resilience in the context of technological change?
20. How can societies ensure that the economic benefits of automation are widely distributed and accessible?

Answers Multiple-choice Questions

Answers:
1. d)
2. a)
3. a)
4. b)
5. c)
6. b)
7. c)
8. c)
9. b)
10. b)

11. b)
12. b)
13. c)
14. c)
15. d)

Review Question:

1. "Roboethics and Societal Implications," delves into the multifaceted aspects of integrating robotics and automation into various industries and the broader society. It explores the ethical considerations, economic impacts, and transformative effects that the advancement of robotics technology brings. This chapter aims to shed light on the profound changes that emerge from the widespread adoption of robots, addressing their implications on the global economy, the workforce, and the equitable distribution of benefits. Furthermore, the chapter highlights the pivotal role of ethics and responsible decision-making in navigating the challenges and opportunities presented by the integration of robotics into our daily lives.
2. The integration of robotics and automation has a profound impact on economies at a global scale. Automation enhances productivity, efficiency, and competitiveness across industries, leading to transformative changes in manufacturing, services, and various sectors. By automating repetitive and labor-intensive tasks, companies can reduce production costs, streamline operations, and maintain consistent product quality. This, in turn, bolsters economic growth and strengthens the position of countries in the global market.
3. The promise of increased productivity is one of the cornerstones of the adoption of robotics and automation. By replacing manual tasks with automated processes, companies can achieve higher output levels and shorter production cycles. Moreover, robots can operate 24/7 without the need for breaks or rest, ensuring continuous production and reduced downtime. This increased productivity contributes to higher output volumes, improved resource utilization, and overall efficiency gains for industries.
4. Labor displacement refers to the phenomenon where automation replaces human workers in performing specific tasks or jobs. As automation technologies advance, routine and repetitive tasks can be efficiently carried out by robots, leading to a decrease in the demand for certain types of manual labor. However, job transformation is also a crucial aspect. Automation creates new roles that involve overseeing, maintaining, and programming robots, along with positions focused on creative and strategic decision-making. Thus, while some jobs may be displaced, others emerge that require different skill sets and expertise.
5. The integration of robotics and automation leads to a significant shift in the demand for skill sets in the workforce. While routine and manual tasks become automated, the demand for skills related to creativity, problem-solving, and innovation increases. Workers need to be adaptable and possess the ability to collaborate with robotic systems. Additionally, proficiency in programming, data analysis, and technical problem-solving becomes essential to ensure effective interaction with automated technologies.
6. Automation plays a pivotal role in fostering economic resilience and competitiveness. Industries that adopt automation are better equipped to respond to market fluctuations, disruptions, and changes in consumer demand. Automation enhances the agility of businesses to adjust production levels, shift focus to different product lines, and optimize resource allocation. This resilience allows companies to maintain their market positions and adapt to changing economic landscapes.
7. Small and Medium Enterprises (SMEs) gain substantial benefits from the integration of automation. Automation technologies enable SMEs to compete on a larger scale by reducing operational costs, enhancing productivity, and expanding production capabilities. As automation solutions become more accessible and customizable, SMEs can adopt robotics to streamline their processes, improve product quality, and increase overall efficiency. This levels the playing field, allowing smaller businesses to remain competitive in a technology-driven market.
8. The integration of robotics impacts traditional value chains and global trade patterns by altering the dynamics of production and distribution. Automation allows for localized manufacturing, reducing the need for extensive global supply chains. This shift can impact the demand for outsourcing and offshoring, potentially leading to increased onshoring or nearshoring of production activities. As a result, global trade patterns may experience adjustments as industries prioritize localized and efficient production methods.
9. The displacement of workers due to automation presents challenges related to job loss, skill gaps, and potential social disruptions. To address these challenges, comprehensive strategies are needed. These strategies may include reskilling and upskilling programs to equip workers with new skills, promoting lifelong learning to adapt to changing technological landscapes, and fostering collaboration between governments, industries, and educational institutions to ensure a smooth transition for affected workers.
10. The integration of robotics can influence wealth distribution and inequality. Automation may lead to the concentration of wealth in industries that heavily adopt robotics and automation, potentially exacerbating income

disparities. This underscores the importance of considering mechanisms to ensure that the benefits of automation are distributed more equitably, addressing disparities and providing opportunities for all segments of society to benefit from technological advancements.

11. Automation contributes significantly to the innovation ecosystem and entrepreneurship. It encourages the development of novel products, services, and business models. By automating routine tasks, entrepreneurs can focus on creative and strategic aspects of their ventures. Moreover, automation technologies themselves become a catalyst for innovation, inspiring entrepreneurs to explore new applications and redefine industries.

12. Governments and industries can support workers facing job displacement due to automation through various measures. These measures include investing in reskilling and upskilling programs, providing financial assistance during transitions, and facilitating opportunities for workers to acquire new skills. Collaboration between governments, industries, and educational institutions is essential to create a supportive ecosystem for workers in the face of changing job landscapes.

13. Bridging the skills gap and providing education are of paramount importance in a changing job landscape. Offering education that aligns with emerging technological trends equips individuals with relevant skills to navigate the new opportunities brought about by automation. Lifelong learning, continuous professional development, and the integration of technology-centric education play vital roles in empowering individuals for sustainable careers.

14. The integration of robotics and automation has a profound impact on global economic dynamics. Automation reshapes industries, supply chains, and production methods, influencing how goods and services are produced, distributed, and consumed. It can lead to shifts in trade patterns, investment priorities, and economic growth trajectories, ultimately redefining the landscape of the global economy.

15. The concept of equitable distribution of benefits is essential to address economic challenges arising from automation. Ensuring that the benefits of automation are accessible to a wide range of stakeholders, including workers, businesses, and communities, promotes inclusivity and prevents the concentration of advantages in specific segments of society. Achieving equitable distribution requires collaboration, policy frameworks, and responsible business practices.

16. Automation technologies democratize opportunities for Small and Medium Enterprises (SMEs) by providing them with access to advanced tools, systems, and capabilities. As automation becomes more accessible and adaptable, SMEs can integrate robotic solutions into their operations, enabling them to enhance productivity, improve product quality, and expand market reach. This democratization levels the playing field, enabling SMEs to compete effectively in the global arena.

17. Innovation and adaptability are crucial in the face of automation. Embracing automation technologies encourages businesses to innovate their processes, products, and services to remain competitive. The ability to adapt to changing technologies and consumer preferences is paramount for sustainable growth and success in a rapidly evolving business landscape shaped by automation.

18. The integration of robotics can impact the on shoring and nearshoring of production activities. Automation technologies enable localized manufacturing, reducing the need for extensive global supply chains. This shift can lead to increased production being brought closer to markets, minimizing logistical complexities and contributing to economic efficiency.

19. Economic resilience is of paramount significance in the context of technological change and automation. Businesses that adopt automation are better equipped to withstand disruptions, market fluctuations, and unexpected challenges. Automation enhances agility, allowing companies to adjust operations, production volumes, and resource allocation to maintain competitiveness and adapt to changing economic scenarios.

20. Ensuring that the economic benefits of automation are widely distributed and accessible requires a multifaceted approach. Governments, industries, and educational institutions need to collaborate to create policies and initiatives that promote equitable distribution. This includes addressing barriers to technology adoption, providing training and upskilling opportunities, and fostering an environment where individuals and businesses can harness the advantages of automation for sustainable growth and shared prosperity.

Online Resources

Online Exam Papers

Link:https://forms.gle/TLsRVMN37gQKrpSU7

QR Code:

YouTube Video Playlist Link

Link:https://www.youtube.com/playlist?list=PLzlh_31VWEFoFKfYGPPrrnR-1txc1_eqR

QR Code:

Chapter 17: Top of Form Robotics in Everyday Life

In recent years, the integration of robotics into everyday life has transformed the way we live, work, and interact with our surroundings. From smart home devices to autonomous vehicles, robotics has become an integral part of modern society, offering enhanced convenience, efficiency, and innovation across various domains. This rapid advancement in robotics technology has paved the way for a future where intelligent machines seamlessly integrate into our daily routines, making tasks easier, safer, and more enjoyable.

One of the most visible applications of robotics in everyday life is within our homes. Smart home devices, driven by robotic technology, have revolutionized the way we manage our living spaces. From voice-activated virtual assistants like Amazon's Alexa and Google Assistant to robotic vacuum cleaners and lawnmowers, these technologies have simplified tasks that were once time-consuming and mundane. These robots can perform tasks such as cleaning floors, adjusting room temperature, turning off lights, and even reminding us of appointments and tasks.

Moreover, the integration of robotics has led to the concept of the connected home, where various devices communicate with each other to create a cohesive and efficient living environment. For instance, smart thermostats can learn user preferences and adjust temperature settings accordingly, while robotic security systems can monitor homes and alert homeowners to potential threats.

Robotics has significantly impacted industries by enhancing efficiency, precision, and safety in manufacturing processes. Factory floors are now filled with robotic arms and machines performing tasks that require speed, accuracy, and consistency. These robots can assemble products, weld components, perform quality checks, and handle hazardous materials, thus reducing the risk to human workers and improving overall production rates.

Collaborative robots, known as cobots, have gained prominence in recent years due to their ability to work alongside human operators. These robots assist workers in tasks that require both human decision-making and robot precision. This collaboration not only enhances productivity but also improves the working conditions and safety of employees.

In the healthcare sector, robotics has brought about revolutionary changes in diagnostics, surgeries, and patient care. Surgical robots, such as the da Vinci Surgical System, enable surgeons to perform minimally invasive procedures with greater precision and control. This results in smaller incisions, reduced recovery times, and improved patient outcomes. Additionally, robotic exoskeletons are being developed to assist patients with mobility impairments, enabling them to regain mobility and independence.

Furthermore, telemedicine robots facilitate remote consultations and diagnosis, especially in underserved or remote areas. These robots allow healthcare professionals to connect with patients, provide medical advice, and monitor their conditions remotely. This approach not only improves access to medical care but also reduces the burden on overcrowded healthcare facilities.

The development of autonomous vehicles is transforming the way we commute and travel. Self-driving cars and drones are becoming more prevalent, promising increased safety, reduced traffic congestion, and enhanced energy efficiency. Autonomous vehicles use a combination of sensors, cameras, and artificial intelligence algorithms to navigate and make decisions on the road. This technology has the potential to drastically reduce accidents caused by human error, while also enabling greater mobility for people who are unable to drive.

In addition to self-driving cars, delivery robots are being tested to transport goods and groceries to people's doorsteps. These robots navigate sidewalks and streets autonomously, using sensors to avoid obstacles and pedestrians. This innovation has the potential to revolutionize the last-mile delivery process, making it more efficient and environmentally friendly.

Robotics has also found its way into the entertainment and leisure industry, enhancing the experiences of people in various ways. Theme parks feature animatronic characters that interact with visitors, creating immersive and engaging environments. Virtual reality (VR) and augmented reality (AR) technologies, often integrated with robotic elements, offer new dimensions to gaming and entertainment experiences.

Moreover, social robots, designed to mimic human behavior and interactions, are being developed to provide companionship to individuals. These robots can engage in conversations, provide emotional support, and even assist in tasks such as medication reminders for the elderly. This fusion of technology and companionship addresses the social isolation and loneliness experienced by some individuals.

While robotics in everyday life offers numerous benefits, there are also challenges and considerations to address. Ethical concerns, such as privacy and security, arise as more devices become connected and collect data. Ensuring that data is protected and used responsibly is crucial to maintaining trust in these technologies. Additionally, concerns about job displacement due to automation and the need for continuous education and training to adapt to technological changes are important considerations.

In conclusion, the integration of robotics into everyday life has ushered in a new era of convenience, efficiency, and innovation. From smart homes to healthcare, manufacturing, transportation, and entertainment, robots are transforming the way we live, work, and interact with the world around us. As robotics technology continues to advance, its potential to enhance our quality of life, drive economic growth, and address pressing societal challenges becomes increasingly evident. However, responsible deployment, ethical considerations, and a balanced approach to addressing potential risks are essential to ensure that robotics continues to serve as a positive force in our everyday lives.

Domestic Robots: Enhancing Daily Living: The integration of robotics and automation into our everyday lives has ushered in a new era of convenience and efficiency. Among the various applications of robotics, domestic robots stand out as innovative solutions that aim to enhance daily living by assisting with household chores, providing entertainment, and offering companionship. These robots are designed to alleviate the burden of repetitive tasks, create a more comfortable living environment, and even improve the overall quality of life for individuals and families.

The Rise of Domestic Robots: Domestic robots, also known as service robots, have witnessed significant advancements in recent years, driven by the convergence of robotics technology, artificial intelligence (AI), and machine learning. These robots are designed to perform tasks that range from cleaning and organizing to entertainment and home security. The goal is to make daily life more convenient, efficient, and enjoyable. One of the most well-known categories of domestic robots is the robotic vacuum cleaner. These autonomous devices use sensors, cameras, and algorithms to navigate through homes, avoiding obstacles and cleaning floors. They not only save homeowners time and effort but also ensure cleaner living spaces.

Household Chores and Automation: One of the primary focuses of domestic robots is to tackle household chores that often consume a significant portion of our time and energy. Robotic vacuum cleaners, as mentioned earlier, are a prime example. They can efficiently clean different types of flooring surfaces and reach areas that are often overlooked. Additionally, robotic mops have been developed to clean hard surfaces, such as tiles and hardwood floors, further reducing the need for manual cleaning. Robotic lawnmowers are another example of domestic robots that have gained popularity. These devices autonomously mow lawns, maintaining the grass at a consistent height and saving homeowners the task of pushing a heavy lawnmower around their yards.

Entertainment and Companionship: Domestic robots extend beyond utilitarian purposes and also contribute to entertainment and companionship. Social robots, designed to interact with humans and mimic human behaviors, offer companionship to individuals who may live alone or seek additional interactions. These robots engage in conversations, respond to voice commands, and even provide emotional support through simple interactions. Moreover, entertainment-focused domestic robots cater to recreational activities within the home. From robots that play music and dance to those that engage in interactive storytelling, these devices contribute to a more enjoyable and dynamic living environment. They can be particularly beneficial for families with children, adding an element of novelty and excitement to daily routines.

Smart Home Integration: The concept of the smart home goes hand in hand with domestic robots, as these devices are often integrated into the larger ecosystem of connected devices. This integration allows for seamless communication and coordination between various devices within a home, creating a cohesive and efficient living environment. For instance, a domestic robot equipped with AI and sensors can be programmed to adjust room temperature, turn off lights, and manage home security systems. Voice-activated virtual assistants can control these robots and other smart home devices, making it easy for homeowners to manage various aspects of their living spaces using natural language commands.

Challenges and Considerations: While domestic robots offer numerous benefits, there are challenges and considerations that need to be addressed. Privacy and data security are paramount, as these robots often collect and transmit data about users and their living spaces. Ensuring that data is protected and that users have control over their personal information is crucial to maintaining trust in these technologies. Furthermore, the cost of domestic robots can be a barrier to adoption for some individuals. High initial costs and ongoing maintenance expenses may limit access to these technologies, especially for those with limited financial resources. Striking a balance between affordability and advanced capabilities is essential for broadening the reach of domestic robots.

Future Prospects: As robotics technology continues to advance, the capabilities of domestic robots are likely to become more sophisticated and tailored to individual needs. AI and machine learning algorithms will play a significant role in enabling these robots to understand user preferences, adapt to changing environments, and perform tasks more autonomously.

Moreover, domestic robots are expected to become more versatile, capable of performing a wider range of tasks beyond cleaning and entertainment. Robots that assist with meal preparation, laundry, and home maintenance are areas of active research and development. These advancements have the potential to significantly impact the lives of individuals, especially those with limited mobility or health challenges.

Domestic robots have the potential to revolutionize daily living by offering convenience, efficiency, and enhanced quality of life. From cleaning and organizing to providing companionship and entertainment, these robots cater to various aspects of our routines. As technology continues to evolve, the integration of AI and automation will drive the development of even more advanced and capable domestic robots. While challenges related to privacy, cost, and accessibility need to be addressed, the future holds the promise of a more automated and intelligent living environment where robots seamlessly contribute to our everyday lives.

Robot-Assisted Learning Environments: In the realm of education, the integration of technology has been steadily transforming the way students learn and teachers instruct. One notable innovation in this field is the development and implementation of robot-assisted learning environments. These environments combine robotics and artificial intelligence (AI) to create interactive and engaging educational experiences that cater to the diverse needs of students. From personalized tutoring to fostering critical thinking skills, robot-assisted learning environments hold the potential to revolutionize education by providing tailored and effective learning opportunities.

The Evolution of Learning Environments: Traditional learning environments often rely on one-size-fits-all approaches, where educators deliver standardized content to a diverse group of students. However, this approach can leave some students behind, failing to address their unique learning styles, paces, and preferences. This is where robot-assisted learning environments come into play. Robot-assisted learning environments leverage AI algorithms to adapt to individual students' needs and learning patterns. By analyzing students' performance, comprehension, and engagement levels, these systems can tailor their interactions and content delivery, ensuring that each student receives the support and challenges they require to thrive academically.

Personalized Learning: One of the key advantages of robot-assisted learning environments is their ability to provide personalized learning experiences. Traditional classrooms often struggle to cater to the diverse needs of students with varying learning speeds and abilities. Robot-assisted learning platforms address this challenge by adapting content, pacing, and instructional strategies to individual learners. For instance, a robot-assisted system can identify areas where a student needs additional support and provide targeted exercises and explanations to reinforce understanding. Conversely, if a student demonstrates proficiency in a certain subject, the system can offer more advanced challenges to keep them engaged and motivated.

Enhanced Engagement: Engagement is a critical factor in effective learning. Robot-assisted learning environments leverage robotics and interactive interfaces to create engaging and immersive educational experiences. These environments often employ humanoid or interactive robots that can engage in conversations, display emotions, and respond to students' questions. Humanoid robots, for instance, can simulate real-life interactions, creating a more dynamic and relatable learning experience. This approach can be particularly effective in subjects like language learning, where students can practice conversational skills with the robot in a low-pressure environment.

Fostering Critical Skills: Robot-assisted learning environments also have the potential to foster critical skills that are highly valuable in today's rapidly evolving world. These environments often emphasize problem-solving, critical thinking, collaboration, and creativity. Through interactive challenges, simulations, and group activities, students can develop these skills in an engaging and practical context. For instance, a robot-assisted learning environment may present students with real-world scenarios that require them to analyze data, make informed decisions, and work together to find solutions. These experiences not only enhance subject-specific knowledge but also prepare students for the challenges they will encounter in their future careers.

Support for Teachers: Robot-assisted learning environments are not meant to replace teachers but rather to complement their efforts. These systems can offer teachers valuable insights into students' progress, strengths, and areas of improvement. By automating certain tasks like grading and assessment, teachers can dedicate more time to providing personalized guidance and support to individual students. Additionally, these systems can assist teachers in creating customized lesson plans and adapting instructional materials to suit different learning styles. This collaborative approach between teachers and technology ensures a holistic and effective learning experience for students.

Challenges and Considerations: While robot-assisted learning environments hold great promise, there are challenges that need to be addressed. Privacy and data security are paramount, as these systems collect and analyze student data to provide personalized learning experiences. Ensuring that data is protected and used responsibly is crucial to maintaining trust in these technologies. Moreover, the integration of technology in education requires teachers to adapt their instructional methods and embrace the role of technology as an educational tool. Proper training and professional development are essential to ensure that teachers are equipped with the skills and knowledge to effectively use robot-assisted learning environments in the classroom.

Future Prospects: The future of robot-assisted learning environments is bright, with continuous advancements in AI, robotics, and educational technology. As these technologies evolve, robot-assisted systems will become even more sophisticated in their ability to adapt to individual students' needs, preferences, and progress. Furthermore, the potential for remote and distance learning is amplified through robot-assisted platforms. Students in remote areas or those with limited access to traditional educational institutions can benefit from high-quality, personalized education delivered through these platforms. The integration of virtual reality and augmented reality could also further enhance the immersive and interactive nature of these learning environments.

Robot-assisted learning environments are transforming the landscape of education by providing personalized, engaging, and adaptive learning experiences. These systems empower students to learn at their own pace, develop critical skills, and engage with educational content in dynamic ways. As technology continues to evolve, the potential for these environments to revolutionize education is vast. By fostering collaboration between teachers, students, and technology, robot-assisted learning environments offer a glimpse into the future of education, where each student's learning journey is tailored for success.

Collaborative Research in Robotics Labs: Collaboration has long been recognized as a driving force behind scientific and technological advancements. In the field of robotics, collaborative research in dedicated labs plays a pivotal role in advancing innovation, expanding knowledge, and solving complex challenges. These labs bring together researchers, engineers, students, and professionals from diverse backgrounds to work collectively on cutting-edge projects that contribute to the evolution of robotics technology and its real-world applications.

The Essence of Collaborative Research Labs: Collaborative research labs in robotics provide an environment where individuals with varied expertise can collaborate on interdisciplinary projects. These labs often operate within academic institutions, research organizations, and industry hubs. Their goal is to harness the collective power of diverse minds to address complex problems that require insights from multiple disciplines, such as computer science, engineering, mathematics, and cognitive science.One of the key characteristics of collaborative robotics labs is the breaking down of traditional silos. Researchers from different disciplines come together to pool their knowledge, skills, and perspectives, resulting in innovative solutions that might not have been possible within the confines of individual research domains.

Interdisciplinary Innovation: The interdisciplinary nature of robotics labs fosters innovation by encouraging researchers to think beyond the boundaries of their respective fields. For example, a collaboration between robotics engineers and neuroscientists might lead to the development of more efficient and biomimetic robotic systems inspired by the human brain's cognitive processes. In addition, these collaborations can lead to the development of novel methodologies and approaches that bridge the gap between theoretical concepts and practical applications. For instance, combining insights from biomechanics, materials science, and robotics can result in the creation of more agile and adaptable robotic platforms for various environments.

Cross-Sectoral Collaboration: Collaborative robotics labs often extend beyond the academic sphere and engage with industry partners. This cross-sectoral collaboration is mutually beneficial. Academia gains access to real-world challenges and resources, while industry partners benefit from cutting-edge research and access to a pool of talented researchers. Industry collaboration also ensures that research conducted within the lab remains relevant and applicable to real-world scenarios. For instance, a robotics lab collaborating with manufacturing companies might focus on developing automation solutions that address specific challenges faced by the industry, thereby accelerating the adoption of robotics in manufacturing processes.

Skill Development and Education: Collaborative robotics labs also play a crucial role in educating the next generation of researchers and professionals. Students who participate in lab activities gain hands-on experience in teamwork, problem-solving, and project management. They are exposed to real-world challenges and contribute to projects that have a tangible impact on society. Furthermore, these labs create an environment that encourages mentorship and knowledge transfer between seasoned researchers and students. This dynamic fosters a culture of continuous learning and skill development, ensuring that the robotics workforce remains well-equipped to tackle emerging challenges.

Translating Research into Impact: Collaborative robotics labs emphasize the translation of research outcomes into practical applications. This focus on translational research ensures that the advancements made in the lab are not confined to academic publications but have a meaningful impact on industries and society as a whole. For instance, a collaborative lab working on healthcare robotics might develop assistive devices that improve the quality of life for individuals with mobility impairments. By actively engaging with end-users, these labs can refine and adapt their solutions to meet real-world needs, ultimately contributing to positive societal outcomes.

Challenges and Considerations: While collaborative robotics labs offer numerous benefits, they also face challenges that need to be addressed. Effective communication and coordination among researchers from diverse disciplines can be complex, requiring strong leadership and clear project management strategies. Additionally, intellectual property rights and concerns may arise when working with industry partners, necessitating well-defined agreements to protect the interests of all parties involved. Maintaining a sustainable funding model is another challenge that collaborative labs often encounter. Securing funding for interdisciplinary research and sustaining long-term projects can be demanding, but it is crucial to ensure the continuity of impactful research.

Future Prospects: The future of collaborative robotics labs is promising, with advancements in technology facilitating even more seamless interdisciplinary collaboration. Virtual collaboration tools, augmented reality, and remote communication platforms have the potential to connect researchers across geographical boundaries, enabling them to work together on shared projects. Furthermore, the integration of artificial intelligence and machine learning can enhance the efficiency of collaborative research by analyzing large datasets, identifying patterns, and providing insights that guide decision-making. These technologies can also facilitate the rapid prototyping and testing of ideas, expediting the research and development process.

Collaborative research labs in robotics represent a dynamic and effective approach to advancing innovation and knowledge. By bringing together experts from various disciplines and industries, these labs foster interdisciplinary innovation, address complex challenges, and accelerate the translation of research into practical applications. Through collaborative efforts, robotics labs contribute to the development of technologies that shape the future of industries, improve quality of life, and drive progress in various sectors. As technology continues to evolve, collaborative research in robotics remains a driving force in shaping a more interconnected and technologically advanced world.

Universities as Hubs for Innovation and Experimentation: In recent years, universities have emerged as dynamic hubs for innovation and experimentation in the field of robotics. These institutions, known for their role in education and research, have increasingly taken on the mantle of driving technological advancements and pushing the boundaries of what robots can achieve. Through their multidisciplinary approach, collaboration with industries, and emphasis on cutting-edge research, universities play a pivotal role in shaping the future of robotics technology and its diverse applications.

Multidisciplinary Synergy: One of the defining features of universities as hubs for innovation is their ability to bring together researchers, students, and experts from various disciplines. Robotics is inherently multidisciplinary, requiring insights from computer science, engineering, mathematics, biology, psychology, and more. Universities provide a unique platform where experts from these diverse fields can collaborate and share their knowledge to create comprehensive solutions.For example, a robotics project aimed at developing robotic assistance for individuals with disabilities may involve engineers designing the hardware, computer scientists developing complex algorithms, psychologists understanding human-robot interaction, and medical professionals providing insights into the specific needs of the target audience. The synergy of these disciplines often leads to breakthroughs that would be challenging to achieve in isolated research environments.

Hands-on Learning and Skill Development: Universities prioritize hands-on learning and skill development, and this approach extends to robotics education and research. Students engaged in robotics programs gain practical experience by building, programming, and testing robots. This experiential learning not only enhances their technical skills but also nurtures creativity, critical thinking, problem-solving abilities, and teamwork – all essential traits for future innovators.Moreover, universities often provide students with access to state-of-the-art labs, equipment, and facilities that are essential for conducting robotics research. This hands-on exposure empowers students to experiment, innovate, and turn their theoretical knowledge into tangible projects that contribute to the field.

Industry Collaboration and Real-world Applications: Universities foster strong ties with industries, creating opportunities for students and researchers to collaborate on real-world projects. Industry partnerships offer several benefits – from access to resources and funding to exposure to practical challenges and current trends. These collaborations often lead to the development of robotics solutions that address industry-specific needs, such as automation in manufacturing, precision agriculture, healthcare robotics, and more.Furthermore, industry partnerships provide valuable insights into the practical applicability of robotics technologies, helping researchers align their work with industry demands. This synergy between academia and industry not only advances the field but also ensures that students are equipped with skills that align with the requirements of the job market.

Incubators and Start-up Ecosystems: Many universities have established innovation centers, incubators, and start-up ecosystems that provide a nurturing environment for entrepreneurial ventures in robotics. These platforms offer aspiring innovators the resources, mentorship, and infrastructure needed to transform their ideas into viable products and services. The collaborative atmosphere of universities encourages students and researchers to explore commercial applications for their robotics projects and contributes to the growth of the robotics industry.Start-ups originating from university incubators often develop groundbreaking technologies that find applications across sectors, ranging from healthcare and agriculture to manufacturing and entertainment. Notable success stories include companies that have developed innovative robotic exoskeletons, autonomous vehicles, and advanced surgical robots.

Addressing Societal Challenges: Universities are uniquely positioned to address some of the pressing societal challenges through robotics research and innovation. For instance, researchers are developing robots that aid in disaster response, search and rescue operations, environmental monitoring, and humanitarian missions. By leveraging robotics technology, universities contribute to finding solutions to complex global issues while providing valuable experiential learning opportunities for students.

Ethical Considerations and Responsible Innovation: As universities drive innovation in robotics, they also address the ethical considerations and societal implications associated with these technologies. Roboethics, a field that explores the ethical challenges posed by robotics and AI, gains prominence within academic institutions. Researchers and scholars in this field investigate issues related to privacy, autonomy, accountability, bias, and human-robot interaction. By engaging in these discussions, universities ensure that technological advancements are aligned with ethical values and contribute to responsible and sustainable innovation.

Public Engagement and Outreach: Universities serve as conduits for public engagement and outreach in robotics. They organize workshops, exhibitions, public lectures, and interactive demonstrations that educate the public about robotics technology and its potential impact on society. These initiatives not only foster interest and curiosity among the general public but also promote the exchange of ideas between researchers and the broader community.

Universities as hubs for innovation and experimentation are essential driving forces in the field of robotics. Through multidisciplinary collaboration, hands-on learning, industry partnerships, start-up incubators, and addressing societal challenges, universities play a crucial role in shaping the future of robotics technology. As they educate the next generation of robotics leaders, researchers, and professionals, universities contribute to advancements that impact industries, society, and everyday life. By embracing an interdisciplinary approach and fostering an ecosystem of innovation, universities ensure that the realm of robotics continues to evolve and redefine the boundaries of technological possibilities.

Exercise

Multiple-choice Questions

1. What is the central focus of "Robotics in Everyday Life"?
a) Historical evolution of robotics technology
b) Ethical considerations in artificial intelligence
c) Economic impact of automation in industries
d) The role of robots in healthcare
2. What is a primary feature of universities as hubs for innovation in robotics?
a) Isolation from other disciplines
b) Lack of collaboration with industries
c) Limited access to state-of-the-art labs
d) Multidisciplinary collaboration
3. How do universities contribute to nurturing creativity and critical thinking in robotics?
a) By providing theoretical lectures only
b) Through isolated research environments
c) By emphasizing hands-on learning and skill development
d) By focusing solely on technical skills
4. What role do universities play in fostering collaboration between students and industries?
a) Universities discourage collaboration with industries
b) Universities limit access to resources and funding
c) Universities establish partnerships that offer practical exposure and industry insights
d) Universities focus solely on theoretical research
5. What benefits do university-industry collaborations provide for robotics students and researchers?
a) Limited access to resources
b) Exposure to practical challenges and trends
c) Isolation from industry demands
d) Solely theoretical research opportunities
6. What type of projects often result from collaborations between universities and industries in robotics?
a) Projects that align with theoretical research goals
b) Projects with no real-world applications
c) Projects that address industry-specific needs
d) Projects with no commercial value

7. How do university innovation centers and incubators support entrepreneurial ventures in robotics?
a) By offering theoretical lectures
b) By limiting access to resources
c) By providing mentorship and infrastructure
d) By focusing solely on research projects
8. What impact do university start-up ecosystems have on the growth of the robotics industry?
a) They hinder the growth of the industry
b) They discourage innovation and experimentation
c) They provide resources for entrepreneurs to transform ideas into viable products
d) They focus solely on theoretical research
9. What societal challenges can universities address through robotics research and innovation?
a) Challenges related to theoretical knowledge
b) Challenges that are unrelated to technology
c) Challenges related to disaster response, humanitarian missions, and environmental monitoring
d) Challenges that are beyond the scope of robotics
10. What is the field that explores ethical challenges posed by robotics and AI?
a) Techno ethics
b) Roboethics
c) Artificial Intelligence Ethics
d) Mechanical Ethics
11. What do universities ensure by engaging in discussions related to robethics?
a) That technological advancements are solely focused on industry needs
b) That ethical considerations are disregarded in robotics research
c) That technological advancements align with ethical values
d) That ethical considerations hinder technological progress
12. What role do universities play in public engagement and outreach in robotics?
a) They discourage public engagement and outreach initiatives
b) They solely focus on theoretical research
c) They organize workshops, exhibitions, public lectures, and interactive demonstrations
d) They exclude the broader community from discussions on robotics
13. What is the significance of universities as conduits for public engagement in robotics?
a) They limit the exchange of ideas between researchers and the public
b) They discourage interest and curiosity among the general public
c) They foster public interest in robotics technology and its societal impact
d) They only cater to the needs of researchers and experts
14. What is the overarching impact of universities as hubs for innovation and experimentation in robotics?
a) Limited impact on the robotics field
b) Negligible influence on technology advancements
c) Significant contributions to shaping the future of robotics technology
d) Sole focus on theoretical research without practical implications
15. What do university start-up ecosystems provide for aspiring innovators?
a) Mentorship and infrastructure
b) Solely theoretical knowledge
c) Limited access to resources
d) Isolation from industry demands

Review Question:

1. What is the overarching focus of "Robotics in Everyday Life"?
2. How do domestic robots contribute to enhancing daily living for individuals and families?
3. Explain the concept of assistive robotics and its significance in improving the quality of life for people with disabilities.
4. What role do educational institutions play in creating robot-assisted learning environments?
5. How do robot-assisted learning environments enhance the educational experience for students?
6. Describe the key features and benefits of collaborative research in robotics labs.
7. How do collaborative research initiatives in robotics labs contribute to advancing the field of robotics?
8. Explain the concept of universities as hubs for innovation and experimentation in robotics.
9. What is the significance of universities in fostering creativity and critical thinking in robotics?

10. How do university-industry collaborations in robotics benefit both students and industries?
11. Describe the role of universities in bridging the gap between research and practical applications in robotics.
12. How do universities address societal challenges through robotics research and innovation?
13. What is the importance of ethical considerations in robotics research and development?
14. Explain how universities engage in discussions related to Roboethics and societal implications of robotics.
15. How do universities contribute to public engagement and outreach in the field of robotics?
16. What is the impact of universities as hubs for innovation and experimentation in robotics?
17. Describe the role of university-based start-up ecosystems in fostering innovation and entrepreneurship in robotics.
18. How do universities nurture creativity among students pursuing robotics education?
19. Explain the concept of democratizing robotics education and its implications.
20. What are the key takeaways from"Robotics in Everyday Life," regarding the integration of robotics into various aspects of society?

Answers Multiple-choice Questions

Answers:
1. b)
2. d)
3. c)
4. c)
5. b)
6. c)
7. c)
8. c)
9. c)
10. b)
11. c)
12. c)
13. c)
14. c)
15. a)

Review Question:

1. "Robotics in Everyday Life," is to explore how robotics technology has become an integral part of our daily lives, transforming various aspects of society and enhancing the way we live, learn, and work. The chapter delves into the diverse applications of robotics, ranging from domestic robots that assist with household tasks to the role of universities in driving innovation and experimentation in the field of robotics. The chapter aims to provide a comprehensive understanding of how robotics has become an everyday phenomenon that impacts multiple facets of our existence.
2. Domestic robots contribute significantly to enhancing daily living by providing assistance and convenience in various tasks. These robots can perform household chores such as cleaning, cooking, and even companionship for elderly individuals. They reduce the burden of repetitive tasks, freeing up time and energy for individuals to focus on other activities. Moreover, domestic robots can be programmed to adapt to specific preferences and schedules, providing a personalized and efficient experience that improves the quality of life for individuals and families.
3. Assistive robotics refers to the utilization of robotic technology to enhance the lives of people with disabilities. These robots are designed to provide assistance in various ways, such as helping with mobility, communication, and everyday tasks. The significance of assistive robotics lies in its potential to empower individuals with disabilities, enabling greater independence and participation in daily life. By bridging the gap between technology and accessibility, assistive robots contribute to a more inclusive and equitable society.
4. Robot-assisted learning environments create educational settings where robots play a role in enhancing the learning experience. These environments utilize robots to facilitate interactive and engaging activities that promote active learning. The integration of robotics in education introduces students to new ways of learning, problem-solving, and collaboration. Robots can provide immediate feedback, adapt to individual learning styles, and create dynamic and immersive learning experiences that go beyond traditional teaching methods.
5. Collaborative research in robotics labs is characterized by the cooperation of researchers from diverse disciplines to address complex challenges in the field of robotics. This collaborative approach brings together experts with different expertise, allowing for the exchange of ideas, perspectives, and innovative solutions. By pooling resources

and knowledge, collaborative research accelerates advancements in robotics and fosters interdisciplinary breakthroughs that may not have been achievable through isolated research efforts.

6. Universities hold a pivotal role as hubs for innovation and experimentation in robotics. They serve as centers where cutting-edge research, development, and testing of robotic technologies take place. Universities facilitate innovation by providing a conducive environment for creative thinking, collaboration, and the exploration of novel concepts. Additionally, universities offer the resources, expertise, and academic freedom required for researchers and students to push the boundaries of robotics, leading to the creation of groundbreaking technologies and solutions.

7. University-industry collaborations in robotics benefit both students and industries in several ways. For students, these collaborations provide opportunities for real-world experiences, exposure to industry practices, and the chance to work on meaningful projects that have practical applications. Industries benefit from the fresh perspectives and innovative ideas that students bring, as well as the potential to identify future talent for their workforce. This collaboration enhances the relevance of academic research and promotes technology transfer from academia to industry.

8. The role of universities in bridging the gap between research and practical applications in robotics is multifaceted. Universities are uniquely positioned to conduct foundational research that underpins advancements in the field. They also offer the space for experimentation, testing, and validation of new robotic technologies. Furthermore, universities contribute to the education and training of the future workforce, ensuring that students are equipped with the skills and knowledge needed to develop, deploy, and maintain robotics solutions in various industries.

9. Universities address societal challenges through robotics research and innovation by leveraging their expertise to develop solutions that address pressing issues. For example, in healthcare, universities collaborate with medical professionals to create robots that assist in surgeries, diagnostics, and patient care. In disaster response, universities work on robots that aid in search and rescue missions. By focusing their efforts on challenges that impact society, universities contribute to positive change and improve the quality of life for individuals and communities.

10. Ethical considerations play a crucial role in robotics research and development. Universities engage in discussions related to Roboethics and societal implications to ensure that advancements in robotics align with ethical standards and values. These discussions encompass a wide range of topics, including the potential consequences of robotics on employment, privacy, safety, and the overall well-being of society. Universities encourage thoughtful reflection on these issues to guide responsible innovation and minimize unintended negative impacts.

11. Universities engage in public engagement and outreach activities related to robotics to foster understanding, awareness, and involvement within the community. These activities can include public lectures, workshops, demonstrations, and collaborative projects with schools and local organizations. By sharing knowledge and showcasing the potential benefits of robotics, universities promote a positive image of the field, inspire future generations, and encourage public participation in shaping the direction of robotics technology.

12. University-based start-up ecosystems play a crucial role in fostering innovation and entrepreneurship in robotics. These ecosystems provide an environment where students, researchers, and faculty members can explore and develop their entrepreneurial ideas. Universities offer access to resources such as mentorship, funding, and infrastructure, enabling aspiring entrepreneurs to transform their robotics-related concepts into viable businesses. This contributes to economic growth, job creation, and the commercialization of robotic technologies.

13. The integration of robotics in everyday life introduces new challenges related to human-robot interaction, privacy, security, and ethical considerations. Ensuring the safety of users, protecting personal data, and preventing potential misuse of robotics technology become paramount concerns. Furthermore, addressing the potential job displacement due to automation and the ethical implications of autonomous decision-making by robots requires careful consideration and well-defined regulations.

14. The economic impact of robotics extends to various sectors, including manufacturing, healthcare, and service industries. The automation of routine and repetitive tasks can lead to increased efficiency, reduced operational costs, and improved product quality. However, the economic transformation brought about by robotics also necessitates measures to address job displacement, reskilling of the workforce, and the equitable distribution of benefits to prevent socio-economic inequalities.

15. The societal implications of robotics encompass both positive and negative aspects. On one hand, robotics technology enhances productivity, convenience, and the quality of life. On the other hand, concerns arise regarding job displacement, ethical dilemmas, and potential threats to privacy. Universities play a vital role in addressing these implications by conducting research, engaging in public dialogue, and guiding the responsible development and deployment of robotics technology to benefit society as a whole.

16. The ethical considerations in robotics are crucial for guiding responsible research and development. Universities encourage discussions about the impact of robotics on human values, social norms, and the overall well-being of

individuals and communities. Ethical considerations encompass areas such as robot autonomy, accountability, transparency, and the potential for biases in decision-making algorithms. By prioritizing ethical considerations, universities contribute to the creation of robotics technologies that align with societal values.

17. Collaborative research in robotics labs facilitates cross-disciplinary collaboration among researchers from different academic fields. This collaborative approach enhances the understanding of complex challenges and encourages the integration of diverse expertise to develop comprehensive solutions. By fostering an environment of knowledge sharing, innovation, and synergy, collaborative robotics labs accelerate advancements and contribute to the holistic development of robotics technology.

18. The economic impact of robotics is multifaceted, encompassing both positive and potentially disruptive effects. While robotics enhances productivity, streamlines operations, and leads to the creation of new job roles, it also raises concerns about job displacement in certain sectors. The transformation of industries due to automation necessitates a strategic approach to reskilling the workforce, promoting innovation, and ensuring equitable distribution of economic benefits.

19. Universities serve as hubs for innovation and experimentation in robotics by providing a fertile ground for creative thinking, research, and development. These institutions create an environment where students and researchers can explore novel ideas, collaborate across disciplines, and push the boundaries of robotics technology. By encouraging experimentation and innovation, universities contribute to the advancement of robotic solutions that address real-world challenges.

20. The collaboration between universities and industries in the field of robotics benefits both parties in various ways. Universities contribute their expertise in research, education, and innovation, while industries offer resources, practical challenges, and opportunities for technology transfer. This collaboration accelerates the translation of research findings into practical applications, stimulates economic growth, and ensures that academia and industry work together to shape the future of robotics technology.

Online Resources

Online Exam Papers

Link:https://forms.gle/sfa7UQ5uuhi25yWf7

QR Code:

YouTube Video Playlist Link

Link:https://www.youtube.com/playlist?list=PLzlh_31VWEFrVeluA8_ZoRDBgI-Y9PWGk

QR Code:

Chapter 18: Top of Form Robotics beyond Earth: Extraterrestrial Exploration

The realm of robotics has extended its reach beyond the confines of Earth, venturing into the vast expanse of space to facilitate extraterrestrial exploration. As technological advancements continue to evolve, robots have emerged as indispensable tools for investigating and navigating distant celestial bodies, unlocking new frontiers of knowledge and understanding. This chapter delves into the captivating domain of "Robotics beyond Earth: Extraterrestrial Exploration," where machines, engineered with ingenuity and precision, play an instrumental role in unraveling the mysteries of the cosmos.

The exploration of space poses numerous challenges that human astronauts alone cannot overcome. Extreme environments, radiation exposure, and vast distances demand innovative solutions, and robotics offers an ideal avenue to address these challenges. Robotic systems can be designed to withstand harsh conditions, operate autonomously in remote locations, and collect data that would otherwise be inaccessible to human explorers. Moreover, robotics reduces the risks associated with human missions, allowing researchers to gather critical information before potentially committing human lives to interplanetary travel.

One of the most significant achievements in extraterrestrial exploration is the deployment of robotic rovers on the surface of Mars. The Mars rovers, such as Spirit, Opportunity, Curiosity, and Perseverance, have revolutionized our understanding of the Martian landscape, geology, and potential habitability. These rovers are equipped with sophisticated instruments that analyze soil samples, study rock formations, and even search for signs of past or present microbial life. The intricate design of these robots enables them to traverse rugged terrains, withstand temperature fluctuations, and communicate with Earth-based mission control centers.

Beyond Mars, other celestial bodies have also captured the attention of robotic explorers. The moon, for instance, has been the target of several robotic missions. Lunar rovers have been deployed to study the moon's surface, analyze its composition, and gather data that could be invaluable for future human missions. These rovers, equipped with cameras and scientific instruments, have expanded our knowledge of Earth's natural satellite and paved the way for further lunar exploration.

Another groundbreaking area of extraterrestrial robotics is the exploration of asteroids and comets. These celestial bodies offer insights into the formation of our solar system and the potential presence of valuable resources. Robotic spacecraft, such as the OSIRIS-REx mission, are designed to collect samples from asteroids and return them to Earth for analysis. Such missions provide a unique opportunity to study materials that have remained relatively unchanged since the birth of our solar system.

The deployment of robotic landers and orbiters on distant planets and moons has also been instrumental in expanding our understanding of the cosmos. These robots provide detailed observations of planetary atmospheres, geological features, and magnetic fields, contributing to the broader field of planetary science. The use of robotics in space exploration has led to groundbreaking discoveries, such as the identification of subsurface oceans on moons like Europa and Enceladus, raising intriguing possibilities for the existence of extraterrestrial life.

Additionally, robotics technology has facilitated the assembly and maintenance of space infrastructure. Space telescopes like the Hubble Space Telescope and the upcoming James Webb Space Telescope are equipped with robotic mechanisms that enable precise adjustments and repairs, ensuring their optimal functionality and longevity. These telescopes provide unprecedented views of the universe, offering insights into distant galaxies, star formation, and the origins of the cosmos.

The success of robotic missions in extraterrestrial exploration can be attributed to meticulous planning, engineering excellence, and effective communication. Mission control centers on Earth operate as nerve centers, remotely guiding robotic spacecraft, rovers, and landers across millions of miles of space. These centers orchestrate complex maneuvers, interpret data streams, and make real-time decisions to ensure the success of each mission.

While robotics has revolutionized our understanding of the cosmos, it also raises ethical considerations. The potential for contamination of extraterrestrial environments by Earth-based microbes carried by robotic spacecraft is a topic of concern. Measures are taken to minimize the risk of contamination and preserve the integrity of potential biospheres on other planets.

In conclusion, the exploration of space has been dramatically transformed by the integration of robotics. From the surface of Mars to the far reaches of our solar system, robotic missions have expanded human knowledge, revealed mysteries of the universe, and laid the foundation for future space exploration. These ingenious machines are the pioneers of the cosmos, extending the reach of human curiosity and enabling us to explore realms that were once deemed unreachable. As technology advances, the relationship between robotics and extraterrestrial exploration will continue to evolve, pushing the boundaries of our understanding and reshaping our perspective of the universe

Rovers on Mars, Moon Landers, and Beyond: The exploration of celestial bodies beyond Earth has been greatly enhanced by the deployment of robotic spacecraft, rovers, and landers. These remarkable machines play a pivotal role in advancing our understanding of the universe, unraveling mysteries of distant planets, moons, asteroids, and comets. This chapter delves into the captivating domain of "Rovers on Mars, Moon Landers, and Beyond," highlighting the achievements, challenges, and significance of robotic exploration in outer space.

Rovers on Mars: Mars, often referred to as the "Red Planet," has captivated human imagination for centuries due to its potential as a habitat for life and its geological similarities to Earth. Robotic rovers have emerged as vital tools for exploring Mars' surface, conducting scientific research, and transmitting valuable data back to Earth. The legacy of Mars rovers began with Sojourner in 1997 and has continued with Spirit, Opportunity, Curiosity, and the latest addition, Perseverance.These rovers are marvels of engineering, equipped with a suite of instruments that enable them to analyze soil samples, study rock formations, measure atmospheric conditions, and even search for signs of past or present microbial life. The success of the Mars rovers lies in their adaptability, as they traverse challenging terrains, overcome steep slopes, and endure temperature fluctuations. Their scientific findings have transformed our understanding of Mars' history, geology, and potential habitability.

Moon Landers: The moon, Earth's natural satellite, has been a focal point of exploration since the Space Age began. While human missions have left a lasting footprint on the lunar surface, robotic landers have also played a crucial role in uncovering its mysteries. Lunar landers provide valuable data about the moon's composition, geology, and potential resources. They have helped scientists understand the moon's formation and its geological evolution.Robotic lunar missions, such as the Change program by China and the Luna program by the Soviet Union, have collected samples, studied moonquakes, and investigated the moon's thin exosphere. These missions have paved the way for human exploration and resource utilization on the moon. Furthermore, advancements in technology have enabled the deployment of landers that can explore regions that were previously inaccessible to human missions.

Beyond Mars and the Moon: Robotic exploration extends far beyond Mars and the moon, reaching distant celestial bodies such as asteroids and comets. These missions provide insights into the origins of our solar system and the potential presence of valuable resources. Notable examples include the OSIRIS-REx mission, which collected samples from asteroid Bennu, and the European Space Agency's Rosetta mission, which orbited comet 67P/Churyumov-Gerasimenko.The study of asteroids and comets not only contributes to our understanding of planetary formation but also offers a glimpse into the conditions that existed in the early solar system. By analyzing the composition and structure of these celestial bodies, scientists can piece together the puzzle of our cosmic origins and gain insights into the building blocks of life.

Challenges and Significance: While robotic exploration has yielded remarkable discoveries, it is not without its challenges. The vast distances between Earth and these celestial bodies introduce communication delays, making real-time control of robots challenging. Autonomous systems are employed to enable robots to make decisions based on pre-programmed instructions and environmental data. However, autonomy must strike a balance between scientific exploration and the need for human intervention when unexpected situations arise.Another challenge is ensuring the sterilization of robotic spacecraft to prevent the contamination of extraterrestrial environments with Earth-based microbes. This is particularly important in the search for potential signs of life on other planets or moons. Stringent protocols are in place to ensure that robotic missions do not compromise the integrity of future scientific investigations.The significance of robotic exploration goes beyond scientific discovery. These missions inspire curiosity, capture the imagination of people worldwide, and foster international collaboration. The data collected by robotic spacecraft contribute to a growing body of knowledge that shapes our understanding of the cosmos and our place within it. Furthermore, the technologies developed for robotic missions have applications in various industries, from aerospace to healthcare.

In conclusion, the deployment of robotic spacecraft, rovers, and landers has revolutionized our understanding of the universe. From Mars to distant comets, these machines have expanded human knowledge, uncovered celestial mysteries, and paved the way for future human exploration. Their ability to withstand harsh conditions, collect data, and transmit information back to Earth has transformed the way we explore and study outer space. As technology continues to advance, robotic exploration will continue to evolve, revealing new insights into the cosmos and inspiring generations to reach for the stars.

Collecting Samples and Analyzing Extraterrestrial Terrain: The quest to understand the composition and history of celestial bodies beyond Earth has driven the development of innovative robotic missions aimed at collecting samples and analyzing extraterrestrial terrain. This endeavor has not only advanced our understanding of the universe but also pushed the boundaries of technology and engineering. In this context, this delves into the fascinating domain of "Collecting Samples and Analyzing Extraterrestrial Terrain," highlighting the significance, challenges, and achievements of these endeavors.

Sample Collection Missions: Collecting samples from extraterrestrial bodies is akin to unearthing treasures from distant worlds. Robotic missions designed for sample collection offer unique insights into the geological, chemical, and physical properties of celestial bodies. These samples serve as time capsules, preserving crucial information about the history and evolution of the solar system. One of the pioneering sample collection missions is NASA's OSIRIS-REx, which targeted the near-Earth asteroid Bennu. The spacecraft successfully collected samples from the asteroid's surface, offering a glimpse into the early solar system's building blocks. Similarly, Japan's Hayabusa2 mission collected samples from the asteroid Ryugu, providing researchers with valuable material to study the formation of planets and the origins of life.

Analyzing Extraterrestrial Terrain: The analysis of extraterrestrial terrain involves studying surface features, geological formations, and compositional characteristics to unravel the mysteries of celestial bodies. Robotic landers, rovers, and orbiters are equipped with a suite of instruments that enable scientists to conduct detailed investigations from afar.The Curiosity rover on Mars exemplifies the success of analyzing extraterrestrial terrain. Equipped with tools such as cameras, spectrometers, and drills, Curiosity has provided valuable insights into Mars' geology, climate history, and potential habitability. By analyzing sedimentary rock layers and mineral compositions, scientists have uncovered evidence of past water presence on the Martian surface, shedding light on its potential as a habitat for life.

Challenges and Technological Innovations: Collecting samples and analyzing extraterrestrial terrain presents a myriad of challenges, from the harsh conditions of space to the limitations of communication. These challenges drive the development of cutting-edge technologies and engineering solutions. The sample collection process, for instance, requires precision and ingenuity. Spacecraft must execute intricate maneuvers to approach and interact with their target bodies. Sophisticated robotic arms, drills, and collectors are designed to secure samples without compromising their integrity. Furthermore, these samples must be carefully preserved and returned to Earth for thorough analysis.Analyzing extraterrestrial terrain from afar demands sophisticated instruments and imaging techniques. Spectrometers, cameras, and other sensors must be optimized for space environments, and data must be transmitted across vast distances. Autonomy is crucial for these missions, as real-time control is often unfeasible due to communication delays.

Scientific Significance and Discoveries: The scientific significance of collecting samples and analyzing extraterrestrial terrain is profound. These missions offer insights into the formation of planets, the origins of life, and the evolution of the solar system. By studying the chemical compositions of asteroids, moons, and planets, researchers can unravel the processes that shaped our cosmic neighborhood. One notable example is the analysis of samples from the moon. The Apollo missions brought back lunar samples that revolutionized our understanding of lunar geology and the moon's history. Similarly, samples from asteroids and comets provide insights into the early solar system's conditions and the building blocks of organic molecules.

Beyond Scientific Inquiry: While the primary focus of these missions is scientific inquiry, their impact extends beyond the realm of research. Collecting samples and analyzing extraterrestrial terrain inspires generations of scientists, engineers, and explorers. These missions captivate public imagination, fostering a sense of wonder and curiosity about the universe. Additionally, the technologies developed for these missions have practical applications on Earth. Innovations in robotics, materials science, imaging, and communication can find uses in various industries, from healthcare to manufacturing. The challenges faced in the extreme environments of space drive the advancement of engineering solutions that have real-world benefits.

The endeavor to collect samples and analyze extraterrestrial terrain represents the epitome of scientific exploration and human curiosity. These missions have redefined our understanding of the universe, uncovering clues about the origins of planets, moons, and asteroids. From spacecraft collecting particles from asteroids to rovers examining the landscapes of Mars, these endeavors have pushed the boundaries of technology and inspired future generations to reach for the stars. As technology continues to evolve, sample collection and terrain analysis will play an increasingly crucial role in

Contributions to Our Understanding of the Cosmos: This topic delves into the remarkable realm of "Contributions to Our Understanding of the Cosmos" through robotic missions beyond Earth. This segment encapsulates the profound significance of these missions, their impact on our comprehension of the universe, and the unprecedented insights they have brought to humanity's understanding of cosmic phenomena.

Revolutionizing Astronomical Observations:

Robotic missions have revolutionized our ability to observe and study celestial bodies, unveiling cosmic wonders that were previously hidden from our view. Telescopes placed in space, such as the Hubble Space Telescope, have captured breathtaking images of distant galaxies, nebulae, and supernovae. These observations have deepened our understanding of the universe's vastness, its evolution, and the forces that shape it.

Additionally, space-based telescopes have the advantage of operating above Earth's atmosphere, which can distort and obscure observations. This enables astronomers to obtain clearer and more detailed images, leading to groundbreaking discoveries about the formation of galaxies, the nature of dark matter and dark energy, and the birth and death of stars.

Probing Exoplanetary Systems: The discovery of exoplanets, planets located beyond our solar system, has been one of the most significant contributions of robotic missions to our understanding of the cosmos. Space telescopes like Kepler and TESS (Transiting Exoplanet Survey Satellite) have identified thousands of exoplanets through the transit method, where a planet passing in front of its host star causes a temporary dimming of the star's light.These discoveries have ignited new avenues of research, including the search for potentially habitable exoplanets and the study of their atmospheres for signs of life. The identification of exoplanets with diverse characteristics challenges our preconceived notions about planetary formation and raises questions about the prevalence of life elsewhere in the universe.

Profound Insights into Cosmic Evolution: Robotic missions to distant galaxies have provided profound insights into the evolution of the cosmos. The Sloan Digital Sky Survey (SDSS), for instance, has mapped millions of galaxies across a large portion of the sky. By studying the distribution of galaxies, astronomers have gained a better understanding of the large-scale structure of the universe, the expansion rate of space, and the nature of dark matter and dark energy.Moreover, robotic surveys have revealed the intricate interplay between galaxies, gas, and cosmic filaments. These observations have led to a deeper appreciation of how galaxies form, evolve, and interact over cosmic timescales. This, in turn, has enhanced our understanding of the underlying physical processes that shape the universe.

Probing the Cosmic Microwave Background: The cosmic microwave background (CMB) radiation is a faint glow of microwave radiation that permeates the universe. It is the afterglow of the Big Bang and provides a snapshot of the universe's state when it was only a few hundred thousand years old. Robotic missions such as the Planck satellite have measured the CMB's temperature fluctuations with unprecedented precision.The analysis of CMB data has led to breakthroughs in cosmology, confirming the inflationary theory of the universe's rapid expansion in its early moments. It has also provided precise measurements of the universe's age, composition, and geometry. By studying the CMB, scientists can glean insights into the conditions of the universe shortly after its birth and refine our understanding of its evolution.

Mapping the Cosmic Web: The cosmic web is a complex structure composed of galaxies, dark matter, and intergalactic gas that forms the framework of the universe. Robotic surveys have contributed to mapping the intricate structure of the cosmic web, revealing how galaxies are interconnected by vast cosmic filaments and clusters.These mappings have enabled astronomers to study the distribution of matter in the universe and uncover its large-scale patterns. The identification of cosmic voids and superclusters has shed light on the underlying forces that drive the universe's expansion and the growth of cosmic structures.

The contributions of robotic missions to our understanding of the cosmos are nothing short of revolutionary. Through the lens of space-based telescopes, the unveiling of exoplanets, the probing of cosmic microwave background radiation, and the mapping of the cosmic web, these missions have provided unprecedented insights into the universe's evolution, structure, and fundamental properties. As technology continues to advance, these missions will undoubtedly continue to reshape our understanding of the cosmos, deepening our connection to the universe and our place within it.

Robotic Repair and Maintenance of Satellites: In the realm of "Robotic Repair and Maintenance of Satellites," a critical aspect of space exploration and satellite operations. This segment highlights the significance of robotic technologies in extending the lifespan of satellites, ensuring their functionality, and mitigating the challenges posed by space debris and other hazards.

Satellites: Pillars of Modern Infrastructure: Satellites have become integral to modern life, serving various functions such as communication, navigation, weather forecasting, and scientific research. These artificial celestial bodies orbit Earth and provide essential services that impact various industries and sectors. However, the harsh conditions of space, including extreme temperatures, radiation, and micrometeoroid impacts, can take a toll on satellite systems over time.

Challenges of Satellite Maintenance: Satellites are complex systems with intricate mechanisms, electronics, and sensitive instruments. Despite meticulous design and engineering, wear and tear, as well as technical malfunctions, can occur due to the harsh space environment. The challenge lies in ensuring the longevity and functionality of satellites, especially those in geostationary and higher orbits, where retrieval and repair missions are prohibitively difficult.

Enter Robotic Repair and Maintenance: Robotic technologies have emerged as promising solutions to address the challenges associated with satellite maintenance. Robotic systems equipped with advanced sensors, manipulators, and propulsion mechanisms can perform intricate tasks in space, including servicing, repairing, and upgrading satellites. These robots can be remotely controlled from Earth or operate autonomously using pre-programmed algorithms.

Robotic Servicing Missions: Robotic servicing missions involve sending specialized robots to rendezvous with malfunctioning or aging satellites in space. Once in proximity, these robots can perform a range of tasks, such as replacing damaged components, deploying solar panels, refueling propulsion systems, and conducting diagnostic tests. These actions extend the operational lifespan of satellites and maximize their efficiency.One notable example is the Robotic Refueling Mission (RRM), developed by NASA. RRM demonstrated the feasibility of refueling satellite propellant tanks in space, a capability that can significantly extend the operational life of satellites and reduce the need for costly replacements.

Dealing with Space Debris: Space debris, consisting of defunct satellites, spent rocket stages, and other fragments, poses a threat to operational satellites. Collisions with even small debris particles can lead to catastrophic damage. Robotic technologies offer a way to mitigate this hazard by removing or redirecting space debris. Concepts like the RemoveDEBRIS mission, which deployed a net to capture debris, exemplify the potential of robotics in reducing space clutter.

Astronaut Assistants: Robotic systems can also serve as assistants to astronauts during extravehicular activities (EVAs) or spacewalks. These robots, often called Extra-Vehicular Robotic Systems (EVRs), are equipped with arms, cameras, and tools that aid astronauts in performing intricate tasks. EVRs can enhance astronaut safety and efficiency during missions, such as repairing solar arrays or conducting maintenance on the International Space Station (ISS).

In-Orbit Manufacturing and Assembly: The concept of in-orbit manufacturing and assembly involves using robotic systems to build and assemble satellites or large structures in space. This approach eliminates the need to launch fully assembled satellites from Earth, reducing launch costs and enabling the creation of more complex and capable systems. It also allows for modular upgrades and repairs, extending the lifespan of space assets.

Ethical Considerations and Challenges: While robotic repair and maintenance of satellites offer promising solutions, they also raise ethical considerations. For instance, there is a debate about whether satellites should be designed with modular components to facilitate repair or whether the focus should be on designing longer-lasting satellites. Additionally, ensuring that robots do not inadvertently create more space debris during repair missions is a critical challenge.

Robotic repair and maintenance of satellites represent a pivotal advancement in space technology. These robotic systems play a vital role in extending the operational lifespan of satellites, ensuring their functionality, and addressing the challenges posed by space debris. As technology continues to evolve, robotic repair and maintenance capabilities will likely become increasingly sophisticated, contributing to a more sustainable and efficient use of space resources.

Building and Assembling Space Structures In-Situ: In the fascinating world of "Building and Assembling Space Structures In-Situ," a cutting-edge concept that holds the potential to revolutionize space exploration and utilization. This segment explores how robotic technologies are paving the way for constructing large-scale space structures directly in space, eliminating the need for launching fully assembled structures from Earth and unlocking new possibilities for space missions and endeavors.

Introduction to In-Situ Space Structure Assembly: The idea of building and assembling structures in space, often referred to as "in-situ construction," is a paradigm shift in space exploration. Traditionally, space missions required launching pre-assembled structures, satellites, and equipment from Earth, which presented significant challenges in terms of launch costs, size limitations, and the complexity of deployment. In-situ assembly aims to address these challenges by leveraging robotic systems and advanced engineering techniques to construct structures directly in space.

Advantages of In-Situ Construction: In-situ space structure assembly offers a range of advantages that can transform the way we approach space missions:

Cost Efficiency: Launching fully assembled structures from Earth is expensive and limits the size and complexity of missions. In-situ construction reduces launch costs by sending compact components and materials to space, which are then assembled into larger structures.

Size and Complexity: In-situ assembly enables the creation of structures that exceed the payload capacity of existing launch vehicles. This opens the door to building large telescopes, habitats, research stations, and other ambitious projects.

Adaptability and Customization: Robots can assemble structures in space using modular components, allowing for adaptability, upgrades, and customization. This is particularly valuable for missions with evolving requirements.

Reduced Launch Risk: Launching fully assembled structures carries inherent risks. In-situ construction eliminates the need for intricate folding mechanisms, deployment systems, and potential points of failure during launch.
Resource Utilization: In-situ construction leverages resources available in space, such as asteroids, lunar soil, or extracted materials, to build structures. This promotes sustainability and reduces the reliance on Earth's resources.
Robotic In-Situ Assembly Techniques: Robotic technologies are at the heart of in-situ space structure assembly. Various methods and approaches are being explored to enable the construction of space structures:
Swarm Robotics: Robotic swarms consist of multiple small robots working collaboratively to perform tasks. These swarms can assemble structures by coordinating their movements, attaching components, and ensuring precise alignment.
3D Printing and Additive Manufacturing: Additive manufacturing techniques allow robots to create structures layer by layer using materials like metal, plastic, or lunar regolith. This approach is being studied for building habitats, tools, and infrastructure.
Teleoperation and Autonomy: Robotic systems can be teleoperated from Earth or programmed with autonomy to perform assembly tasks. Advanced AI algorithms enable robots to adapt to changing conditions and unforeseen challenges.
Modular Assembly: Components are sent to space in compact form and assembled into larger structures using robotic arms or specialized mechanisms. This modular approach facilitates flexibility and adaptability.
Utilization of Local Resources: Robots can mine resources from celestial bodies like the Moon or asteroids to create construction materials. This reduces the need to transport materials from Earth and enhances sustainability.
Potential Applications and Impact: The possibilities enabled by in-situ space structure assembly are vast:
Space Habitats: In-situ construction can facilitate the creation of habitable structures on the Moon, Mars, or other celestial bodies. These habitats can support long-duration missions and colonization efforts.
Telescopes and Observatories: Large-scale telescopes and observatories can be built in space, free from Earth's atmospheric interference, enabling unprecedented insights into the cosmos.
Research Stations: In-situ assembly allows for the construction of research stations in unique locations, such as lunar lava tubes or the surface of asteroids, enabling scientific exploration.
Resource Utilization: The ability to extract and process resources in space for construction materials opens avenues for sustainable resource utilization and reduced reliance on Earth's resources.
Challenges and Considerations: In-situ space structure assembly also presents challenges:
Robust Robotics: Developing reliable and robust robotic systems capable of working in the harsh space environment is essential for successful in-situ construction.
Navigation and Coordination: Ensuring precise navigation, coordination, and communication among robotic systems is critical to avoid collisions and maintain alignment.
Materials and Adaptability: Identifying suitable construction materials and adapting to unforeseen challenges during assembly are key considerations.
Regulatory and Ethical Considerations: The regulatory landscape for in-situ assembly needs to be defined, and ethical considerations regarding space resource utilization must be addressed.
In-situ space structure assembly is poised to revolutionize space exploration and utilization. The integration of robotic technologies, additive manufacturing techniques, and innovative engineering approaches opens the door to ambitious missions, habitats, telescopes, and more. As researchers and engineers continue to push the boundaries of space technology, in-situ construction holds the promise of transforming our understanding of space and our ability to create a sustainable presence beyond Earth.
Reducing Space Debris and Enhancing Satellite Lifetimes: We will explore the critical topic of "Reducing Space Debris and Enhancing Satellite Lifetimes," focusing on the challenges posed by space debris and the innovative ways robotics and technology are being employed to address this growing concern. This also delves into the strategies for prolonging the operational lifespan of satellites, ensuring the sustainability and safety of space activities for current and future generations.
Introduction to Space Debris and its Consequences:
The proliferation of space activities, including satellite launches and missions, has led to the accumulation of space debris in Earth's orbit. Space debris comprises defunct satellites, spent rocket stages, fragments from collisions, and other discarded objects. As this debris population continues to increase, it poses significant risks to operational satellites, spacecraft, and even the International Space Station (ISS). Collisions with debris can result in catastrophic damage, threatening space missions and exacerbating the problem further.
Challenges and Dangers of Space Debris: The presence of space debris in Earth's orbit raises several challenges and dangers:

Collision Risks: Space debris poses a collision risk to operational satellites and spacecraft, potentially causing mission failures and disruptions to vital services such as communication, navigation, and weather forecasting.

Kessler Syndrome: The Kessler Syndrome refers to a theoretical scenario in which the density of space debris becomes so high that collisions create a self-perpetuating cascade of debris, making certain orbital regions unusable for future missions.

Sustainability Concerns: As the space industry continues to grow, the accumulation of debris raises sustainability concerns, as it limits the safe and accessible regions of space for future exploration and exploitation.

Robotic Solutions for Space Debris Mitigation: Robotic technologies play a crucial role in addressing the challenges posed by space debris:

Debris Removal and Capture: Robotic missions can be designed to capture and remove larger pieces of debris using robotic arms, nets, harpoons, or other mechanisms. These missions aim to deorbit the debris, allowing it to re-enter Earth's atmosphere and burn up.

Active Debris Removal: Satellite operators can design spacecraft to include propulsion systems that enable controlled reentry at the end of their operational lives, reducing the risk of adding to the space debris problem.

Debris Tracking and Surveillance: Robotic systems equipped with advanced sensors and cameras can track and catalog space debris, helping to predict potential collisions and enabling safer navigation for operational satellites.

Sustainable Satellite Design: Robotics can facilitate the development of satellites with more robust and resilient designs, minimizing the creation of additional debris through collisions or breakups.

Enhancing Satellite Lifetimes: Prolonging the operational lifespan of satellites is essential for reducing the frequency of satellite replacements and minimizing the accumulation of space debris. Robotics and technology contribute to enhancing satellite lifetimes in various ways:

Orbit Adjustments: Robotic systems can extend the operational life of satellites by periodically adjusting their orbits to counteract the effects of atmospheric drag and gravitational perturbations.

Refueling and Maintenance: Robotic servicing missions can refuel satellites, replace worn-out components, and perform repairs to extend their operational lifespan.

End-of-Life Disposal: Robotics can facilitate controlled deorbiting at the end of a satellite's mission, reducing the chances of it becoming a piece of space debris.

Upgradability: Modular designs and robotic systems can enable upgrades and repairs to satellites in orbit, prolonging their usefulness and reducing the need for replacements.

Collaborative Efforts and Policy Considerations:

Addressing the challenges of space debris and enhancing satellite lifetimes require collaboration among governments, space agencies, satellite operators, and the private sector:

International Cooperation: The international community must work together to establish guidelines, regulations, and best practices for responsible space activities that reduce the creation of space debris.

Transparency: Collaboration among satellite operators and agencies is essential for sharing data on spacecraft locations and intentions, allowing for collision avoidance maneuvers.

Policy Frameworks: Governments and space agencies can implement policies that encourage sustainable satellite design, responsible mission planning, and debris mitigation efforts.

The challenges of space debris and the need for sustainable space activities underscore the importance of robotics and technology in safeguarding the space environment. As the space industry continues to expand, the integration of robotic solutions for debris removal, enhanced satellite lifetimes, and responsible space practices becomes paramount. By leveraging the capabilities of robotics, we can mitigate the risks posed by space debris, extend the operational lives of satellites, and contribute to the long-term sustainability of space exploration and utilization.

Exercise

Multiple-choice Questions

1. What is the primary advantage of using robots in extraterrestrial exploration?

a) They require less power

b) They are immune to cosmic radiation

c) They can operate in harsh environments

d) They are cheaper than human missions

2. Which planet in our solar system has been the primary focus of robotic exploration missions?

a) Mars

b) Jupiter

c) Venus

d) Mercury

3. What is the main challenge of using robots in deep-space exploration missions?

a) Limited communication bandwidth
b) High radiation levels
c) Lack of robotic technology
d) Low gravity environments
4. Which of the following is an example of a rover used for extraterrestrial exploration?
a) Curiosity
b) Hubble
c) Voyager
d) Kepler
5. What is the purpose of the robotic arm on the Mars rovers?
a) To take selfies
b) To collect soil samples
c) To measure cosmic radiation
d) To communicate with Earth
6. The Hubble Space Telescope is an example of which type of robotic technology?
a) Rover
b) Lander
c) Orbiter
d) Telescope
7. Which moon of Jupiter is considered a prime target for robotic exploration due to its subsurface ocean?
a) Titan
b) Europa
c) Io
d) Ganymede
8. What is the primary propulsion method used by robotic spacecraft for interplanetary travel?
a) Wind power
b) Solar sails
c) Nuclear fusion
d) Rocket engines
9. Which organization successfully landed a robotic rover named "Zhurong" on Mars in 2021?
a) NASA
b) ESA
c) Roscosmos
d) CNSA
10. Which planet's atmosphere is often considered challenging for robotic landings due to its thick and hazy composition?
a) Venus
b) Mars
c) Saturn
d) Uranus
11. What type of energy source is commonly used to power robots on extraterrestrial surfaces?
a) Coal
b) Natural gas
c) Solar panels
d) Nuclear reactors
12. Which mission successfully collected samples from an asteroid and returned them to Earth using a robotic spacecraft?
a) Voyager
b) New Horizons
c) Hayabusa2
d) Cassini
13. What is the term for the phenomenon where communication signals to and from Earth are temporarily blocked due to planetary alignment?
a) Radio blackout
b) Cosmic interference
c) Solar eclipse

d) Communication lag
14. In addition to exploration, what other potential application does robotic technology have in extraterrestrial environments?
a) Agriculture
b) Artistic expression
c) Telepathy
d) Time travel
15. Which region of Mars has been of particular interest for astrobiological studies using robotic missions?
a) Olympus Mons
b) Valles Marineris
c) Hellas Planitia
d) Jezero Crater

Review Questions:

1. What are some advantages of using robotic systems for extraterrestrial exploration compared to human missions?
2. Describe the challenges that robots face in terms of communication during deep-space missions.
3. How do robotic rovers collect and analyze samples on other planets?
4. Provide examples of robotic missions that have explored celestial bodies other than Earth.
5. What role does a robotic arm play on a planetary rover, and how is it utilized?
6. Explain the significance of the Hubble Space Telescope in the context of extraterrestrial exploration.
7. Why is Europa considered a potential target for robotic exploration, and what features make it intriguing?
8. Compare and contrast the propulsion methods used by robotic spacecraft for interplanetary travel.
9. Discuss the main sources of power used by robots on extraterrestrial surfaces and their advantages and limitations.
10. How does the atmosphere of Venus pose challenges to robotic landings and exploration?
11. Describe a specific mission where a robotic spacecraft successfully collected samples from an asteroid. What were the objectives of this mission?
12. Explain the concept of the "Goldilocks Zone" in the context of extraterrestrial exploration and its relevance to habitability.
13. How does the lack of a strong magnetic field on Mars affect the capabilities of robotic systems?
14. Discuss the potential applications of robotic technology beyond exploration in extraterrestrial environments.
15. Explain the significance of the Jezero Crater on Mars in the search for signs of past life.
16. Describe the achievements and challenges of the Rosetta mission, including the role of the robotic lander Philae.
17. What is the importance of teleoperation and autonomous navigation in robotic exploration?
18. How does robotic technology enable the study of lunar water and its implications for future space missions?
19. Compare the challenges faced by robots in exploring the surfaces of the Moon and Mars.
20. Explain how stellar navigation can be utilized by robotic systems to orient themselves in extraterrestrial environments.

Answers Multiple-choice Questions

Answers:
1. c)
2. a)
3. a)
4. a)
5. b)
6. d)
7. b)
8. d)
9. d)
10. a)
11. c)
12. c)
13. a)
14. a)
15. d)

Review Questions:

1. Robotic systems offer several advantages over human missions in extraterrestrial exploration. Firstly, robots can withstand extreme conditions, such as high radiation levels, low temperatures, and harsh terrains that would be

hazardous to humans. They don't require life support systems, food, or water, making them more adaptable for long-duration missions. Robots can be designed for specialized tasks, such as collecting samples, conducting repetitive measurements, or performing precise manipulations, which may not be feasible for humans. Additionally, the cost of sending robots is generally lower compared to human missions, reducing the financial burden. Remote operation allows scientists to control robots from a safe distance, enabling real-time decision-making based on data received from the robots. Overall, robotic systems play a crucial role in gathering data, conducting experiments, and exploring challenging extraterrestrial environments.

2. Communication during deep-space missions is challenging due to the vast distances involved. The time it takes for a signal to travel between Earth and a spacecraft increases with the distance. This delay, known as signal latency, can range from minutes to hours, making real-time control and decision-making difficult. Additionally, the limited available bandwidth for data transmission restricts the amount of information that can be sent back to Earth. This limitation necessitates careful data compression and prioritization. The alignment of planets and celestial bodies can lead to communication blackout periods when the spacecraft is blocked from direct communication with Earth. To address these challenges, mission planners use complex scheduling algorithms to optimize communication windows, and autonomous systems on board the spacecraft allow it to perform tasks and make decisions without continuous human intervention.

3. Robotic rovers collect and analyze samples from extraterrestrial surfaces using a combination of tools and instruments. A robotic rover typically carries a suite of scientific instruments, including cameras, spectrometers, drills, and sensors. To collect samples, the rover's robotic arm is used to manipulate tools such as drills, scoops, or corers to gather soil, rock, or ice samples from the surface. Once collected, these samples can be analyzed on-board using the rover's instruments or stored for future return to Earth. The instruments can determine the composition, mineralogy, and physical properties of the collected samples. For example, spectroscopy can identify the presence of certain minerals, while microscopic imaging can reveal the sample's texture and structure. By analyzing these samples, scientists can gain insights into the geological history, potential habitability, and past presence of water or organic materials on other planets and moons.

4. The robotic arm on a planetary rover serves as a versatile tool for interacting with the environment and conducting scientific investigations. Equipped with various instruments, cameras, and tools, the arm enables the rover to perform tasks such as collecting samples, digging trenches, and manipulating objects. For sample collection, the arm can use drills, coring devices, or scoops to gather soil or rock samples from the surface. The arm can also position scientific instruments close to interesting features for analysis, such as using spectrometers to identify mineral compositions. Cameras on the arm provide detailed visual information, helping mission operators plan routes, avoid obstacles, and select targets of interest. The arm's dexterity allows it to investigate locations that are inaccessible to other parts of the rover, contributing to a more comprehensive exploration and scientific understanding of the extraterrestrial environment.

5. The Hubble Space Telescope has been a transformative tool for extraterrestrial exploration, revolutionizing our understanding of the universe. Launched in 1990, Hubble orbits Earth, above the atmosphere, which can distort and block incoming light. This unobstructed view has allowed Hubble to capture incredibly detailed images of distant galaxies, nebulae, stars, and planets. Hubble's observations have led to numerous discoveries, such as the precise measurement of the rate of the universe's expansion and the observation of exoplanets outside our solar system. By studying the light emitted by various celestial objects, Hubble's instruments can analyze their compositions, temperatures, and movements. This has contributed to our understanding of the origins of the universe, the life cycles of stars, and the formation of galaxies. In the context of robotic exploration, Hubble demonstrates the power of remote sensing and the importance of space-based observatories in advancing our knowledge of the cosmos.

6. Europa, a moon of Jupiter, has garnered significant interest for robotic exploration due to its potential to host a subsurface ocean beneath its icy crust. This ocean might contain more than twice the water volume of Earth's oceans and could provide a suitable environment for life as we know it. Robotic missions to Europa aim to investigate the moon's surface and subsurface to understand its geology, chemistry, and potential habitability. A robotic lander or probe could analyze surface materials and search for signs of organics or other biomarkers. Subsurface exploration might involve deploying penetrators or melting probes to reach the ocean beneath the ice. Such missions could provide insights into the conditions necessary for life and the possibilities of extraterrestrial ecosystems. However, the challenges include the thick ice crust, radiation from Jupiter, and the need for technology capable of operating in extreme environments. Despite these challenges, the potential discoveries on Europa make it a compelling target for future robotic exploration.

7. Robotic spacecraft utilize various propulsion methods for interplanetary travel, each with its advantages and limitations. Chemical rockets, commonly used for launching from Earth, provide high thrust but are limited by the amount of propellant they can carry. Electric propulsion systems, such as ion or Hall-effect thrusters, offer much

higher efficiency by accelerating ions to generate thrust. These systems provide continuous low thrust over longer durations, allowing spacecraft to reach higher speeds. Solar sails use radiation pressure from sunlight to propel the spacecraft, which offers unlimited fuel and gradual acceleration. Nuclear propulsion, though not yet widely used, has the potential to provide high thrust and efficiency. Different missions require different propulsion strategies based on mission objectives, payload mass, travel distance, and time constraints. Choosing the right propulsion method involves balancing trade-offs between efficiency, thrust, mission duration, and technology readiness.

8. Robotic and human space missions each have distinct advantages and challenges. Robotic missions excel in exploring extreme environments, collecting data from hazardous locations, and conducting repetitive tasks with high precision. They can operate for extended periods without the need for life support systems, food, or oxygen. Additionally, robotic missions are cost-effective and can be launched more frequently, enabling broader exploration. However, robotic systems lack the ability to make complex decisions in unpredictable situations and cannot provide the same level of adaptability and creativity as humans. Human missions, on the other hand, offer the ability to respond to unexpected discoveries, troubleshoot issues in real-time, and perform intricate tasks that require human intelligence and judgment. Humans can also directly experience the environment, which can lead to a deeper understanding of the challenges and discoveries. Nonetheless, human missions are significantly more resource-intensive and come with additional risks. Thus, a combination of robotic and human missions can provide a comprehensive approach to exploring and understanding extraterrestrial environments.

9. Robotic exploration of the lunar and Martian surfaces presents unique challenges due to the distinct characteristics of these environments. On the Moon, challenges include the extreme temperature variations between day and night, the presence of abrasive lunar dust, and the lack of a protective atmosphere to shield against cosmic radiation and micrometeoroid impacts. On Mars, challenges include the thin atmosphere, which limits aerodynamic braking during descent, as well as the potential for dust storms that can obscure solar panels and affect visibility. Both environments require careful consideration of mobility and navigation, as well as the development of systems to cope with limited resources and communication delays. Addressing these challenges involves designing robust landing systems, developing reliable power sources, optimizing thermal management, and ensuring the resilience of communication systems. Advances in materials, technology, and autonomous decision-making are essential for successful robotic exploration of these surfaces.

10. The capabilities of robotic technology extend beyond exploration and can play significant roles in various applications. One such application is the establishment of extraterrestrial habitats, where robots could assist in building structures, preparing habitats, and performing maintenance tasks in advance of human missions. In-situ resource utilization (ISRU) is another key application, where robots could mine, extract, and process resources from the environment to produce materials, propellants, or even life support systems. Robots can also serve as precursors for human missions, scouting potential landing sites, conducting surveys, and testing equipment. Furthermore, robots can support scientific investigations by autonomously conducting experiments, collecting data, and transmitting results back to Earth. In the long term, the development of self-replicating or adaptable robotic systems could pave the way for more sustainable and self-sufficient exploration efforts. Overall, robotic technology's potential extends beyond exploration to facilitate the establishment of human presence and utilization of resources in extraterrestrial environments.

11. The Hayabusa2 mission, launched by the Japan Aerospace Exploration Agency (JAXA), is a prime example of a successful robotic mission that collected samples from an asteroid. The primary target of Hayabusa2 was the near-Earth asteroid Ryugu. The mission aimed to study Ryugu's composition, physical characteristics, and potential insights into the early solar system. The spacecraft utilized various innovative techniques to achieve its objectives. It deployed rovers to the asteroid's surface to study its texture and structure, and it used an impactor to create an artificial crater to access pristine material beneath the surface. Hayabusa2's robotic arm collected samples from the crater, employing a horn-like sampler to collect particles ejected by the impact.

12. The "Goldilocks Zone," also known as the habitable zone, is a region around a star where conditions are suitable for the existence of liquid water on the surface of a planet. This zone is defined by the distance from the star where the planet's surface temperature allows water to be in a liquid state, neither boiling away nor freezing solid. The concept is crucial in the context of extraterrestrial exploration as it indicates the potential for a planet to support life as we know it. If a planet orbits too close to its star, any water present would evaporate due to excessive heat, while a planet too distant would have water frozen solid due to extreme cold. The habitable zone identifies the range where conditions might be conducive to the development and sustenance of life forms similar to those on Earth, making it a key consideration when searching for exoplanets with the potential for habitability.

13. Mars lacks a strong global magnetic field like Earth's, which has several implications for robotic systems and exploration. Earth's magnetic field protects our planet from the solar wind and cosmic rays by deflecting charged particles. On Mars, robotic systems are exposed to higher levels of cosmic and solar radiation due to the absence of

this protective shield. This radiation can interfere with the operation of electronic components, potentially leading to malfunctions or data corruption. The thin atmosphere on Mars does not provide sufficient protection from these high-energy particles. Robotic systems exploring Mars must be designed with radiation-hardened components and shielding to mitigate the effects of radiation exposure, ensuring their reliability and longevity on the Martian surface.

14. Robotic technology offers a wide range of potential applications beyond exploration in extraterrestrial environments. One significant application is the construction of habitats and infrastructure for future human missions. Robots could assist in assembling habitats, preparing landing sites, and constructing protective shielding. In-situ resource utilization (ISRU) is another valuable application, where robots mine, extract, and process local resources to produce materials, propellants, and life support systems. Additionally, robots can conduct experiments autonomously, remotely controlled by scientists on Earth, enabling continuous data collection. Robotic technology also has potential in medical research, where low-gravity environments may provide insights into human physiology and the development of new treatments. Overall, robotic technology can facilitate the establishment of sustainable human presence and utilization of resources on other planets.

15. The Jezero Crater on Mars is of significant importance in the search for signs of past life due to its geological features and the potential presence of a former lake. This crater is thought to have once contained a lake, making it an excellent candidate for preserving evidence of ancient microbial life. The sediments that accumulated in the lakebed could contain organic molecules and other bio signatures that could indicate the presence of life in Mars' distant past. The Perseverance rover, which landed in the Jezero Crater, is equipped with scientific instruments designed to analyze rocks and sediments, potentially providing insights into the history of Mars' climate and habitability. By studying Jezero Crater, scientists hope to shed light on the conditions that may have supported life billions of years ago and contribute to our understanding of the potential for life beyond Earth.

16. The Rosetta mission, led by the European Space Agency (ESA), achieved groundbreaking success by orbiting the comet 67P/Churyumov-Gerasimenko and deploying a robotic lander named Philae. The mission's primary goal was to study the comet's composition and gain insights into the early solar system's formation. Rosetta used a complex series of gravity-assist maneuvers to rendezvous with the comet, and it orbited the comet for over two years, studying its surface, structure, and activity. Philae, the robotic lander, successfully touched down on the comet's surface in November 2014. However, due to an issue with its harpoons, it bounced multiple times and eventually landed in a partially shaded region, limiting its solar power supply. Despite these challenges, Philae still managed to transmit valuable data before its batteries depleted. The Rosetta mission revolutionized our understanding of comets, offering insights into their composition and the processes that shaped our solar system.

17. Teleoperation and autonomous navigation are critical aspects of robotic exploration, enabling robots to navigate and interact with their environment effectively. Teleoperation involves controlling robots remotely from a distance, allowing human operators to manipulate the robots and respond to real-time situations. This is particularly important in situations where robots encounter unexpected obstacles, challenging terrain, or complex tasks that require human judgment and decision-making. Autonomous navigation, on the other hand, allows robots to make decisions independently based on pre-programmed instructions, onboard sensors, and real-time data analysis. This capability is crucial when communication delays make teleoperation impractical, such as in deep space missions. Autonomous systems enable robots to execute tasks, avoid hazards, and adapt to changing conditions, enhancing their efficiency and effectiveness in conducting scientific investigations and exploring extraterrestrial environments.

18. Robotic technology plays a crucial role in studying lunar water and its implications for future space missions. The presence of water on the Moon, in the form of ice within permanently shadowed craters at the poles, has significant implications for future human exploration and sustainable presence. Robotic landers and rovers equipped with scientific instruments are used to detect and characterize these water deposits. They can analyze the composition of the ice, study its distribution, and assess its potential usability as a resource for producing water, oxygen, and hydrogen for rocket fuel. This discovery is transformative, as it reduces the need to transport these resources from Earth, making future lunar missions more cost-effective and sustainable. Robotic technology enables detailed mapping, analysis, and planning for utilizing lunar water, which has the potential to revolutionize the way we approach space travel and exploration beyond Earth.

19. Exploring the surfaces of the Moon and Mars presents unique challenges due to the distinct characteristics of these environments. On the Moon, challenges include extreme temperature fluctuations between day and night, ranging from scorching hot to freezing cold, which can affect the performance of equipment and materials. The abrasive lunar regolith, or dust, poses risks to mechanical systems and can hinder optical instruments. On Mars, challenges include the thin atmosphere that provides limited aerodynamic braking during descent, requiring precise landing techniques. The presence of dust storms can reduce solar panel efficiency and impede visibility. Additionally, both environments have low-gravity conditions that impact mobility, stability, and the behavior of loose surface material. Navigating and overcoming these challenges necessitates the development of robust robotic

systems, advanced materials, thermal management solutions, and precise navigation techniques tailored to the unique conditions of each environment.

20. Stellar navigation is a technique employed by robotic systems to orient themselves in extraterrestrial environments by observing the positions of stars. This method relies on the consistent positions of stars in the night sky, which change predictably as a spacecraft moves. By comparing the observed positions of stars to their known positions, a robotic system can accurately determine its own orientation in space. Stellar navigation is particularly valuable in deep space missions and on planetary bodies where other navigation methods, such as GPS, are not available or accurate. This technique allows robotic systems to determine their attitude and heading independently, aiding in precision landing, accurate pointing of scientific instruments, and aligning with desired trajectories. Stellar navigation enhances the autonomy and reliability of robotic systems in extraterrestrial exploration, contributing to their ability to navigate and conduct tasks with accuracy and efficiency.

Online Resources

Online Exam Papers

Link:https://forms.gle/iFGFyhp8mXhgSHfK7

QR Code:

YouTube Video Link

YouTube Video Playlist Link

Link:https://www.youtube.com/playlist?list=PLzlh_31VWEFp3CTNFEX0g_Y_ukYzopvhJ

QR Code:

Chapter 19: Human-Centric Robotics and Assistive Technologies

Human-centric robotics and assistive technologies are driving transformative advancements in various fields, including healthcare, rehabilitation, and daily life support. These technologies are designed with a primary focus on improving the well-being, quality of life, and independence of individuals, especially those with disabilities or limitations. The integration of robotics and assistive technologies into various applications offers promising solutions that empower individuals to overcome challenges and participate more actively in society. This article explores the concept of human-centric robotics, discusses the key components of assistive technologies, and highlights their impact on individuals and society.

Human-centric robotics refers to the design, development, and application of robotic systems that are specifically tailored to interact effectively and harmoniously with humans. Unlike traditional industrial robots that operate in controlled environments, human-centric robots are designed to work alongside, assist, or interact directly with humans in real-world settings. These robots prioritize safety, adaptability, and user-friendliness, aiming to enhance human capabilities and augment daily activities. They are equipped with sensors, artificial intelligence (AI), and advanced control systems that enable them to understand human gestures, speech, and intentions, facilitating seamless collaboration and communication.

Assistive technologies encompass a broad range of devices, systems, and solutions designed to assist individuals with disabilities or limitations in performing daily tasks, communicating, and engaging in social interactions. These technologies aim to bridge the gap between an individual's abilities and the demands of their environment. Key components of assistive technologies include:

Mobility Aids: Mobility-enhancing technologies such as wheelchairs, exoskeletons, and prosthetics provide individuals with mobility impairments the ability to move independently and engage in various activities.

Communication Devices: Speech generating devices, communication apps, and eye-tracking systems empower individuals with speech disorders or communication challenges to express themselves effectively.

Sensory Aids: Technologies like hearing aids, cochlear implants, and vision enhancement devices help individuals with sensory impairments regain or enhance their auditory and visual capabilities.

Cognitive Support Systems: These include memory aids, reminder systems, and cognitive training software that assist individuals with cognitive impairments in managing daily tasks and maintaining cognitive function.

Home Automation: Smart home technologies offer voice-activated control of appliances, lighting, security systems, and more, enhancing the accessibility and convenience of living spaces.

Wearable Assistive Devices: Wearable technologies such as smart glasses, smartwatches, and wearable sensors provide real-time assistance and monitoring for various needs.

Impact on Individuals and Society:

Enhanced Independence: Human-centric robotics and assistive technologies enable individuals with disabilities to perform tasks independently, enhancing their self-esteem, confidence, and overall quality of life.

Inclusion and Participation: These technologies promote social inclusion by reducing barriers to participation in various activities, education, employment, and community engagement.

Improved Healthcare: Assistive technologies play a pivotal role in healthcare settings, aiding in medical diagnostics, rehabilitation, and therapeutic interventions. They facilitate personalized care and support telemedicine initiatives.

Empowerment: By granting individuals greater control over their environment and activities, assistive technologies empower them to actively pursue personal and professional goals.

Economic Opportunities: The development and deployment of assistive technologies create economic opportunities by spurring innovation, driving research and development, and generating jobs in the technology and healthcare sectors.

Despite their potential, human-centric robotics and assistive technologies face challenges such as cost, accessibility, interoperability, and ethical considerations related to privacy and data security. To address these challenges, collaboration among researchers, engineers, healthcare professionals, policymakers, and end-users is crucial. The future of these technologies holds exciting possibilities, including more advanced AI and machine learning integration, improved natural language processing, greater customization, and expanded applications in fields such as education and elderly care. Human-centric robotics and assistive technologies are transforming lives, offering solutions that empower individuals with disabilities to lead more independent, fulfilling, and connected lives. By focusing on the needs, preferences, and aspirations of users, these technologies embody a human-centered approach that places individuals at the forefront of innovation. As technology continues to evolve, the potential for positive impact on individuals and society at large is vast. The collaboration between technology developers, researchers, healthcare professionals, and policymakers is essential in unlocking the full potential of these technologies and creating a more inclusive and accessible world for everyone.

Robotic Prosthetics and Exoskeletons: In the realm of healthcare and assistive technologies, the fields of robotic prosthetics and exoskeletons have witnessed remarkable advancements in recent years. These cutting-edge technologies are designed to enhance the mobility and quality of life for individuals with limb loss or impaired mobility. By merging robotics, engineering, and medical science, these solutions are transforming the lives of those facing physical challenges. This article delves into the world of robotic prosthetics and exoskeletons, exploring their design principles, technological innovations, benefits, and future prospects.

Understanding Robotic Prosthetics: Robotic prosthetics involve the development of artificial limbs that mimic the form, function, and appearance of natural limbs. Unlike traditional passive prosthetics, which are static and lack mobility, robotic prosthetics are equipped with advanced sensors, actuators, and electronics that replicate the intricate movements of the human body. These prosthetics are often controlled using myoelectric signals, which are generated by the user's muscle contractions. By harnessing these signals, individuals can perform a wide range of actions, from grasping objects to walking with a more natural gait.

Key Components and Design Considerations: Robotic prosthetics consist of several key components that enable their functionality and integration with the human body:

Sensors: Advanced sensors embedded within the prosthetic limb provide real-time feedback on the user's movements, pressure, and interactions with the environment. These sensors enable precise control and coordination.

Actuators: Actuators, such as motors and pneumatics, replicate the movements of muscles, tendons, and joints. These components facilitate lifelike motions, allowing users to perform tasks with greater ease and efficiency.

Control Systems: Myoelectric control systems interpret the electrical signals generated by the user's muscles and translate them into specific movements of the prosthetic limb. Machine learning algorithms enhance the accuracy and responsiveness of these control systems over time.

Materials and Comfort: Lightweight and durable materials ensure the prosthetic limb's comfort and longevity. The design also considers factors such as weight distribution, socket fit, and alignment to enhance user comfort and minimize fatigue.

Benefits and Impact: Robotic prosthetics offer a range of benefits that significantly impact the lives of individuals with limb loss:

Restored Functionality: Robotic prosthetics restore a higher degree of functionality, enabling users to perform tasks that were previously challenging or impossible.

Enhanced Quality of Life: With improved mobility, individuals experience enhanced independence, self-esteem, and overall quality of life.

Natural Movements: The integration of sensors and advanced control systems allows for smoother, more natural movements that closely resemble human actions.

Reduced Phantom Limb Pain: Some advanced robotic prosthetics incorporate sensory feedback, helping alleviate phantom limb pain by providing sensory input to the residual limb.

Exoskeletons: Augmenting Human Mobility:Exoskeletons are wearable robotic devices that provide external support to the user's body, enhancing their physical capabilities and mobility. These devices can be designed to assist individuals with mobility impairments, aid rehabilitation efforts, or even enhance the performance of able-bodied users. Exoskeletons typically consist of a lightweight frame, actuators, sensors, and control systems that work in tandem to augment human movements.

Types of Exoskeletons:

Rehabilitation Exoskeletons: These exoskeletons are employed in physical therapy and rehabilitation settings to assist individuals in relearning movements and regaining muscle strength after injuries or surgeries.

Assistive Exoskeletons: Designed to enhance mobility, assistive exoskeletons provide powered support to individuals with mobility impairments. They enable users to stand, walk, and navigate their environment more effectively.

Industrial and Military Exoskeletons: These exoskeletons find applications in industries and military contexts, where they augment human strength and endurance for tasks such as heavy lifting or carrying heavy loads.

Technological Innovations and Challenges: Robotic prosthetics and exoskeletons continue to advance through innovations in materials, control systems, and AI:

Mind-Controlled Prosthetics: Some prosthetics use brain-computer interfaces (BCIs) to interpret neural signals, enabling users to control their prosthetic limbs through their thoughts.

Soft Exoskeletons: Emerging designs focus on creating lightweight and flexible exoskeletons that closely resemble the human body's natural movements.

Autonomous Adaptation: AI-driven exoskeletons can adapt to users' movements and intentions in real time, providing seamless and intuitive support.

Benefits and Challenges of Exoskeletons: Exoskeletons offer a range of benefits, including:

Rehabilitation: Exoskeletons aid in physical therapy, helping individuals regain strength and mobility after injuries or surgeries.

Enhanced Mobility: Assistive exoskeletons empower individuals with mobility impairments to stand, walk, and navigate their surroundings.

Reduced Strain: Industrial exoskeletons reduce strain on workers by supporting their posture and reducing the risk of musculoskeletal injuries.

However, exoskeletons also face challenges, such as:

Cost: Developing and manufacturing exoskeletons can be expensive, limiting accessibility for some individuals and healthcare facilities.

Weight and Comfort: Ensuring that exoskeletons are lightweight and comfortable for extended use is crucial for user acceptance.

Adaptation: Exoskeletons must be **adaptable** to various body sizes, shapes, and mobility needs.

Future Prospects: The future of robotic prosthetics and exoskeletons holds promising possibilities:

Neural Integration: Advancements in brain-computer interfaces could enable direct neural control of prosthetics and exoskeletons.

Personalization: Customized designs and fit will become more prevalent, ensuring optimal comfort and performance for individual users.

Medical Integration: These technologies could be integrated more closely with the medical field, promoting collaborative efforts between engineers, clinicians, and researchers.

Robotic prosthetics and exoskeletons are exemplars of technological innovation converging with healthcare and rehabilitation. These solutions transcend the boundaries of disability and empower individuals to regain their independence, mobility, and quality of life. As research continues and technology evolves, the future of these fields promises even greater advancements, driving positive change for individuals facing physical challenges and society as a whole.

Enhancing Mobility and Independence: The fields of robotic prosthetics and exoskeletons have revolutionized the way individuals with mobility impairments regain their independence and mobility. These innovative technologies not only provide physical support but also offer a renewed sense of empowerment, allowing users to engage in daily activities and participate actively in society. This article explores how robotic prosthetics and exoskeletons are enhancing mobility and independence, the technological advancements that underpin their capabilities, their impact on individuals' lives, and the potential future directions for these transformative technologies.

Technological Advancements in Robotic Prosthetics: Robotic prosthetics have evolved significantly over the years, thanks to advancements in materials, sensors, and control systems. These technological innovations have propelled prosthetic limbs from being static replacements to dynamic and responsive devices that closely mimic the functions of natural limbs.

Myoelectric Control Systems: One of the most significant breakthroughs in robotic prosthetics is the development of myoelectric control systems. These systems enable users to control their prosthetic limbs using electrical signals generated by their remaining muscles. Electrodes placed on the skin pick up these signals, allowing individuals to perform intricate movements with their prosthetic limbs, such as opening and closing fingers or bending the elbow. The integration of machine learning algorithms has improved the accuracy and efficiency of myoelectric control, making these prosthetics more intuitive and responsive to the user's intentions.

Sensor Integration: Advanced sensors are a cornerstone of modern robotic prosthetics. These sensors provide real-time feedback about the user's movements, pressure, and interaction with the environment. By incorporating tactile sensors, users can gain a sense of touch and pressure, enhancing their ability to handle objects delicately. Proprioceptive sensors provide information about the prosthetic limb's position and orientation, enabling smoother and more coordinated movements.

Bionic Limbs: The concept of bionic limbs has become a reality, with prosthetics that can replicate natural movements in a remarkably lifelike manner. The integration of sophisticated actuators and artificial muscles allows for fluid and dynamic movements that closely resemble human motions. Individuals with robotic prosthetics can now perform tasks with a level of dexterity and finesse that was once thought unattainable.

Impact on Mobility and Independence: The impact of robotic prosthetics on mobility and independence is profound, offering individuals with limb loss or congenital limb differences the opportunity to lead more fulfilling lives:

Restored Functionality: Robotic prosthetics restore a significant degree of functionality, enabling users to perform a wide range of activities that were previously challenging or impossible. From picking up objects and typing on a keyboard to participating in sports and recreational activities, these prosthetics empower users to engage in various pursuits.
Enhanced Independence: Perhaps the most significant contribution of robotic prosthetics is the restoration of independence. Users can carry out tasks without relying heavily on assistance, boosting their self-esteem and confidence. The ability to perform daily activities independently has a cascading effect, positively impacting mental and emotional well-being.
Social Integration: Robotic prosthetics facilitate social integration by reducing the visible difference between individuals with prosthetic limbs and those with natural limbs. This can lead to increased social interactions, reduced stigma, and a broader sense of belonging.
Technological Innovations in Exoskeletons: Exoskeletons represent another innovative approach to enhancing mobility and independence, particularly for individuals with mobility impairments due to spinal cord injuries, neuromuscular disorders, or aging-related conditions.
Powered Assistance: Exoskeletons provide powered assistance to the user's movements, enabling them to stand, walk, and navigate their environment. Advanced actuators and motors generate the necessary torque to support the body's weight and facilitate movements, reducing the effort required by the user.
Customization and Adaptation: Modern exoskeletons can be customized to accommodate different body shapes and sizes. This adaptability ensures a comfortable fit and optimal support for individual users. The ability to adjust the exoskeleton's settings allows users to tailor the device to their specific needs and capabilities.
Hybrid Control Systems: Hybrid control systems combine user-initiated commands with automated responses. This synergy between human intention and robotic assistance creates a seamless and natural walking experience. Some exoskeletons even incorporate machine learning algorithms that adapt to the user's gait and movements, further enhancing comfort and efficiency.
Impact on Mobility and Independence: Exoskeletons have a transformative impact on mobility and independence, offering users the opportunity to regain their ability to walk and engage in physical activities:
Walking Rehabilitation: For individuals recovering from spinal cord injuries or stroke, exoskeletons play a crucial role in rehabilitation. By providing weight-bearing support and promoting correct walking patterns, exoskeletons aid in retraining the nervous system and regaining muscle strength.
Enhanced Mobility: Exoskeletons empower individuals with mobility impairments to stand up, walk, and navigate their surroundings. This newfound mobility enhances users' physical capabilities, enabling them to participate in activities that were previously beyond their reach.
Psychological Well-being: The ability to stand and walk has significant psychological benefits. Exoskeleton users often report improvements in mood, self-esteem, and overall well-being. The act of standing eye-to-eye with others can foster a sense of equality and social inclusion.
Challenges and Future Directions: While the advancements in robotic prosthetics and exoskeletons are impressive, several challenges remain:
Accessibility and Affordability: Cost remains a significant barrier to access for many individuals. Both prosthetics and exoskeletons can be expensive, and insurance coverage may not always be sufficient.
Comfort and Wear ability: Ensuring that these devices are comfortable for extended wear is critical. Users should be able to wear prosthetics and exoskeletons for extended periods without discomfort or fatigue.
Neural Integration: Exploring ways to integrate robotic prosthetics and exoskeletons with the user's nervous system could lead to more intuitive control and a greater sense of embodiment.
The fields of robotic prosthetics and exoskeletons are at the forefront of enhancing mobility and independence for individuals with mobility impairments. These technologies not only restore physical capabilities but also inspire confidence, empowerment, and a renewed sense of self. As research continues and technology evolves, the potential for further advancements in design, control systems, and neural integration holds the promise of even greater transformation in the lives of those who rely on these remarkable innovations. Ultimately, the convergence of engineering, medical science, and human resilience is driving the creation of a more inclusive and accessible world.
User-Centered Design and Personalization: The integration of user-centered design and personalization has emerged as a cornerstone in the development of robotic prosthetics and exoskeletons. These advanced technologies, designed to enhance mobility and improve the quality of life for individuals with mobility impairments, are increasingly being tailored to individual users' needs and preferences. The marriage of cutting-edge engineering, medical science, and human-centric design principles is shaping the future of assistive technologies. This article delves into the significance of user-centered design and personalization in the realm of robotic prosthetics and exoskeletons, highlighting their impact, challenges, and potential for transforming lives.

User-Centered Design: A Human-Centric Approach:User-centered design places the individual user at the heart of the design and development process. This approach emphasizes understanding users' needs, preferences, and challenges to create solutions that align seamlessly with their requirements. In the context of robotic prosthetics and exoskeletons, user-centered design goes beyond functional considerations to encompass factors such as comfort, aesthetics, ease of use, and emotional well-being.

Comprehensive Needs Assessment: Effective user-centered design begins with a comprehensive needs assessment that involves direct engagement with individuals who will benefit from the technology. This assessment encompasses physical capabilities, daily routines, desired activities, and psychological factors. The insights gained from these interactions inform the design process and guide engineers and designers in creating solutions that cater to specific user needs.

Iterative Prototyping: User-centered design involves an iterative approach, with multiple rounds of prototyping and user testing. This iterative process allows designers to gather feedback, make refinements, and iteratively improve the technology's functionality, usability, and fit. It ensures that the final product aligns closely with user requirements and preferences.

Inclusive and Accessible Design: Inclusive design principles are integral to user-centered design. These principles ensure that the technology is accessible to a diverse range of users, including those with varying abilities and needs. Designers take into account factors such as universal usability, clear user interfaces, and compatibility with assistive devices.

Personalization: Tailoring Solutions to Individual Users:Personalization takes user-centered design a step further by customizing robotic prosthetics and exoskeletons to suit individual users' unique characteristics, preferences, and requirements. Personalized solutions acknowledge that no two individuals have the same needs, and they strive to create technologies that adapt seamlessly to each user's circumstances.

Customized Fit and Functionality: Personalization enables the customization of the technology's fit, form, and functionality. Prosthetic limbs and exoskeletons can be tailored to match the user's specific residual limb size, body structure, and mobility goals. This level of customization enhances comfort, usability, and overall effectiveness.

Intuitive Control Interfaces: Personalized control interfaces take into consideration the user's physical abilities, cognitive capabilities, and personal preferences. For example, individuals who have lost their limbs may prefer control mechanisms that align with their residual muscle movements, enabling intuitive control of their prosthetic devices.

Emotional Connection: Personalization extends beyond functionality to evoke an emotional connection between the user and the technology. Customization options, aesthetic choices, and even the incorporation of personalized designs or colors contribute to a sense of ownership and identity.

Impact on Mobility and Quality of Life: The incorporation of user-centered design and personalization in robotic prosthetics and exoskeletons yields a range of transformative benefits:

Enhanced Usability: By tailoring the technology to individual needs and preferences, user-centered design and personalization enhance the usability and effectiveness of robotic prosthetics and exoskeletons. This results in more intuitive control, increased comfort, and improved overall functionality.

Improved Quality of Life: Personalized solutions contribute to an improved quality of life for individuals with mobility impairments. The technology not only supports physical mobility but also fosters psychological well-being, self-esteem, and a greater sense of agency.

Participation and Social Inclusion: User-centered design and personalization facilitate greater participation and social inclusion. When individuals can seamlessly integrate robotic prosthetics or exoskeletons into their daily routines and engage in activities they previously enjoyed, they experience a higher level of integration and belonging in society.

Challenges and Considerations: While user-centered design and personalization offer significant benefits, they also present challenges and considerations:

Technological Complexity: Developing personalized solutions requires a deep understanding of both engineering and user needs. The integration of sensors, actuators, control systems, and personalized interfaces demands multidisciplinary collaboration.

Accessibility and Affordability: Achieving widespread accessibility and affordability of personalized technologies remains a challenge. Customization may increase production costs, potentially limiting access for some individuals.

Ethical Considerations: Ensuring that personalization aligns with ethical considerations, such as data privacy and informed consent, is critical. Personalized technologies often involve collecting sensitive user data, necessitating robust safeguards.

Future Prospects and Innovations: The future of user-centered design and personalization in robotic prosthetics and exoskeletons holds promising possibilities:

Neurofeedback Integration: Advancements in neurofeedback technology could enable direct integration between the user's nervous system and the robotic technology, creating a seamless and intuitive connection.
Smart Learning Algorithms: Personalized solutions could incorporate smart learning algorithms that adapt to the user's movements and preferences over time, enhancing usability and comfort.
Interconnected Ecosystems: Personalized technologies could be integrated into broader ecosystems, collaborating with other assistive devices, smart home technologies, and healthcare systems to provide a holistic approach to mobility and well-being.
User-centered design and personalization represent pivotal shifts in the field of robotic prosthetics and exoskeletons. By prioritizing individual needs, preferences, and aspirations, these technologies are transcending mere functionality to become vehicles of empowerment, independence, and improved quality of life. The integration of advanced engineering, human-centric design, and personalized solutions is propelling the evolution of assistive technologies, creating a future where mobility challenges are met with innovation, compassion, and transformative change. As research and development continue, the potential for even more personalized, accessible, and impactful solutions on the horizon is both promising and inspiring.
Socially Assistive Robots: Enhancing Emotional Well-Being: The emergence of socially assistive robots represents a paradigm shift in how technology can enhance emotional well-being and human interaction. These robots are designed not only to perform tasks but also to engage with individuals on a social and emotional level, offering companionship, support, and a bridge to improved mental health. As technology becomes more integrated into our daily lives, the potential of socially assistive robots to address loneliness, promote social engagement, and enhance emotional well-being is gaining recognition. This article explores the significance of socially assistive robots, their impact on emotional well-being, their design principles, challenges, and the evolving landscape of human-robot interaction.
Understanding Socially Assistive Robots: Socially assistive robots are a subset of assistive robotics that focus on providing emotional and social support to individuals, particularly those who may experience isolation, loneliness, or other emotional challenges. Unlike industrial or task-oriented robots, socially assistive robots prioritize human interaction, communication, and emotional connection.
Emotional Engagement: What sets socially assistive robots apart is their ability to establish emotional engagement with users. These robots are designed to read and respond to human emotions, utilizing sensors and artificial intelligence to recognize facial expressions, voice tone, and other non-verbal cues.
Companionship and Support: Socially assistive robots provide companionship and support in various settings, such as elderly care facilities, hospitals, schools, and even private homes. They engage in conversations, offer encouragement, provide reminders for medication or appointments, and even participate in recreational activities.
Mental Health Enhancement: These robots have the potential to contribute to mental health enhancement by providing a non-judgmental space for individuals to express their thoughts and emotions. Through regular interactions, socially assistive robots can offer emotional relief and reduce feelings of loneliness.
Design Principles of Socially Assistive Robots: Creating robots that effectively enhance emotional well-being involves a blend of design principles from both robotics and psychology. The following principles guide the development of socially assistive robots:
Empathy and Emotional Intelligence: Socially assistive robots are designed to exhibit empathy and emotional intelligence. This entails recognizing human emotions, responding appropriately, and providing comfort or support when needed.
Personalization and Adaptation: To be effective, these robots must adapt to individual users' preferences, communication styles, and emotional needs. Personalization enhances the sense of companionship and fosters a stronger emotional connection.
Respectful Interaction: Robots should engage with users in a respectful and supportive manner. This includes maintaining appropriate boundaries, using positive language, and respecting users' privacy and personal space.
Impact on Emotional Well-Being: The impact of socially assistive robots on emotional well-being is substantial and multifaceted:
Alleviating Loneliness: Loneliness is a widespread issue, particularly among the elderly and individuals with limited social interactions. Socially assistive robots provide companionship and reduce feelings of isolation, addressing this significant emotional challenge.
Providing Emotional Support: For individuals dealing with stress, anxiety, or depression, socially assistive robots offer a safe and non-judgmental outlet for expressing emotions. Sharing thoughts with a robot can alleviate emotional burden and contribute to a more positive outlook.

Enhancing Social Engagement: These robots encourage social engagement by initiating conversations, suggesting activities, and facilitating interactions with others. By fostering connections, they contribute to a sense of belonging and community.

Challenges and Considerations: As promising as socially assistive robots are, they also raise several challenges and ethical considerations:

Ethical Boundaries: Establishing ethical boundaries in human-robot interactions is crucial. While robots can provide emotional support, it's essential to recognize their limitations and avoid creating unrealistic expectations.

Privacy and Data Security: Interactions with socially assistive robots involve sharing personal thoughts and emotions. Ensuring robust data security and respecting users' privacy are paramount.

Overreliance on Robots: While these robots can enhance emotional well-being, they should not replace human relationships. Encouraging users to maintain connections with real people remains vital.

Human-Robot Interaction: The Evolving Landscape:The interaction between humans and socially assistive robots is evolving rapidly, driven by advancements in artificial intelligence, machine learning, and natural language processing:

Natural Language Understanding: Improvements in natural language understanding enable robots to comprehend human speech, respond appropriately, and engage in meaningful conversations.

Facial and Emotion Recognition: Robots are becoming adept at recognizing facial expressions and emotional cues, allowing them to respond sensitively to users' moods.

Emotional Context Interpretation: Robots are learning to interpret emotional context through dialogue and behavior, enabling more nuanced and empathetic responses.

Future Prospects and Innovations: The future of socially assistive robots holds several exciting possibilities:

Mental Health Monitoring: Advanced robots could monitor users' emotional states and provide early interventions for individuals at risk of mental health challenges.

Social Skills Development: Socially assistive robots could be used as tools to help individuals, including children on the autism spectrum, develop and refine social skills.

Multimodal Interaction: Future robots might incorporate multiple modes of interaction, including gestures, touch, and even scent, to create richer and more immersive experiences.

Socially assistive robots represent a transformative convergence of technology and emotional well-being. These robots go beyond functional assistance, offering companionship, emotional support, and a unique bridge to improved mental health. As the field continues to advance, the potential for positive impact on individuals' lives is immense. However, it is crucial to approach the integration of these robots with sensitivity, ethical considerations, and a clear understanding of their role in the broader context of human relationships. Through careful design, continuous research, and responsible implementation, socially assistive robots have the potential to become valuable allies in the journey toward enhanced

Robotic Rehabilitation: Restoring Functionality: The field of robotic rehabilitation has emerged as a revolutionary approach to restoring functionality and improving the quality of life for individuals recovering from injuries, surgeries, or neurological conditions. By integrating robotics, artificial intelligence, and advanced sensing technologies, robotic rehabilitation offers personalized and effective therapy solutions that accelerate recovery, enhance mobility, and enable individuals to regain their independence. This article delves into the significance of robotic rehabilitation, its impact on restoring functionality, the technological advancements driving its effectiveness, challenges faced, and the potential future directions of this transformative field.

Understanding Robotic Rehabilitation: Robotic rehabilitation involves the use of robotic devices to assist and enhance the process of physical rehabilitation. These devices are designed to guide patients through targeted exercises and movements, with the goal of improving motor skills, strength, coordination, and overall functional abilities. Robotic rehabilitation is particularly beneficial for individuals who have experienced stroke, spinal cord injuries, traumatic brain injuries, and other neurological disorders.

Personalized Therapy: One of the key advantages of robotic rehabilitation is its ability to provide personalized therapy. Robotic devices can be tailored to an individual's specific needs, adjusting the intensity, range of motion, and resistance levels based on their current abilities and progress.

Precise and Repetitive Movements: Robotic devices ensure consistent and precise movements during rehabilitation sessions. This precision is essential for neuromuscular retraining and promoting neuroplasticity, which are critical factors in the recovery process.

Real-Time Feedback: Robotic rehabilitation devices incorporate sensors that provide real-time feedback to patients and therapists. This feedback helps patients understand their progress and make necessary adjustments to their movements.

Impact on Restoring Functionality: Robotic rehabilitation has a profound impact on restoring functionality and improving the quality of life for individuals undergoing recovery:

Accelerated Recovery: The focused and intensive nature of robotic rehabilitation accelerates the recovery process. Patients often experience quicker improvements in motor skills and functional abilities compared to traditional therapy methods.

Enhanced Motor Control: Robotic devices help individuals regain motor control by engaging specific muscle groups and promoting precise movements. This is particularly beneficial for patients with conditions affecting coordination and muscle strength.

Increased Range of Motion: Robotic rehabilitation can gradually increase the range of motion in joints, preventing stiffness and promoting flexibility. This is crucial for individuals recovering from surgeries or injuries that limit their mobility.

Technological Advancements Driving Effectiveness:Robotic rehabilitation owes its effectiveness to continuous advancements in technology:

Robotic Interfaces: Robotic devices are equipped with user-friendly interfaces that allow therapists to customize therapy plans, monitor progress, and adjust settings in real time.

Haptic Feedback: Haptic feedback technology provides patients with sensory cues, such as vibrations or pressure that simulate the sense of touch. This feedback enhances the patient's awareness of their movements and contributes to improved motor control.

Gamification and Virtual Reality: Robotic rehabilitation is increasingly incorporating gamification and virtual reality elements to make therapy sessions engaging and motivating. These technologies encourage active participation and compliance with therapy routines.

Challenges in Robotic Rehabilitation: While robotic rehabilitation holds great promise, it also presents challenges that need to be addressed:

Cost and Accessibility: Robotic rehabilitation devices can be costly, making them inaccessible to some individuals or healthcare facilities. Efforts are needed to make these technologies more affordable and widely available.

Clinical Integration: Integrating robotic rehabilitation into clinical practice requires training for therapists and ensuring that devices align with established rehabilitation protocols.

Personalized Treatment Plans: Creating personalized treatment plans requires a deep understanding of patients' needs and progress. Developing algorithms that adapt to individual patient profiles is a complex endeavor.

Future Directions and Innovations: The future of robotic rehabilitation holds several promising directions:

Tele-Rehabilitation: Advancements in telehealth and remote monitoring could enable patients to engage in robotic rehabilitation sessions from the comfort of their homes, expanding access to therapy.

Brain-Computer Interfaces: Incorporating brain-computer interfaces could allow patients to control robotic devices directly through their neural signals, enhancing the precision and efficiency of rehabilitation.

Wearable Rehabilitation: Wearable robotic devices could provide continuous support and therapy throughout the day, promoting ongoing recovery and functional improvements.

Robotic rehabilitation is revolutionizing the field of physical therapy by harnessing technology's power to restore functionality, improve mobility, and enhance quality of life for individuals recovering from injuries and neurological conditions. The integration of precision, personalization, and real-time feedback is reshaping the rehabilitation landscape and accelerating the recovery process. As research and development continue, the potential for even more advanced robotic rehabilitation solutions holds the promise of transforming the lives of countless individuals, allowing them to reclaim their independence and regain their functional abilities.

Exercise

Multiple-choice Questions

1. What is the primary focus of human-centric robotics and assistive technologies?
a) Industrial automation
b) Enhancing human interaction
c) Environmental monitoring
d) Space exploration
2. What is the main goal of socially assistive robots?
a) Industrial production
b) Enhancing emotional well-being
c) Space exploration
d) Environmental conservation
3. What distinguishes robotic prosthetics from traditional passive prosthetics?
a) They are made of different materials

b) They are controlled using brain-computer interfaces
c) They are more cost-effective
d) They lack mobility
4. Myoelectric control systems in robotic prosthetics are based on:
a) Voice commands
b) Brain signals
c) Muscle contractions
d) Eye movements
5. What is the primary purpose of exoskeletons in the context of human-centric robotics?
a) Space exploration
b) Enhancing human strength
c) Environmental monitoring
d) Autonomous navigation
6. Which field is the "Goldilocks Zone" concept relevant to in extraterrestrial exploration?
a) Robotic prosthetics
b) Socially assistive robots
c) Planetary habitability
d) Human-robot interaction
7. What does the term "teleoperation" refer to in robotic exploration?
a) Remote control of robots from a distance
b) Autonomous navigation in outer space
c) Communication between robots and humans
d) Robotic learning algorithms
8. How do socially assistive robots contribute to emotional well-being?
a) By performing physical tasks for users
b) By providing companionship and emotional support
c) By monitoring environmental conditions
d) By analyzing data from space missions
9. What role does artificial intelligence play in the field of robotic rehabilitation?
a) It helps robots perform tasks autonomously
b) It enhances emotional well-being
c) It enables remote control of robots
d) It personalizes therapy plans and adapts to user needs
10. What is the primary advantage of gamification and virtual reality in robotic rehabilitation?
a) Increasing the cost of therapy
b) Making therapy sessions boring
c) Enhancing patient engagement and motivation
d) Reducing the effectiveness of therapy
11. What distinguishes user-centered design from traditional design approaches?
a) It focuses on aesthetics over functionality
b) It prioritizes the needs and preferences of users
c) It only considers technological advancements
d) It excludes customization options
12. What does the "Goldilocks Zone" refer to in the context of space exploration?
a) A region of optimal temperature and conditions for life
b) A zone with gold deposits on asteroids
c) A region with strong magnetic fields
d) An area with high radiation levels
13. What is the purpose of teleoperation in the field of robotics?
a) Enhancing communication between robots
b) Enabling robots to navigate autonomously
c) Allowing remote control of robots by humans
d) Providing emotional support to users
14. Which technology enables individuals to control robotic prosthetics using their muscle contractions?
a) Brain-computer interfaces
b) Virtual reality headsets

c) Voice recognition software
d) Myoelectric control systems
15. What are exoskeletons primarily designed to assist with?
a) Enhancing human strength and endurance
b) Space exploration on other planets
c) Controlling robots remotely
d) Environmental monitoring

Review Questions:

1. What is the primary objective of human-centric robotics and assistive technologies?
2. How do socially assistive robots differ from traditional industrial robots?
3. Explain the concept of "Goldilocks Zone" in the context of extraterrestrial exploration.
4. How do myoelectric control systems enhance the functionality of robotic prosthetics?
5. What role do sensors play in robotic prosthetics and exoskeletons?
6. Describe the potential impact of socially assistive robots on emotional well-being.
7. What are the advantages of using gamification and virtual reality in robotic rehabilitation?
8. How does user-centered design influence the development of assistive technologies?
9. What is the significance of haptic feedback in human-robot interaction?
10. Explain the concept of teleoperation and its relevance in robotic exploration.
11. How do robotic prosthetics contribute to restoring limb functionality?
12. Describe the potential applications of robotic technology beyond exploration.
13. How does the lack of a strong magnetic field on Mars affect robotic systems?
14. What is the significance of the Jezero Crater on Mars in the search for signs of past life?
15. Discuss the achievements and challenges of the Rosetta mission, including the role of the robotic lander Philae.
16. Explain the importance of personalization in the context of robotic prosthetics and exoskeletons.
17. How can socially assistive robots alleviate feelings of loneliness and isolation?
18. Compare the challenges faced by robots exploring the Moon and Mars.
19. How do wearable robotic devices contribute to ongoing rehabilitation?
20. What are the potential future directions for the field of robotic rehabilitation?

Answers

Multiple-choice Questions

Answers:
1. b)
2. b)
3. c)
4. c)
5. d)
6. c)
7. c)
8. b)
9. d)
10. c)
11. a)
12. a)
13. c)
14. d)
15. a)

Review Questions:

1. The primary objective of human-centric robotics and assistive technologies is to create robotic systems and devices that enhance human well-being, improve quality of life, and provide assistance in various domains. Unlike traditional robotics that focus on industrial automation, human-centric robotics prioritizes human interaction, emotional well-being, and personalized support. These technologies aim to bridge the gap between humans and machines by fostering meaningful relationships, promoting independence, and addressing specific needs of individuals, such as those with mobility impairments, cognitive challenges, or emotional needs.
2. Socially assistive robots are a distinct category of robots designed to provide emotional support, companionship, and cognitive assistance to individuals. Unlike traditional industrial robots, which are mainly used for tasks

requiring precision and efficiency, socially assistive robots focus on human interaction and well-being. These robots use sensors to recognize human emotions, engage in conversations, and even monitor users' emotional states. By offering companionship and emotional support, socially assistive robots play a vital role in enhancing emotional well-being, reducing feelings of loneliness, and providing a sense of connection to individuals who may be isolated or in need of social interaction.

3. The "Goldilocks Zone" concept, often referred to as the habitable zone, pertains to the region around a star where conditions are neither too hot nor too cold to support the presence of liquid water on a planet's surface. This zone is crucial for planetary habitability as water is a fundamental requirement for life as we know it. In the context of extraterrestrial exploration, understanding the "Goldilocks Zone" helps scientists identify planets with the potential to host life and directs efforts towards studying their atmospheres and surface conditions. The concept is significant in the search for exoplanets that could potentially harbor life beyond Earth.

4. Myoelectric control systems are a technology used in robotic prosthetics to enable individuals with limb loss to control their prosthetic devices using muscle contractions. Electrodes placed on the surface of the user's residual muscles pick up electrical signals generated when the user attempts to move their missing limb. These signals are then translated into specific movements of the prosthetic limb, allowing the user to control the device in a more intuitive and natural manner. Myoelectric control systems enhance the functionality and usability of robotic prosthetics, providing users with greater control over their movements and improving their quality of life.

5. Sensors play a pivotal role in both robotic prosthetics and exoskeletons by providing real-time feedback and control. In the context of robotic prosthetics, sensors embedded within the device can detect the user's muscle contractions or nerve signals, enabling precise and coordinated movements of the prosthetic limb. Similarly, in exoskeletons, sensors can monitor the user's posture, balance, and gait, allowing the device to provide assistance and adjust its movements accordingly. By continuously collecting data and providing feedback, sensors enhance the interaction between humans and robotic devices, contributing to better functionality, safety, and user experience.

6. The potential impact of socially assistive robots on emotional well-being is profound. These robots offer companionship, emotional support, and cognitive engagement, particularly to individuals who may experience loneliness, isolation, or cognitive challenges. Studies have shown that interactions with socially assistive robots can lead to improved mood, reduced feelings of loneliness, and increased overall emotional well-being. These robots provide a non-judgmental space for individuals to express their thoughts and emotions, fostering a sense of connection and belonging. By addressing emotional needs, socially assistive robots contribute to a holistic approach to well-being and quality of life.

7. Gamification and virtual reality are two innovative approaches used in robotic rehabilitation to enhance patient engagement and motivation. Gamification involves incorporating game-like elements, such as challenges, rewards, and competition, into rehabilitation exercises. This approach makes therapy sessions more engaging and enjoyable, encouraging patients to actively participate and complete their exercises with enthusiasm. Virtual reality takes this a step further by immersing patients in simulated environments where they can interact with virtual objects and scenarios. This not only adds an element of novelty but also provides a distraction from the repetitive nature of rehabilitation exercises, leading to improved compliance and better outcomes.

8. User-centered design is a design approach that prioritizes the needs, preferences, and experiences of users when creating products, systems, or technologies. In the context of assistive technologies, user-centered design ensures that devices are tailored to the specific requirements of individuals with disabilities or challenges. This approach involves actively involving end-users in the design process, conducting thorough needs assessments, and iterating on designs based on user feedback. The goal is to create technologies that are intuitive, effective, and closely aligned with users' needs, resulting in solutions that genuinely enhance their quality of life and promote independence.

9. Haptic feedback is a technology that provides users with tactile sensations through the use of vibrations, pressure, or other physical cues. In human-robot interaction, haptic feedback enhances the sensory experience by simulating the sense of touch. For example, a prosthetic limb equipped with haptic feedback can provide users with a sense of pressure when gripping an object, allowing them to better perceive their interactions with the environment. This technology enhances the user's awareness of their movements and interactions, contributing to improved motor control and a more intuitive connection with the robotic device.

10. Teleoperation is the practice of controlling robots remotely from a distance. In robotic exploration, teleoperation is used to navigate robots in environments where it might be challenging or dangerous for humans to operate directly. Teleoperation allows human operators to guide robotic systems using remote control interfaces, ensuring that tasks are performed accurately and safely. This approach is particularly valuable in scenarios such as space exploration, where robots are deployed to distant planets or environments with harsh conditions. Teleoperation enables humans to explore and gather data from these environments while minimizing risks to human operators.

11. Robotic prosthetics contribute to restoring limb functionality in individuals who have experienced limb loss. These prosthetics are designed to mimic the natural movements and actions of missing limbs, allowing users to perform everyday tasks and activities with greater ease. Advanced technologies, such as myoelectric control systems, enable users to control the prosthetic limb using muscle contractions, enhancing precision and coordination. By restoring limb functionality, robotic prosthetics empower individuals to regain their independence, engage in physical activities, and improve their overall quality of life.
12. Robotic technology has applications beyond exploration and can impact various aspects of human life. In addition to space missions, robotic systems are used in healthcare for surgeries and rehabilitation, in manufacturing for automation, in entertainment for interactive experiences, in education for interactive learning, and in disaster response for remote interventions. Robotic technology enhances productivity, safety, and efficiency across diverse domains, leading to advancements in various fields and improving the overall human experience.
13. The lack of a strong magnetic field on Mars affects robotic systems in several ways. Earth's magnetic field plays a vital role in shielding the planet from harmful solar radiation and cosmic rays. Mars' weak magnetic field exposes its surface to higher levels of radiation, which can damage electronic components of robotic systems. This radiation exposure poses challenges for the durability and functionality of robotic explorers on Mars. Additionally, a weak magnetic field affects the accuracy of navigation systems that rely on magnetic cues for orientation, requiring robotic systems to rely more on other navigation methods such as visual or satellite-based navigation.
14. The Jezero Crater on Mars holds significant importance in the search for signs of past life. This crater is believed to have once contained a lake, making it a prime location to search for ancient microbial life or evidence of past habitability. The lake's sediments could have preserved traces of microbial life or organic molecules, providing valuable insights into Mars' past and its potential for supporting life. The Perseverance rover's mission in the Jezero Crater aims to collect samples that could potentially be returned to Earth for further analysis, advancing our understanding of Mars' history and the possibility of life beyond Earth.
15. The Rosetta mission, led by the European Space Agency (ESA), aimed to study the comet 67P/Churyumov-Gerasimenko and its interactions with the Sun. The mission included the robotic lander Philae, which was designed to land on the comet's surface and perform experiments. However, Philae faced challenges during its landing, as it bounced multiple times before settling in an area with limited sunlight. Despite these challenges, Philae managed to transmit valuable data back to Rosetta and contributed to our understanding of cometary composition and behavior. The mission as a whole provided insights into the early solar system and the origin of water on Earth.
16. Socially assistive robots have the potential to significantly impact emotional well-being by providing companionship, emotional support, and cognitive engagement. These robots are designed to interact with users in ways that foster emotional connections and alleviate feelings of loneliness and isolation. Through regular conversations, active listening, and even simulated empathy, socially assistive robots create a space for individuals to express their thoughts and emotions without fear of judgment. By offering a non-judgmental outlet for communication and emotional expression, these robots can contribute to improved mood, reduced stress, and an overall enhanced sense of emotional well-being.
17. Wearable robotic devices are an innovative approach in rehabilitation that aims to provide continuous support and therapy throughout the day. These devices are designed to be worn by users and can assist in mobility, posture, and movement-related activities. By offering ongoing support, wearable robotic devices enable users to engage in daily tasks and activities more independently, promoting functional recovery and overall well-being. These devices can be particularly beneficial for individuals recovering from injuries, surgeries, or conditions that affect their mobility and motor skills.
18. Robotic prosthetics contribute to restoring limb functionality by replicating the natural movements and actions of missing limbs. These prosthetics are designed with advanced sensors, actuators, and control systems that mimic the biomechanics of human limbs. By analyzing the user's muscle contractions or nerve signals, robotic prosthetics can translate these signals into corresponding movements of the prosthetic limb. This enables users to perform a wide range of tasks, such as grasping objects, walking, and even more intricate movements. By providing users with enhanced motor control and the ability to perform daily activities, robotic prosthetics empower individuals to regain their independence and improve their quality of life.
19. The potential impact of socially assistive robots on reducing feelings of loneliness is significant. Loneliness is a widespread issue, particularly among the elderly, individuals with limited social interactions, or those facing cognitive challenges. Socially assistive robots provide companionship, engage in conversations, and offer emotional support to individuals in need. These robots can initiate interactions, listen actively, and provide a sense of connection, thereby reducing feelings of isolation and loneliness. By creating opportunities for social engagement and emotional interaction, socially assistive robots contribute to an improved sense of well-being and a more positive outlook on life.

20. The primary focus of exoskeletons in the context of human-centric robotics is to enhance human mobility and strength. Exoskeletons are wearable robotic devices designed to provide physical assistance to users by augmenting their movements and capabilities. These devices are particularly valuable for individuals with mobility impairments or muscle weakness. By providing support to joints and muscles, exoskeletons enable users to walk, stand, and perform activities that might otherwise be challenging. The goal is to enhance the user's overall mobility, promote independence, and improve their quality of life by restoring functional abilities and enabling greater participation in daily activities.

Online Resources

Online Exam Papers

Link:https://forms.gle/bk7JGoZai5972qGK6

QR Code:

YouTube Video Playlist Link

Link:https://www.youtube.com/playlist?list=PLzlh_31VWEFoOm9xh3oyE8AJ5l_HjDTX1

QR Code:

Chapter 20: Top of Form Entrepreneurship and Innovation in Robotics

Entrepreneurship and innovation have always been driving forces behind technological advancements and economic growth. In the realm of robotics, these factors play a pivotal role in shaping the landscape of industries, transforming business models, and creating new opportunities for both startups and established enterprises. This article explores the intricate relationship between entrepreneurship, innovation, and robotics, highlighting how these elements intersect to foster groundbreaking developments and drive economic progress.

Robotics, once confined to industrial settings, has undergone a remarkable evolution fueled by innovation. From assembly line automation to sophisticated humanoid robots, technological advancements have expanded the scope of robotics into numerous domains, including healthcare, agriculture, logistics, and even space exploration. This expansion is propelled by a synergy between entrepreneurship and innovation, where visionary minds identify opportunities, develop novel solutions, and bring them to the market.

Entrepreneurs with a vision to bridge technological gaps and address real-world challenges drive innovation in robotics. These individuals recognize untapped market niches, envision the potential of robotic solutions, and embark on journeys to create products that disrupt industries. Their ability to envision the future of robotics and integrate it into diverse sectors drives innovation and creates new avenues for economic growth.

Innovation in robotics often stems from cross-disciplinary collaboration. Entrepreneurs with diverse backgrounds, such as engineering, computer science, and design, bring varied perspectives to the table. This convergence of knowledge and skills fuels the creation of robotic solutions that are not only technologically advanced but also cater to end-users' needs and preferences.

The integration of cutting-edge technologies, such as artificial intelligence, machine learning, and sensors, is a hallmark of innovation in robotics. Entrepreneurs leverage these technologies to create robots that can learn, adapt, and interact intelligently with their environments. The fusion of technologies from various fields pushes the boundaries of what robotics can achieve, resulting in innovative applications across industries.

The innovation ecosystem plays a critical role in nurturing entrepreneurship in the robotics sector. An environment conducive to idea generation, resource availability, and collaboration enables entrepreneurs to bring their robotic innovations to fruition.

Research institutions and universities contribute to the innovation ecosystem by conducting cutting-edge research in robotics. Entrepreneurs often collaborate with these institutions to access knowledge, expertise, and potential partnerships that accelerate the development of novel robotic solutions.

Incubators and accelerators provide startups with the necessary resources, mentorship, and funding to transform ideas into market-ready products. In the field of robotics, these programs offer access to specialized labs, prototyping facilities, and industry experts who guide entrepreneurs in refining their innovations.

Venture capitalists, angel investors, and government grants play a crucial role in fueling innovation in robotics. Their financial support empowers entrepreneurs to overcome technological challenges, conduct research, and bring their robotic solutions closer to commercialization.

Collaborative workspaces and innovation hubs create environments where entrepreneurs can interact, share ideas, and collaborate on projects. These spaces foster a sense of community, enabling cross-pollination of ideas and the formation of partnerships that drive innovation.

Numerous entrepreneurial success stories underscore the transformative impact of innovation in robotics:

Companies like Intuitive Surgical have pioneered the field of robotic-assisted surgery with their da Vinci Surgical System. This technology enables surgeons to perform minimally invasive procedures with precision, reducing patient trauma and recovery time.

Entrepreneurs have harnessed drone technology for various applications, from aerial photography to package delivery. Companies like DJI have revolutionized industries by developing drones that are accessible, easy to operate, and equipped with advanced features.

Robotics companies are creating AI-powered assistants that enhance productivity and convenience. Pepper, a humanoid robot developed by SoftBank Robotics, can interact with humans, recognize emotions, and provide information in retail and hospitality settings.

Startups like Waymo are driving innovation in autonomous vehicles, aiming to reshape transportation systems. Through a combination of robotics and AI, these companies are working toward safer and more efficient transportation solutions.

Creating innovative robotic solutions requires deep technical expertise and multidisciplinary collaboration. Entrepreneurs must navigate the complexity of hardware design, software development, and integration of advanced technologies.

The introduction of new robotic technologies often raises regulatory questions related to safety, ethics, and liability. Entrepreneurs must work closely with regulatory bodies to ensure compliance and gain market acceptance.

Securing funding for robotics ventures can be challenging, particularly for startups working on cutting-edge innovations. Investors may hesitate due to the high development costs and longer timelines associated with bringing robotic products to market.

As robots become more integrated into human lives, ethical considerations arise, including questions about job displacement, privacy, and the potential impact on social dynamics. Entrepreneurs must address these concerns to build public trust in their innovations.

Ensuring seamless and intuitive human-robot interaction remains a challenge. Entrepreneurs need to design robots that are user-friendly, capable of understanding natural language, and able to adapt to users' preferences and needs.

Entrepreneurship and innovation are driving forces that continue to shape the robotics landscape, transforming industries, and creating new economic opportunities. The symbiotic relationship between visionary entrepreneurs and technological innovators fuels the development of robots that enhance human lives, improve business processes, and push the boundaries of what is possible. As robotics technology continues to advance, entrepreneurs will play a pivotal role in identifying unmet needs, driving innovation, and navigating the challenges of bringing cutting-edge solutions to the global market. The future of robotics promises even more groundbreaking developments, driven by the passion, creativity, and ingenuity of entrepreneurial minds.

Robotic Startups: Driving Technological Innovation: The world of robotics has witnessed a transformative evolution over the past few decades, with technological advancements changing the way we live, work, and interact. One of the driving forces behind this evolution is the emergence of robotic startups that are fueled by innovation, creativity, and a bold entrepreneurial spirit. These startups have played a pivotal role in pushing the boundaries of robotics technology, introducing groundbreaking solutions, and disrupting traditional industries. This article delves into the realm of robotic startups, exploring their significance in driving technological innovation and reshaping various sectors.

The Rise of Robotic Startups: Robotic startups have emerged as dynamic players in the field of technology, harnessing the potential of robotics to address real-world challenges and create transformative solutions. Unlike established corporations, startups are characterized by their agility, willingness to take risks, and disruptive thinking. They are driven by a passion to create products that challenge the status quo, offer unique value propositions, and capitalize on emerging trends.

Innovating Through Disruption: Robotic startups thrive on disruption. They identify gaps in the market, areas where traditional solutions fall short, and untapped opportunities to revolutionize industries through robotics technology. By offering novel approaches and fresh perspectives, startups challenge established norms and introduce solutions that can transform the way businesses operate and individuals interact with technology.

Navigating Emerging Technologies:Startups have the flexibility to embrace emerging technologies and integrate them into their solutions. From artificial intelligence and machine learning to advanced sensors and computer vision, startups leverage a wide array of technologies to create cutting-edge robotic systems that are smarter, more intuitive, and capable of addressing complex tasks.

Agile Development and Iteration:Agility is a hallmark of startups. They have the ability to iterate quickly, experiment with different approaches, and pivot based on feedback and market demand. This iterative process enables startups to refine their products, optimize functionalities, and align with customer needs more effectively than larger, more bureaucratic organizations.

Impact across Industries: Robotic startups are not confined to a single industry; their impact is felt across diverse sectors, redefining the way we perceive and interact with technology.

Healthcare and Medical Robotics:Startups in this domain have developed robotic solutions for surgery, diagnostics, rehabilitation, and even mental health support. These technologies enhance precision, minimize invasiveness, and offer personalized treatment options, ultimately improving patient outcomes.

Agriculture and AgTech:Startups are driving innovation in agriculture by developing robotic systems for tasks such as planting, harvesting, and monitoring crops. These solutions enhance efficiency, reduce resource consumption, and contribute to sustainable farming practices.

Logistics and Warehousing:Robotic startups are revolutionizing the logistics and warehousing industry by introducing autonomous robots for material handling, order fulfillment, and inventory management. These robots optimize workflows, reduce human labor, and improve operational efficiency.

Manufacturing and Industry 4.0:Startups are at the forefront of the Industry 4.0 revolution, introducing collaborative robots (cobots) that work alongside humans, smart factories powered by data analytics, and innovative solutions for predictive maintenance. These advancements enhance productivity, minimize downtime, and drive process optimization.

Retail and Customer Engagement:Robotic startups are changing the retail landscape with interactive robots that guide customers, provide product information, and create engaging in-store experiences. These robots bridge the gap between online and offline shopping, enhancing customer engagement and loyalty.

Education and EdTech:Startups are leveraging robotics to enhance educational experiences by introducing robots that teach coding, engineering concepts, and problem-solving skills. These robots make learning more interactive, engaging, and accessible to students of all ages.

Challenges and Opportunities: While robotic startups have the potential to drive innovation, they also face challenges that require strategic navigation.

Funding and Investment:Securing funding is a common challenge for startups, especially those operating in capital-intensive fields like robotics. Investors may be cautious due to the longer development timelines and complexities associated with robotics technology.

Technological Complexities:Developing robotic systems involves intricate technical challenges, including hardware design, software integration, and algorithm development. Startups must overcome these complexities to create reliable and user-friendly products.

Regulatory Landscape:Navigating regulatory frameworks for robotic technologies can be complex and time-consuming. Startups must ensure compliance with safety standards, ethical considerations, and industry-specific regulations.

Market Acceptance:Introducing innovative robotic solutions requires market education and acceptance. Startups must communicate the value of their products, address potential skepticism, and build trust among potential customers and stakeholders.

Future Directions and Transformative Potential: The future of robotic startups is promising, with numerous opportunities to further drive technological innovation and reshape industries.

Customization and Personalization:Startups are poised to create robotic solutions that cater to individual needs, preferences, and contexts. Customization and personalization will become key factors in enhancing user experiences and driving widespread adoption.

Collaboration and Integration:Collaboration between startups and established corporations will become more prevalent, enabling startups to access resources, expertise, and distribution channels. This collaboration will accelerate the integration of robotic technologies into existing ecosystems.

Ethical Considerations:As robotic solutions become more integrated into daily life, startups will play a critical role in addressing ethical considerations, including job displacement, privacy, and equitable access to technology.

Robotic startups are at the forefront of driving technological innovation and transforming industries. Their entrepreneurial spirit, commitment to disruption, and ability to leverage emerging technologies have the potential to reshape the way we interact with technology and tackle societal challenges. As these startups continue to push boundaries, collaborate with stakeholders, and overcome challenges, they will undoubtedly play a pivotal role in shaping the future of robotics and its impact on various sectors.

Robotic Innovation Ecosystems: Incubators and Accelerators: Innovation in the field of robotics is a dynamic and rapidly evolving process, driven by advancements in technology, interdisciplinary collaboration, and a strong entrepreneurial spirit. Within this ecosystem, incubators and accelerators play a pivotal role in nurturing early-stage robotic startups and propelling them towards success. These innovation hubs provide startups with essential resources, mentorship, networking opportunities, and funding, creating an environment where groundbreaking ideas can flourish and transform into market-ready products. This article explores the significance of robotic innovation ecosystems, focusing on the role of incubators and accelerators in fostering entrepreneurial growth and driving technological innovation.

The Role of Incubators and Accelerators: Incubators and accelerators are designed to provide startups with the support and resources necessary to thrive in the competitive landscape. While both share the goal of helping startups succeed, they operate at different stages of a company's development.

Incubators: Incubators focus on nurturing early-stage startups and providing them with a nurturing environment to develop their ideas and products. Incubator programs typically offer a range of resources, including office space, mentorship, access to experts, and often seed funding. The primary goal of incubators is to guide startups through the ideation and initial development phases, helping them refine their business models, test their concepts, and prepare for further growth.

Accelerators: Accelerators, on the other hand, cater to startups that have a validated product and are ready to scale their operations. Accelerator programs are more intensive and time-limited, typically lasting a few months. During this period, startups receive mentorship, business development support, access to potential investors, and sometimes additional funding in exchange for equity. The goal of accelerators is to fast-track a startup's growth, assist in market entry, and enable them to secure follow-up funding or achieve other key milestones.

Advantages of Robotic Incubators and Accelerators: Robotic incubators and accelerators offer several advantages that contribute to the growth and success of startups in the field of robotics:

Access to Expertise: Incubator and accelerator programs often provide startups with access to a network of industry experts, seasoned entrepreneurs, and mentors with deep knowledge of the robotics domain. This access to expertise helps startups overcome technical challenges, navigate regulatory hurdles, and make informed strategic decisions.

Networking Opportunities: Participating in incubator or accelerator programs exposes startups to a diverse community of fellow entrepreneurs, investors, potential customers, and industry partners. This networking environment fosters collaboration, partnership opportunities, and access to valuable connections that can propel the startup's growth.

Tailored Support: Robotic incubators and accelerators understand the unique challenges faced by startups in the field of robotics. As a result, their programs often offer tailored support that addresses technical complexities, prototyping needs, and the integration of cutting-edge technologies.

Validation and Credibility: Being accepted into a reputable incubator or accelerator program can validate a startup's concept, providing credibility that attracts investors, customers, and collaborators. The association with a respected program can enhance a startup's reputation and increase its chances of success.

Funding Opportunities: Many incubator and accelerator programs provide startups with access to funding, either through direct investment or by facilitating introductions to potential investors. This financial support is crucial for startups to further develop their products, conduct market validation, and achieve key milestones.

Business Development Resources: Startups often require guidance in developing their business strategies, marketing plans, and go-to-market approaches. Incubators and accelerators offer business development resources, workshops, and mentorship to help startups refine their value propositions and create sustainable growth strategies.

Accelerated Learning Curve: Participating in an incubator or accelerator program compresses the learning curve for startups. The structured curriculum, mentorship, and exposure to industry trends enable startups to rapidly develop skills, adapt to market dynamics, and refine their offerings.

Case Studies: Robotic Incubators and Accelerators: Several prominent robotic incubators and accelerators have demonstrated the impact of their programs on startups in the field:

Mass Robotics: Based in Boston, Mass Robotics is a robotics-focused innovation hub that provides startups with co-working spaces, prototyping facilities, and access to a network of experts. Their programs facilitate collaboration, knowledge sharing, and access to resources that accelerate the development of robotic technologies.

Robot Launch: Robot Launch is a global competition for startups in the robotics and automation space. It provides participants with exposure to industry experts, potential investors, and mentors. The competition acts as a platform for startups to showcase their innovations and gain valuable feedback from a diverse panel of judges.

NVIDIA Inception: NVIDIA Inception is an accelerator program for AI and deep learning startups, which often intersect with robotics. The program offers support in the form of technical assistance, mentorship, and access to NVIDIA's GPU hardware and software resources. It empowers startups to leverage AI technologies and accelerate their development cycles.

Y Combinator: While not exclusively focused on robotics, Y Combinator is one of the most renowned startup accelerators globally. Many robotics startups have successfully graduated from Y Combinator's accelerator program, benefiting from its comprehensive support, mentorship, and access to a vast network of alumni and investors.

Challenges and Future Directions: While robotic incubators and accelerators offer numerous advantages, they also face challenges that must be addressed for continued success:

Industry-Specific Expertise: The field of robotics is diverse and multidisciplinary, encompassing areas such as hardware design, software development, AI, and mechanical engineering. Incubators and accelerators need to provide a range of expertise to effectively support startups across these domains.

Sustainable Funding Models: Incubators and accelerators themselves require funding to operate. Developing sustainable funding models that balance the needs of startups, investors, and program administrators is essential for the longevity and effectiveness of these programs.

Global Reach: With the growth of the robotics industry on a global scale, there is a need for incubators and accelerators to reach startups beyond their local regions. Ensuring that startups from various geographical locations can access and benefit from these programs is crucial for fostering innovation on a broader scale.

Rapid Technological Evolution: The field of robotics is characterized by rapid technological advancements. Incubators and accelerators must stay up-to-date with these advancements to offer startups relevant and timely support.

Robotic incubators and accelerators are instrumental in shaping the future of the robotics industry by nurturing startups and driving technological innovation. Through their tailored support, access to expertise, and networking opportunities, these programs empower startups to overcome challenges, develop cutting-edge solutions, and transform their ideas into market-ready products. As the field of robotics continues to evolve, incubators and accelerators will play a crucial role in fostering entrepreneurial growth, enabling collaboration, and propelling the development of groundbreaking robotic technologies that shape industries and improve lives.

Intellectual Property in Robotics: Patents and Innovations: In the ever-evolving world of robotics, intellectual property (IP) plays a pivotal role in safeguarding innovations, promoting creativity, and driving technological advancements. As the field of robotics continues to expand across industries and domains, protecting novel ideas, inventions, and designs becomes crucial for fostering innovation and ensuring a fair competitive landscape. This article delves into the realm of intellectual property in robotics, focusing on patents as a means to secure innovations, encourage research and development, and shape the future of the robotics industry.

The Significance of Intellectual Property in Robotics: Intellectual property encompasses various forms of legal protection that allow creators, inventors, and innovators to safeguard their intellectual creations. In the context of robotics, intellectual property protection is essential for several reasons:

Encouraging Innovation: By providing creators with exclusive rights to their innovations, intellectual property protection encourages inventors to invest time, resources, and effort into developing new robotic technologies. This encourages continuous research and development, fostering innovation and driving technological progress.

Attracting Investment: Investors are more likely to support startups and companies that have secured their intellectual property. Intellectual property rights provide a level of certainty that the innovations developed by these entities are protected from unauthorized use or reproduction, increasing the attractiveness of investment opportunities.

Fair Competition: Intellectual property protection ensures that inventors and creators can compete fairly in the marketplace. Without such protection, larger companies could easily copy or replicate innovative technologies, undermining the efforts of smaller startups and individual inventors.

Knowledge Sharing: While intellectual property provides exclusivity, it also encourages knowledge sharing. Innovators disclose their inventions in patent applications, contributing to the global body of knowledge and enabling further advancements based on those ideas.

Economic Growth: A robust system of intellectual property protection stimulates economic growth by fostering innovation, attracting investment, and creating a conducive environment for startups to flourish. This growth, in turn, leads to the creation of jobs and the development of new industries.

Patents and Their Role in Robotic Innovations: Patents are a crucial aspect of intellectual property in the field of robotics. A patent grants inventors the exclusive right to make, use, sell, and license their invention for a specific period, typically 20 years from the filing date. This exclusive right provides inventors with a competitive edge and encourages them to disclose their innovations to the public.

Types of Patents in Robotics: In the realm of robotics, several types of patents are relevant:

Utility Patents: These cover the functional aspects of inventions, including methods, processes, machines, and articles of manufacture. In robotics, utility patents can protect innovative algorithms, sensor technologies, mechanical designs, and control systems.

Design Patents: Design patents protect the ornamental appearance of an invention, rather than its functionality. In robotics, this could include the aesthetic design of robotic devices, user interfaces, and other visual elements.

Software Patents: Software is a significant component of robotic systems. Software patents protect novel algorithms, code, and software processes that contribute to the functionality of a robotic invention.

Method Patents: These cover specific methods or processes involved in achieving a particular outcome. In robotics, method patents can protect novel techniques for robot navigation, object recognition, or task execution.

Patent Benefits for Robotic Innovators: For innovators in the robotics field, obtaining patents offers several advantages:

Exclusivity: Patents grant inventors the exclusive right to their inventions, preventing competitors from using, making, or selling the patented technology without permission.

Monetization: Patent holders can license their inventions to other companies, generating revenue through royalties. This can be particularly lucrative when the patented technology is widely applicable.

Market Advantage: Patented technologies give companies a competitive edge in the marketplace, attracting customers, partners, and investors who value innovative solutions.

Leveraging Partnerships: Patented technologies can serve as valuable assets in forming strategic partnerships or collaborations with other companies.

Challenges in Patenting Robotics Innovations: While patents offer significant benefits, there are challenges specific to the field of robotics:

Interdisciplinary Nature: Robotics involves a wide range of disciplines, including engineering, computer science, mechanics, and more. Patent applications need to adequately cover these interdisciplinary aspects, which can be complex.

Rapid Technological Evolution: The pace of technological advancement in robotics is rapid. Patent applications need to be strategically timed to ensure protection for inventions while considering potential future developments.

Overlap with Prior Art: Given the vast amount of research and development in robotics, there is a risk of overlap with prior art. Inventors need to thoroughly research existing patents and publications to ensure their invention is novel.

Robotics Patents and Innovations: Several robotics patents have made a significant impact on the field, shaping industries and influencing the direction of technological advancements:

iRobot's Roomba Patents: iRobot's Roomba vacuum cleaning robot revolutionized the household cleaning industry. The company's patents covering navigation algorithms, obstacle detection, and autonomous cleaning contributed to the Roomba's success and established iRobot as a leader in consumer robotics.

Intuitive Surgical da Vinci Surgical System: Intuitive Surgical patents on the da Vinci Surgical System, a robotic-assisted surgical platform, transformed the field of medical robotics. These patents cover innovative mechanisms for precise surgical movements and instrument control.

Boston Dynamics' Quadruped Robot Patents: Boston Dynamics' patents on quadruped robots, such as Spot and Big Dog, have paved the way for advancements in autonomous navigation, dynamic stability, and load-carrying capabilities. These patents have influenced the development of robots for various applications, including search and rescue and industrial inspection.

Tesla's Autonomous Vehicle Patents: While not exclusively a robotics company, Tesla's patents related to autonomous driving technology have had a significant impact on the robotics and automotive industries. These patents cover innovations in sensors, machine learning, and autonomous control systems.

Future Directions and Emerging Trends: The landscape of intellectual property in robotics is evolving alongside technological advancements and shifts in industry priorities:

Collaboration and Cross-Licensing: As robotics becomes more interdisciplinary, companies may explore cross-licensing agreements to access each other's patented technologies. This approach encourages collaboration, accelerates innovation, and reduces the risk of legal disputes.

Ethical and Regulatory Considerations: As robotics technologies become more integrated into society, ethical and regulatory concerns arise. Innovators must consider the ethical implications of their inventions and navigate evolving regulations related to data privacy, safety, and AI ethics.

Open Source and Collaboration: Some innovators in the robotics field opt for open-source approaches, sharing their innovations with the community to foster collaboration and rapid development. Balancing open-source principles with the need for intellectual property protection is a challenge that requires careful consideration.

Intellectual property, particularly patents, is a cornerstone of innovation in the field of robotics. It fosters a culture of creativity, encourages investment, and rewards inventors for their contributions to technological progress. As robotics continues to shape industries, transform daily life, and redefine the way humans interact with technology, intellectual property protection will remain essential for promoting innovation, ensuring fair competition, and shaping the future of the robotics industry.

Exercise

Multiple choice Questions

1. What role do entrepreneurship and innovation play in the field of robotics?
a) They have no impact on robotics.
b) They hinder the progress of robotics.
c) They drive technological advancements and economic growth in robotics.
d) They are only relevant in other industries.

2. Which of the following best describes the concept of an entrepreneurial mindset in robotics?

a) Avoiding risks and maintaining the status quo.
b) Embracing challenges and pursuing innovative solutions.
c) Ignoring market trends and established practices.
d) Focusing solely on short-term profits.
3. What is the primary goal of a startup in the robotics industry?
a) To maintain the existing technological landscape.
b) To create disruption and challenge established norms.
c) To solely focus on financial gains.
d) To imitate successful products and solutions.
4. How do robotics startups differ from established corporations?
a) They avoid taking risks.
b) They have limited access to funding.
c) They are more agile and willing to take risks.
d) They are less innovative.
5. Which of the following is a common challenge faced by robotic startups?
a) Lack of innovation.
b) Excessive government regulations.
c) Limited access to funding.
d) Reluctance to collaborate.
6. What is the purpose of innovation ecosystems such as incubators and accelerators in robotics?
a) To stifle innovation.
b) To restrict the growth of startups.
c) To provide resources, mentorship, and funding for startups.
d) To discourage collaboration between startups.
7. Which phase of a startup's development is an accelerator program designed to support?
a) Early-stage idea generation.
b) Scaling and rapid growth.
c) Established market presence.
d) Exit strategy implementation.
How do incubators differ from accelerators?
a)Incubators focus on established startups, while accelerators support early-stage ideas.
b) Accelerators provide funding, while incubators offer mentorship.
c) Incubators offer short-term programs, while accelerators offer long-term programs.
d) Accelerators focus on refining and scaling startups, while incubators provide initial support.
9. What advantage do robotic startups gain from participating in incubator or accelerator programs?
a) Increased competition.
b) Reduced access to resources.
c) Access to expertise, mentorship, and networking opportunities.
d) Limited exposure to potential investors.
10. Why is intellectual property (IP) protection significant in robotics?
a) It limits innovation.
b) It discourages collaboration.
c) It safeguards inventions, promotes creativity, & drives innovation.
d) It encourages infringement of patents.
11. How does IP protection encourage innovation in robotics?
a) By stifling creativity.
b) By limiting access to technology.
c) By providing inventors with exclusive rights and rewards.
d) By restricting access to patents.
12. What is the duration of a typical patent's exclusivity in robotics?
a) 5 years.
b) 10 years.
c) 20 years.
d) 50 years.
13. Which type of patent protects the ornamental appearance of a robotic invention?
a) Utility patent.

b) Design patent.
c) Software patent.
d) Method patent.
14. How do software patents contribute to the field of robotics?
a) They limit the development of robotic software.
b) They encourage open-source collaboration.
c) They protect novel algorithms and software processes.
d) They restrict access to coding.
15. What advantage do patent holders have in the robotics market?
a) Increased competition from non-patented products.
b) Reduced access to funding.
c) Market advantage and monetization opportunities.
d) Limited access to expertise.

Review Questions:

1. What is the role of entrepreneurship and innovation in the field of robotics?
2. How does an entrepreneurial mindset contribute to innovation in robotics?
3. What distinguishes startups in the robotics industry from established corporations?
4. How do incubators and accelerators support the growth of robotic startups?
5. What is the primary goal of an accelerator program for startups in robotics?
6. Explain the difference between utility patents and design patents in the context of robotics.
7. How do intellectual property rights encourage innovation in the robotics industry?
8. What benefits do robotic innovators gain from obtaining patents for their creations?
9. Describe the concept of cross-licensing in the context of intellectual property in robotics.
10. How do robotic startups benefit from networking opportunities provided by innovation ecosystems?
11. What are the challenges faced by robotic startups, and how do incubators address these challenges?
12. How do robotic incubators and accelerators contribute to fostering collaboration among startups?
13. How does intellectual property protection strike a balance between innovation and competition in robotics?
14. Explain the importance of an open-source approach in robotics and its impact on innovation.
15. What ethical considerations should innovators in robotics address while protecting their intellectual property?
16. Compare and contrast the roles of incubators and accelerators in supporting startups.
17. How does intellectual property protection impact investment decisions in the robotics industry?
18. Describe the concept of disruptive innovation and its significance in robotics entrepreneurship.
19. What factors contribute to the success of a robotic startup in terms of intellectual property strategy?
20. How do interdisciplinary challenges in robotics affect the patenting process and innovation?

Answers Multiple-choice Questions

Answers:
1. c)
2. b)
3. b)
4. c)
5. c)
6. c)
7. b)
8. d)
9. c)
10. c)
11. c)
12. c)
13. b)
14. c)
15. c)

Review Questions:

1. Entrepreneurship and innovation are pivotal forces shaping the trajectory of the robotics industry. Entrepreneurial endeavors involve identifying market gaps, developing novel solutions, and implementing strategic approaches to transform ideas into reality. In the context of robotics, entrepreneurs harness innovation to create cutting-edge technologies, drive economic growth, and address complex challenges. By fostering an entrepreneurial mindset,

individuals and organizations in the robotics domain seek out new opportunities, embrace risks, and push the boundaries of what is achievable. Innovation complements entrepreneurship by providing the creative and technical foundation for groundbreaking advancements. Together, entrepreneurship and innovation contribute to the rapid evolution of robotic technologies, enabling industries to be reshaped, enhancing human capabilities, and paving the way for a technologically advanced future.

2. An entrepreneurial mindset in robotics encapsulates the willingness to embrace challenges, take calculated risks, and cultivate innovation to drive technological advancement. In the robotics context, this mindset encourages individuals and teams to think beyond conventional boundaries and to approach problems with a solution-oriented perspective. Entrepreneurs in robotics focus on creating value by identifying unmet needs, conceptualizing innovative solutions, and navigating the complexities of transforming those concepts into market-ready products. This mindset encompasses qualities such as resilience, adaptability, open-mindedness, and the drive to continually learn and iterate. By fostering an entrepreneurial mindset, the robotics industry encourages a culture of experimentation, rapid prototyping, and collaboration that fuels the development of cutting-edge robotic technologies.

3. Robotics startups stand apart from established corporations due to their agility, risk-taking mentality, and capacity to drive innovation. Unlike established companies, startups are characterized by their lean operational structure, focus on disruptive solutions, and their ability to pivot swiftly based on market feedback. Startups in robotics often emerge with the intent to challenge conventional norms, introduce novel technologies, and address unmet needs within industries. This contrasts with larger corporations, which may have established processes, resources, and infrastructure but might be less inclined to take high-risk ventures. Robotics startups leverage their nimbleness to explore emerging technologies, experiment with novel applications, and create solutions that reshape industries, foster economic growth, and address complex societal challenges.

4. Incubators and accelerators play critical roles in supporting the growth and success of robotic startups. Incubators focus on nurturing early-stage startups by providing resources such as mentorship, workspace, and access to industry experts. These programs help startups refine their concepts, develop prototypes, and validate their business models. Accelerators, on the other hand, target startups with proven products, aiming to rapidly scale their operations. Accelerator programs provide intensive mentorship, networking opportunities, and potential funding in exchange for equity. Both incubators and accelerators contribute to the entrepreneurial ecosystem by providing startups with crucial resources, guidance, and a network of professionals who can offer expertise, connections, and funding to accelerate their growth and enhance their chances of success.

5. An entrepreneurial mindset offers a plethora of benefits in the robotics industry. It drives individuals and organizations to identify opportunities, create innovative solutions, and adapt to changing market dynamics. This mindset fosters a culture of collaboration, creativity, and continuous learning, enabling the development of technologies that address complex challenges. An entrepreneurial approach in robotics promotes resilience in the face of obstacles, encourages experimentation, and leads to the creation of products with practical applications. By embracing an entrepreneurial mindset, robotics professionals can leverage their technical expertise to push boundaries, create value, and contribute to advancements that reshape industries, enhance human lives, and redefine the possibilities of technology.

6. Incubators and accelerators are vital components of the robotics innovation ecosystem, providing startups with tailored support to accelerate their growth and success. Incubators are primarily focused on nurturing early-stage startups by offering mentorship, resources, and networking opportunities. They guide startups through the validation of their ideas, refinement of their business models, and development of their prototypes. Accelerators, on the other hand, target startups with validated products and the potential for rapid scaling. These programs offer a concentrated period of mentorship, access to investors, and intensive business development support. Both incubators and accelerators contribute to the evolution of the robotics industry by nurturing innovative ideas, fostering collaboration, and providing startups with the tools they need to succeed in a competitive market.

7. Intellectual property (IP) protection is a cornerstone of innovation in the robotics field. It grants inventors exclusive rights to their creations, motivating them to invest time, resources, and effort into developing new technologies. By providing legal protection against unauthorized use or reproduction, IP rights encourage innovators to disclose their inventions to the public, contributing to the advancement of knowledge in the field. This protection also attracts investment, as investors are more likely to support ventures with secured IP. Intellectual property safeguards innovation, fosters fair competition, and promotes the development of cutting-edge robotic technologies that have the potential to transform industries, drive economic growth, and improve the quality of human life.

8. Innovation serves as the lifeblood of robotics entrepreneurship, driving the creation of novel solutions, disrupting traditional norms, and shaping the future of industries. Innovation in robotics involves identifying opportunities to enhance existing technologies, developing groundbreaking concepts, and translating these ideas into practical

applications. Entrepreneurial ventures in robotics rely on innovation to create value, solve complex challenges, and provide solutions that address unmet needs. By fostering a culture of innovation, robotics entrepreneurs continuously push the boundaries of technological possibilities, creating products and services that drive economic growth, transform industries, and contribute to a more technologically advanced society.

9. Cross-licensing in the context of intellectual property involves the exchange of licensing rights between two or more entities. In the robotics industry, cross-licensing plays a role in promoting collaboration and innovation. When companies hold patents that complement each other's technologies, cross-licensing agreements allow them to access each other's intellectual property. This promotes the sharing of knowledge, reduces legal disputes, and accelerates the development of innovative solutions. By fostering collaboration and reducing barriers to entry, cross-licensing contributes to a more open and dynamic ecosystem where companies can build on each other's innovations to create advanced robotic technologies.

10. Innovation ecosystems, such as incubators and accelerators, provide robotic startups with valuable networking opportunities. These opportunities extend to industry experts, mentors, fellow entrepreneurs, potential investors, and strategic partners. Networking allows startups to access diverse perspectives, gain insights into market trends, and forge valuable connections that can accelerate their growth. Through interactions with experts and peers, startups can receive guidance, feedback, and mentorship that helps refine their business strategies, navigate challenges, and optimize their innovations for market success.

11. Robotic startups encounter several challenges that impact their growth trajectory. Limited access to funding is a common challenge, as innovative robotics projects often require significant resources for research, development, and prototyping. Overcoming technical complexities and interdisciplinary barriers can also be challenging, particularly in a field as diverse as robotics. Additionally, navigating regulatory frameworks and addressing ethical considerations can pose obstacles to innovation. Incubators and accelerators play a crucial role in addressing these challenges by providing startups with resources, mentorship, and support to navigate these complexities and drive successful outcomes.

12. Robotic incubators and accelerators foster collaboration among startups by creating environments conducive to knowledge sharing, idea exchange, and interdisciplinary interaction. These programs bring together startups from diverse backgrounds, encouraging cross-pollination of ideas and expertise. Collaborative interactions with mentors, industry professionals, and fellow entrepreneurs enable startups to gain insights, identify opportunities, and refine their innovations through collective problem-solving. This collaborative ethos not only accelerates the development of individual startups but also contributes to the growth of the broader robotics ecosystem by nurturing a culture of cooperation and mutual support.

13. Intellectual property protection strikes a delicate balance between fostering innovation and promoting healthy competition in the robotics industry. While patents provide exclusive rights to inventors, they also require public disclosure of innovations. This disclosure contributes to the advancement of knowledge and encourages further innovation by allowing others to build upon existing ideas. At the same time, patent protection ensures that inventors can benefit from their creations without undue competition. This balance prevents monopolies, encourages collaboration, and enables a fair competitive landscape where multiple players can contribute to the advancement of robotics technologies.

14. The open-source approach in robotics involves sharing technological advancements, code, and knowledge with the broader community. In contrast to proprietary models, open-source initiatives encourage collaboration, transparency, and collective innovation. In robotics, open-source projects allow developers to access and build upon existing solutions, accelerating the development of technologies and encouraging a culture of knowledge exchange. While some might argue that open-source initiatives could undermine revenue opportunities, proponents highlight that they foster a community-driven ecosystem that enables rapid prototyping, experimentation, and collective problem-solving for the betterment of the field.

15. While intellectual property protection is essential for fostering innovation, it also raises ethical considerations in the robotics industry. Striking the right balance between IP protection and ethical responsibilities is crucial. Innovators must consider the potential societal impacts of their inventions, ensuring that their creations are used responsibly and do not harm individuals or communities. Additionally, there is a need to address potential biases, privacy concerns, and unintended consequences that could arise from the deployment of new technologies. Ethical considerations guide innovators in making informed decisions that prioritize both technological advancement and the well-being of society.

16. Incubators and accelerators serve distinct stages of a startup's development journey in the robotics industry. Incubators are tailored to support early-stage startups, providing resources such as mentorship, workspace, and access to expertise. They assist startups in refining their ideas, validating their concepts, and building initial prototypes. On the other hand, accelerators target startups with validated products and aim to expedite their growth.

Accelerator programs offer intensive mentorship, networking opportunities, and potential funding in exchange for equity. Both play crucial roles in nurturing innovation, supporting entrepreneurs, and fostering the growth of the robotics ecosystem.

17. Intellectual property protection significantly influences investment decisions in the robotics industry. Investors seek startups with secured IP rights as it provides assurance that their investments are protected from potential imitations or infringements. IP protection enhances a startup's credibility, signaling that its innovations are valuable and unique. This makes the startup a more attractive investment opportunity, increasing the likelihood of securing funding for research, development, and scaling operations. Intellectual property protection provides startups with a competitive edge, facilitating partnerships, and accelerating their journey towards market success.

18. Disruptive innovation refers to the process by which new technologies, often emerging from startups, redefine industries by providing novel solutions that challenge existing norms. In robotics entrepreneurship, disruptive innovation involves introducing technologies that revolutionize how tasks are performed, create new markets, and replace traditional methods. Such innovations often target unmet needs, leverage emerging technologies, and offer more efficient, cost-effective, or user-friendly alternatives. Disruptive innovations have the potential to reshape industries, transform business models, and drive significant economic growth by addressing challenges and capitalizing on new opportunities.

19. In the field of robotics, a successful intellectual property (IP) strategy requires a multi-faceted approach that balances protection, collaboration, and innovation. Startups should identify key technologies and inventions that offer a competitive advantage and seek appropriate IP protection for these innovations. Simultaneously, they can engage in cross-licensing agreements to access complementary technologies from other players in the industry. Additionally, startups should consider ethical implications, collaborate with experts, and align their IP strategy with their long-term business goals. A well-crafted IP strategy positions startups to navigate the competitive landscape, attract investment, and contribute to the broader innovation ecosystem.

20. Innovation is the driving force behind the success of robotics entrepreneurship. It encompasses the creation of novel solutions, the introduction of disruptive technologies, and the transformation of industries. In the realm of robotics, entrepreneurs leverage innovation to identify unmet needs, develop cutting-edge technologies, and bring ideas to life. Innovation extends beyond technical advancements; it involves creative problem-solving, business model innovation, and addressing ethical and societal considerations. By fostering an innovation-centric approach, entrepreneurs in robotics contribute to economic growth, reshape industries, and improve the quality of life by harnessing the power of technology to address challenges and create meaningful impact.

Online Resources

Online Exam Papers

Link:https://forms.gle/6XrMehV2JCnhjcFy7

QR Code:

YouTube Video Playlist Link

Link:https://www.youtube.com/playlist?list=PLzlh_31VWEFrb3qC4QAagtXZJfW9vXED1

QR Code:

Chapter 21: Top of Form Beyond the Horizon: Speculating on Future Robotics

The future of robotics holds a tantalizing promise of transformative advancements that will reshape industries, redefine human interaction with technology, and unlock unprecedented possibilities. As we venture into uncharted territories, the horizons of robotics extend beyond the limitations of our current understanding, beckoning us to explore realms where imagination and innovation intersect. Speculating on the future of robotics is an exercise that combines visionary thinking, technological trends, and the convergence of various disciplines to envision a world enriched by robotic marvels.

At the heart of this speculation lies the concept of robotic intelligence. Beyond the confines of programmed algorithms, future robots are expected to exhibit a form of intelligence akin to human cognition. Artificial General Intelligence (AGI) is a frontier that researchers and experts are ardently pursuing, envisioning robots that can learn, reason, and adapt across diverse tasks and environments. The rise of AGI could potentially revolutionize industries such as healthcare, manufacturing, and exploration by enabling robots to perform complex tasks, make autonomous decisions, and interact seamlessly with humans.

One of the most compelling prospects on the horizon is the symbiotic relationship between humans and robots. As robots evolve to possess greater cognitive capabilities, they will be capable of collaborating with humans in ways that were once relegated to science fiction. This collaboration extends beyond mere task execution; robots may become companions, mentors, and partners in creative endeavors. From healthcare assistants providing personalized care to elderly individuals to robotic tutors tailoring education to individual learning styles, the integration of robots into our lives could foster a new era of human-robot companionship.

The advancement of robotics is intricately linked with breakthroughs in materials science and engineering. Soft robotics, inspired by the pliability of biological organisms, could usher in a new paradigm of robots that are safer, adaptable, and capable of intricate movements. These robots might seamlessly interact with humans in close proximity, revolutionizing fields such as medical procedures, disaster response, and exploration of hazardous environments. The fusion of soft robotics, bioengineering, and advanced materials holds the potential to give rise to entirely new forms of robotic life.

The frontier of robotics extends far beyond Earth's boundaries. Space exploration is poised to benefit from autonomous robotic systems that can navigate distant planets, moons, and asteroids. Robotic missions could become precursors to human exploration, paving the way for colonization and resource utilization on celestial bodies. Furthermore, the idea of self-replicating robots constructing infrastructure on distant worlds has captured the imagination of scientists and visionaries, presenting a future where robots are the vanguard of humanity's expansion into the cosmos.

Ethical considerations loom large in the vista of future robotics. As robots assume more sophisticated roles in society, questions of responsibility, accountability, and the potential impact on employment become paramount. Striking a balance between innovation and ensuring that robots are developed and deployed ethically is crucial. Roboticists, ethicists, policymakers, and society at large must collaboratively establish guidelines that govern the development and use of advanced robotic technologies to ensure a future that is both innovative and humane.

In the realm of healthcare, robotics holds the promise of transforming medical diagnosis, treatment, and surgery. Miniaturized robots could navigate the human body to deliver targeted therapies, perform intricate surgeries, and diagnose ailments at their earliest stages. Remote telemedicine robots could bridge geographical gaps, providing medical expertise to underserved regions and ensuring access to quality healthcare for all. The fusion of robotics with biotechnology could lead to personalized medical interventions tailored to an individual's genetic makeup, further enhancing the precision and efficacy of treatments.

The vision of future robotics also extends into the domain of environmental conservation. Autonomous robotic systems could revolutionize how we monitor and address ecological challenges, from monitoring endangered species and habitats to cleaning up plastic waste from oceans. Drones, robotic sensors, and AI-driven analytics might collaborate to offer real-time insights into environmental changes and contribute to the preservation of our planet's biodiversity.

In conclusion, speculating on the future of robotics is an endeavor that transcends the boundaries of imagination and delves into the realms of technological possibilities. From the integration of AGI to the symbiotic relationship between humans and robots, from soft robotics to space exploration and ethical considerations, the trajectory of future robotics is marked by innovation, collaboration, and ethical introspection. As we embark on this journey beyond the horizon, we are poised to witness the transformation of industries, the enhancement of human capabilities, and the realization of a future where robots are not just tools but companions in our quest for progress.

Quantum Computing and Quantum Sensing in Robotics: The intersection of robotics and quantum technology represents a remarkable frontier that holds the potential to revolutionize various aspects of our technological landscape. Quantum computing and quantum sensing, two key components of quantum technology, are poised to bring unprecedented computational power and sensing capabilities to the field of robotics. As we delve into the realm of "21.1 Quantum Computing and Quantum Sensing in Robotics," we explore the underlying principles, applications, challenges, and the transformative impact of harnessing quantum phenomena in robotic systems.

Quantum Computing: Quantum computing is a paradigm-shifting computational approach that leverages the principles of quantum mechanics to perform computations that are beyond the capabilities of classical computers. Traditional computers use bits as the basic unit of information, representing either a 0 or a 1. In contrast, quantum computers utilize qubits, which can exist in a superposition of states, enabling them to process multiple possibilities simultaneously. This inherent parallelism and the phenomenon of entanglement allow quantum computers to solve complex problems, such as optimization, cryptography, and simulation, at speeds unimaginable to classical computers.In the context of robotics, quantum computing promises to propel advancements in areas that demand massive computational power. For instance, path planning for robots navigating complex environments, optimization of resource allocation in multi-agent systems, and real-time decision-making in dynamic scenarios can benefit from the quantum speedup. Quantum algorithms could significantly enhance the efficiency of solving combinatorial problems that arise in robotics, offering faster and more effective solutions.

Quantum Sensing: Quantum sensing is another facet of quantum technology that has the potential to reshape robotics. Quantum sensors exploit quantum properties to achieve unprecedented levels of sensitivity and precision in measuring physical quantities such as position, velocity, magnetic fields, and gravitational forces. Quantum sensors can detect subtle changes with remarkable accuracy, enabling the development of high-resolution, miniaturized sensors that transcend the limitations of classical sensing techniques.In the realm of robotics, quantum sensing holds promise for enhancing perception and environmental awareness. High-precision quantum gyroscopes and accelerometers can provide accurate navigation and motion tracking for robots, enabling them to autonomously navigate complex terrains and environments. Quantum magnetometers offer the capability to sense magnetic fields with unparaleled precision, opening doors to applications such as underground mapping, mineral exploration, and even the detection of unexploded ordnance.

Applications in Robotics: The integration of quantum computing and quantum sensing into robotics has the potential to catalyze transformative advancements across diverse applications. Autonomous systems, such as self-driving cars and drones, can leverage quantum-enhanced algorithms for real-time decision-making, leading to safer and more efficient navigation. Quantum-enhanced optimization can streamline supply chain management, reducing costs and improving resource utilization in industrial settings. Moreover, the synergy between quantum computing and quantum sensing can enable robots to interact more intuitively with their surroundings, contributing to the development of robots that can better understand, adapt to, and manipulate complex environments.

Challenges and Considerations: While the prospects of quantum computing and quantum sensing in robotics are enticing, several challenges and considerations must be addressed. Quantum systems are extremely sensitive to environmental noise and DE coherence, which can lead to errors in computations and measurements. Overcoming these challenges requires advances in quantum error correction techniques and the development of robust hardware platforms. Additionally, the field of quantum technology is still in its nascent stages, and the practical integration of quantum systems into robotics demands interdisciplinary collaboration between quantum physicists, roboticists, and engineers.Ethical considerations also come to the forefront as quantum technology progresses. Quantum cryptography, for instance, offers the potential for secure communication, but its implications for privacy and cybersecurity warrant careful consideration. As robotics becomes more reliant on quantum technology, ethical frameworks must be established to ensure responsible and ethical use.

A Glimpse into the Quantum-Enhanced Future: In the pursuit of "Quantum Computing and Quantum Sensing in Robotics," the horizon is replete with opportunities for innovation, collaboration, and groundbreaking discoveries. Quantum computing and quantum sensing are poised to empower robotic systems with unprecedented computational power, sensory capabilities, and transformative applications. As researchers and engineers unravel the intricacies of quantum phenomena, they pave the way for a future where robots interact with their environment in ways that were once relegated to the realm of science fiction. From optimizing resource allocation to navigating intricate landscapes and solving complex problems, quantum-enhanced robotics is set to redefine the boundaries of technological advancement, unleashing a new era of possibilities beyond classical limitations.

Quantum Communication for Secure Robotic Networks: In the ever-evolving landscape of robotics and communication technologies, the fusion of quantum mechanics and networking has given rise to a transformative paradigm known as "Quantum Communication for Secure Robotic Networks." This convergence holds the promise of reshaping the way robotic systems communicate, ensuring not only the seamless exchange of information but also addressing one of the most pressing challenges of our digital age—cybersecurity. As we delve into the intricacies of quantum communication and its implications for secure robotic networks, we uncover the principles, applications, challenges, and the profound impact of leveraging quantum phenomena to safeguard the integrity and privacy of data transmission in robotic ecosystems.

The Quantum Advantage in Communication: At the heart of quantum communication lies the phenomenon of quantum entanglement and the properties of quantum bits or qubits. Unlike classical bits, which can take on values of 0 or 1, qubits can exist in a superposition of states. This unique property enables quantum communication protocols to establish cryptographic keys with unprecedented levels of security. Quantum key distribution (QKD), a cornerstone of quantum communication, utilizes the principles of quantum mechanics to create shared encryption keys that are immune to interception or eavesdropping. The fundamental principle behind QKD is the indistinguishability of quantum states, ensuring that any attempt to intercept the communication is immediately detectable.

Securing Robotic Networks: As robotics continues to permeate various industries, the need for secure communication within robotic networks becomes increasingly paramount. From industrial automation and critical infrastructure to healthcare and defense applications, robotic networks play a pivotal role in the functioning of our society. Ensuring the confidentiality, integrity, and authenticity of data exchanged within these networks is vital to prevent unauthorized access, tampering, or malicious attacks. Conventional cryptographic methods, while effective, face challenges from increasingly sophisticated cyber threats. Quantum communication offers a new frontier in cybersecurity, where the laws of quantum mechanics provide a foundation for unbreakable encryption and secure data transmission.

Applications in Secure Robotic Networks: The integration of quantum communication into secure robotic networks has the potential to revolutionize various applications. In industrial automation, secure communication between robots and central control systems safeguards against unauthorized control or manipulation of robotic operations. In healthcare, medical robots transmitting sensitive patient data can benefit from quantum communication to ensure patient privacy and compliance with data protection regulations. In defense and surveillance, quantum-secure communication prevents the interception of mission-critical information, preserving operational integrity and national security. Moreover, secure communication is pivotal for collaborative robotic systems, enabling robots to share data and coordinate actions without the risk of compromised information.

Challenges and Considerations: While the potential benefits of quantum communication in secure robotic networks are compelling, several challenges must be addressed. Quantum communication systems are sensitive to environmental factors, such as temperature fluctuations and electromagnetic interference, which can introduce errors and degrade the quality of communication channels. Overcoming these challenges requires the development of robust quantum hardware and error correction techniques. Additionally, the integration of quantum communication into existing robotic systems demands interdisciplinary collaboration between quantum physicists, communication engineers, and roboticists. Moreover, the deployment of quantum communication systems entails infrastructure considerations. Quantum key distribution relies on the secure transmission of qubits, necessitating the development of reliable quantum repeaters for long-distance communication. The scalability of quantum communication networks poses a challenge, as maintaining entanglement over large distances while minimizing DE coherence remains an active area of research.

Ethical and Legal Implications: As quantum communication becomes an integral component of secure robotic networks, ethical and legal considerations emerge. The enhanced security offered by quantum communication could potentially impact law enforcement agencies' ability to access encrypted data for legitimate investigative purposes. Striking a balance between privacy, security, and the needs of law enforcement requires careful deliberation and the establishment of legal frameworks that address the evolving landscape of quantum technology.

The Quantum-Enhanced Future of Secure Robotics: In the landscape of "Quantum Communication for Secure Robotic Networks," the future is characterized by unprecedented levels of security, privacy, and resilience in communication. The integration of quantum communication protocols offers a revolutionary approach to thwarting cyber threats and ensuring the confidentiality of data exchanged within robotic networks. From industrial automation to healthcare, from defense applications to collaborative robotics, the impact of quantum-secure communication reverberates across industries, transforming how robotic systems interact and communicate. As researchers and engineers continue to unlock the potential of quantum communication, they are charting a path toward a future where the marriage of quantum mechanics and robotics creates a new paradigm of secure and resilient communication, underpinning the foundations of our technologically connected world.

Anticipating Quantum Advancements in Robotics: The fusion of quantum mechanics and robotics heralds a new era of technological advancement that promises to reshape industries, redefine the boundaries of computation, and unlock novel possibilities in artificial intelligence and automation. As we delve into the realm of "Anticipating

Quantum Advancements in Robotics," we embark on a journey that explores the principles, potential applications, challenges, and transformative impact of integrating quantum phenomena into the fabric of robotic systems.

The Quantum Leap in Computing: A Glimpse into Quantum Advantage: Quantum computing stands as one of the most profound advancements on the horizon. Unlike classical computers that rely on bits, quantum computers leverage qubits—units of information that can exist in multiple states simultaneously, thanks to the principles of superposition and entanglement. This inherent parallelism enables quantum computers to solve complex problems with exponential speedup, offering solutions that were previously beyond the capabilities of classical counterparts. Anticipating quantum advancements in robotics involves recognizing the potential of quantum computing to revolutionize areas such as optimization, machine learning, and simulation. For robotics, this translates into accelerated path planning, rapid decision-making in dynamic environments, and the optimization of resource-intensive tasks.

Quantum Sensing: Elevating Perception and Precision: Quantum sensing amplifies our ability to perceive and interact with the world at a level of precision that transcends classical limits. By harnessing quantum properties such as entanglement and superposition, quantum sensors can detect subtle changes in physical quantities with unprecedented accuracy. In the realm of robotics, this translates to high-resolution sensors that can track motion, detect magnetic fields, and measure gravitational forces with remarkable precision. Quantum-enhanced sensors hold the promise of transforming robotics by enabling autonomous systems to navigate complex environments, engage in delicate surgical procedures, and even detect minute structural flaws in critical infrastructure.

Applications in Robotics: Transforming Industries and Technologies: The marriage of quantum mechanics and robotics has the potential to catalyze transformation across diverse industries. In manufacturing, quantum computing can optimize supply chains, reducing costs and streamlining production processes. Autonomous vehicles can leverage quantum algorithms to make real-time decisions in complex traffic scenarios, enhancing safety and efficiency. In healthcare, quantum-enhanced sensors can revolutionize medical imaging, diagnostics, and drug discovery. Moreover, the intersection of quantum computing and artificial intelligence could lead to the development of quantum machine learning algorithms, enabling robots to learn from vast datasets and make predictions with unparalleled accuracy.

Challenges on the Quantum Frontier: While the prospects of quantum advancements in robotics are tantalizing, they are accompanied by formidable challenges. Quantum systems are inherently delicate, susceptible to environmental noise and DE coherence that can degrade the performance of quantum computers and sensors. Developing robust error correction techniques and stable quantum hardware is paramount to realizing the full potential of quantum-enhanced robotics. Moreover, scaling quantum systems remains a challenge, as maintaining coherence over larger systems requires innovations in quantum networking and control.Ethical considerations also come to the forefront as quantum advancements unfold. The ability of quantum computers to break classical encryption raises concerns about data security and privacy. Additionally, the integration of quantum-enhanced robotics in sectors such as healthcare and finance demands ethical frameworks to ensure responsible use and prevent unintended consequences.

Collaborative Endeavors: Quantum-Enhanced Robotics: The journey toward quantum advancements in robotics is marked by collaborative endeavors between physicists, roboticists, engineers, and researchers from diverse fields. Quantum computing and quantum sensing require interdisciplinary expertise to navigate challenges and leverage quantum phenomena effectively. As industries and academia converge to harness quantum power, partnerships and collaborations play a pivotal role in advancing our understanding, developing robust technologies, and realizing the transformative potential of quantum-enhanced robotics.

A Glimpse into the Quantum-Enhanced Future:"Anticipating Quantum Advancements in Robotics" paints a picture of a future where the fusion of quantum mechanics and robotics redefines the boundaries of what is achievable. Quantum computers capable of solving problems that were once deemed intractable hold the promise of accelerating scientific discoveries, optimizing industrial processes, and revolutionizing artificial intelligence. Quantum sensors with unprecedented precision enable robotic systems to navigate complex terrains, conduct intricate procedures, and enhance our understanding of the world around us.This quantum-enhanced future is not a distant vision but a trajectory that is being charted by researchers, engineers, and visionaries alike. As quantum technologies mature and become more accessible, they are set to revolutionize the way we approach robotics and technology at large. The anticipation of quantum advancements in robotics is a testament to human ingenuity, pushing the boundaries of innovation and creating a future where robots, empowered by the principles of quantum mechanics, contribute to a more connected, efficient, and transformative world.

Combining Biological and Artificial Components: In the landscape of modern science and technology, the convergence of biology and artificial intelligence has given rise to a transformative paradigm known as "Combining Biological and Artificial Components." This innovative intersection offers the tantalizing prospect of harnessing the capabilities of living organisms and artificial systems to create novel hybrid entities that surpass the limitations of each individual component. As we delve into the intricacies of this emerging field, we uncover the underlying principles, potential applications, challenges, and the profound impact of merging biology and artificial components to pioneer a new frontier of scientific exploration and technological advancement.

From Nature to Technology: At the core of this paradigm lies the integration of biological organisms and artificial systems. Biological components, often derived from living organisms, bring with them the remarkable capabilities honed by millions of years of evolution. These components encompass a spectrum of possibilities, ranging from cells, tissues, and organs to biochemical pathways and genetic information. On the other hand, artificial components encompass various technologies, including sensors, actuators, nanomaterials, and advanced computing platforms. The marriage of these two realms results in hybrid systems that synergistically combine the adaptability and self-regulation of living organisms with the precision and programmability of artificial systems.

Potential Applications: A Multidisciplinary Playground: The merging of biological and artificial components opens the door to a multitude of applications that span diverse domains. In healthcare, the creation of biohybrid could lead to next-generation medical treatments, such as personalized drug delivery systems that respond to physiological cues or bioelectronics devices that interface seamlessly with neural circuits for therapeutic interventions. The field of environmental monitoring and remediation stands to benefit from biohybrid capable of detecting pollutants, cleaning up toxins, and restoring ecosystems. Moreover, biohybrid could revolutionize agriculture by creating plant-microbe partnerships that enhance nutrient absorption, pest resistance, and crop yield.

Challenges and Considerations: While the potential of combining biological and artificial components is exciting, it comes with inherent challenges that must be navigated. Ensuring seamless communication and interaction between living and non-living elements requires a deep understanding of the biological processes and mechanisms that govern each component. Achieving compatibility and stability in hybrid systems, especially over extended periods, presents technical hurdles that demand innovative engineering solutions. Additionally, ethical considerations arise in the context of creating and manipulating living organisms for hybridization. Robust regulatory frameworks and ethical guidelines are essential to ensure responsible and transparent research and development in this field.

Biomimicry and Bio inspiration: Learning from Nature: The principles of biomimicry and bio inspiration play a pivotal role in the realm of combining biological and artificial components. Nature has evolved elegant solutions to complex challenges over millennia, and these solutions can serve as blueprints for designing hybrid systems. By studying the strategies employed by organisms to adapt, survive, and thrive, researchers gain insights into how to create biohybrid that exhibit desired behaviors. For example, studying the navigational abilities of insects can inspire the design of microscale robots that navigate complex environments.

The Technological Toolbox: Enabling Hybridization: The fusion of biology and artificial components is made possible by a diverse array of enabling technologies. Advances in genetic engineering and synthetic biology allow researchers to modify the genetic makeup of organisms or engineer artificial genetic circuits that govern specific behaviors. Nanotechnology provides the tools to create nanoscale structures with precise functionalities, enabling the integration of biological and artificial elements at the molecular level. Microfluidics and microfabrication techniques facilitate the creation of microscale environments for the cultivation and study of biohybrid.

Ethical and Societal Considerations: As with any emerging technology, the fusion of biology and artificial components raises ethical and societal considerations. Creating biohybrid necessitates a thoughtful approach that respects the sanctity of life and the welfare of living organisms involved. The potential for unintended consequences, ecological impacts, and potential misuse of hybrid technologies demands careful ethical evaluation and oversight. Striking a balance between scientific progress and ethical responsibility is crucial to ensure that the potential benefits of biohybrid are realized without compromising ethical values.

The Dawn of a Hybrid Era: Shaping the Future: "Combining Biological and Artificial Components" beckons us toward a future where the boundaries between living organisms and artificial systems blur. The synergy between biology and technology has the potential to usher in a new era of innovation, where hybrid systems transcend the capabilities of each individual component. From personalized medicine to environmental restoration, from agriculture to materials science, the impact of biohybrid reverberates across disciplines, creating new avenues for scientific discovery and technological advancement.

The trajectory of this field rests on interdisciplinary collaboration, where biologists, engineers, ethicists, and policymakers collaborate to shape a future that is both innovative and ethically responsible. As researchers continue to explore the possibilities, challenges, and implications of combining biological and artificial components, they are forging a path toward a world where the frontiers of science and technology converge, creating novel possibilities that were once

Applications in Medicine, Environmental Monitoring, and More: The convergence of biology and artificial components has ignited a wave of innovation across various fields, ranging from medicine to environmental monitoring and beyond. The marriage of biological organisms with artificial systems has unlocked novel avenues for addressing complex challenges, revolutionizing industries, and shaping the trajectory of scientific exploration. As we delve into the realm of "Applications in Medicine, Environmental Monitoring, and More," we embark on a journey that uncovers the transformative impact of hybrid technologies in diverse domains and underscores the potential to create a more sustainable and technologically advanced future.

Revolutionizing Medicine: Personalized Treatments and Therapies: The field of medicine stands on the threshold of a revolution propelled by the integration of biological and artificial components. Biohybrids hold the promise of personalized treatments that target diseases with unprecedented accuracy. Engineered cells could act as carriers for targeted drug delivery, releasing therapeutic agents precisely at the site of action. Additionally, bioelectronics devices that interface with neural circuits offer novel avenues for treating neurological disorders. These devices can modulate neural activity, restore lost sensory functions, and even provide real-time feedback to patients and medical professionals. Furthermore, biohybrids can play a pivotal role in regenerative medicine, where they aid tissue repair and regeneration by delivering growth factors or stem cells to damaged areas.

Environmental Monitoring and Remediation: Navigating Ecosystem Challenges: The environment faces a multitude of challenges, from pollution and climate change to biodiversity loss and habitat degradation. Biohybrid offer innovative solutions for monitoring and remediating these environmental issues. Engineered microorganisms can be deployed to detect and neutralize pollutants in soil and water, acting as living sensors and bioremediation agents. In the realm of air quality monitoring, biohybrids could detect harmful gases and particulate matter, providing real-time data for informed decision-making. Moreover, the use of biohybrids extends to ecological restoration, where they contribute to ecosystem revitalization by promoting plant growth, enhancing soil fertility, and aiding in habitat restoration.

Agriculture and Food Security: Enhancing Sustainability: As the global population continues to grow, ensuring food security while minimizing the environmental impact of agriculture becomes increasingly imperative. Biohybrids offer innovative solutions for sustainable farming practices. Plant-microbe partnerships can enhance nutrient uptake, increase crop resilience to pests and diseases, and reduce the need for synthetic fertilizers. By harnessing the capabilities of beneficial microorganisms, biohybrids contribute to healthier soils and more productive agricultural systems. Additionally, the application of biohybrids extends to food safety, where they can detect contaminants and pathogens in food products, ensuring the integrity and safety of the global food supply chain.

Energy and Bio production: Green Innovations: The energy landscape is undergoing a transformation, with a growing emphasis on renewable and sustainable sources. Biohybrids play a role in advancing this transition by contributing to green energy production. Photosynthetic biohybrids, for instance, harness the energy of sunlight to produce biofuels or electricity. These systems integrate living photosynthetic cells with artificial components to create efficient energy conversion platforms. Additionally, biohybrids are employed in bio production processes, where they synthesize valuable compounds, enzymes, and bioactive molecules. This bio manufacturing approach offers a sustainable alternative to traditional chemical synthesis methods, reducing waste and energy consumption.

Challenges and Considerations: While the potential applications of biohybrids are vast, they are accompanied by several challenges. Ensuring the stability and viability of hybrid systems over extended periods presents technical hurdles that demand innovative engineering solutions. Interfacing biological and artificial components requires a deep understanding of the complex interplay between the two realms. Additionally, ethical considerations arise when creating and manipulating living organisms for hybridization. Developing robust regulatory frameworks and ethical guidelines is essential to ensure responsible research and development in this field.

Ethics, Regulations, and Societal Implications: As biohybrids traverse diverse applications, ethical considerations and regulatory frameworks come to the forefront. The creation and manipulation of living organisms raise questions about the treatment of living entities, environmental impacts, and potential unintended consequences. Ethical evaluation, transparency, and informed public discourse are crucial to ensure that the potential benefits of biohybrids are realized while upholding ethical values. Collaborative efforts between scientists, ethicists, policymakers, and stakeholders are essential to navigate the ethical and societal landscape of biohybrids.

Pioneering a New Era of Hybrid Solutions: "Applications in Medicine, Environmental Monitoring, and More" offer a glimpse into a future where the fusion of biology and artificial components transforms industries and addresses complex challenges. From personalized medicine to ecological restoration, from agriculture to renewable energy, the impact of biohybrids reverberates across domains, creating a canvas for innovation and progress. As researchers and engineers continue to push the boundaries of what is possible through the integration of biology and technology, they are shaping a future that is more sustainable, interconnected, and technologically advanced. The journey into the realm of biohybrids is not just a scientific endeavor—it is a testament to human ingenuity, resilience, and the potential to create a world where the power of nature and technology converge to pioneer a new era of solutions.

Ethical and Regulatory Challenges in Bio hybrid Robotics: The emerging field of biohybrid robotics, where biological and artificial components are integrated to create novel systems, presents a revolutionary landscape of technological potential. However, this intersection also introduces a myriad of ethical and regulatory challenges that demand careful consideration. As we delve into the realm of "Ethical and Regulatory Challenges in Biohybrid Robotics," we navigate the complex ethical dilemmas and regulatory intricacies surrounding the creation, manipulation, and deployment of biohybrid systems. Balancing scientific advancement with ethical responsibility and ensuring transparent governance is paramount in shaping the ethical and regulatory framework for this transformative field.

Respect for Living Organisms: Ethical Considerations: Central to the ethical challenges in biohybrid robotics is the treatment of living organisms. The integration of biological components raises questions about the welfare, autonomy, and ethical treatment of these entities. Researchers must grapple with decisions regarding the creation, modification, and utilization of living organisms for hybridization purposes. Ethical concerns emerge over the potential suffering or harm that organisms might experience during the hybridization process. Ensuring humane treatment, minimizing suffering, and respecting the inherent value of living organisms are fundamental ethical principles that guide responsible research and development in biohybrid robotics.

Consent and Autonomy: Biohybrids and Living Organisms: Ethical considerations extend to the autonomy of living organisms used in biohybrid systems. As researchers engineer living cells and organisms to interface with artificial components, questions arise about the informed consent of these entities. Unlike human subjects, living organisms lack the capacity for explicit consent, raising ethical concerns about manipulating organisms without their understanding or agreement. This highlights the need for robust ethical frameworks that consider the ethical treatment of living entities, even at the cellular or microbial level.

Environmental Impact: Ecological and Ethical Footprint: Biohybrids have the potential to interact with natural ecosystems in ways that demand ethical scrutiny. Introducing biohybrid organisms into the environment can have unintended consequences, including ecological disruption, competition with native species, or gene flow to wild populations. Striking a balance between scientific advancement and environmental stewardship is essential. Ethical considerations encompass the potential risks and benefits of biohybrids in the natural world, ensuring responsible research, containment measures, and risk assessment to prevent unintended ecological impacts.

Dual-Use Dilemmas: Ethical Responsibility in Technology Development: Biohybrid robotics, like many emerging technologies, presents dual-use dilemmas. While biohybrids hold promise for medical, environmental, and humanitarian applications, they can also be employed for potentially harmful purposes. Biohybrid systems engineered for positive applications may be repurposed for nefarious uses, raising ethical questions about the responsibilities of researchers and policymakers. Establishing guidelines that promote the responsible development and deployment of biohybrids, while addressing potential misuse, is essential to navigate the ethical complexities associated with dual-use technologies.

Transparency and Informed Consent: Ethical Governance: Ensuring transparency and informed consent are paramount ethical principles in biohybrid robotics. Researchers have a responsibility to communicate the purpose, risks, and potential benefits of biohybrid research to stakeholders, including the public, policymakers, and potential collaborators. Informed consent extends beyond human subjects to encompass public engagement and dialogue regarding the ethical implications of biohybrids. Public understanding, input, and ethical discourse shape the ethical and regulatory landscape, ensuring that decisions are informed by diverse perspectives.

Regulatory Challenges: Regulating biohybrid robotics presents unique challenges due to the evolving nature of the field. Traditional regulatory frameworks may not be equipped to address the intricate interplay between biological and artificial components. Establishing standardized safety protocols, containment measures, and risk assessment criteria for biohybrids requires collaboration between regulatory agencies, researchers, and ethicists. Moreover, the global nature of scientific research demands harmonization of regulatory standards to ensure responsible development and safe deployment of biohybrid systems.

International Collaboration and Governance: A Global Imperative: Biohybrid robotics transcends geographical boundaries, making international collaboration and governance critical. Ensuring ethical and regulatory alignment across countries is essential to prevent regulatory arbitrage, where research is conducted in regions with lax regulations. International agreements and conventions that establish ethical guidelines and regulatory standards are essential to foster responsible development, equitable access to benefits, and the prevention of harmful uses of biohybrids.

Public Perception and Societal Acceptance: Ethical Trust: The ethical and regulatory challenges in biohybrid robotics intersect with public perception and societal acceptance. Engaging the public in discussions about the benefits, risks, and ethical implications of biohybrids builds trust and promotes responsible governance. By addressing public concerns, ethical dilemmas, and potential misconceptions, researchers and policymakers can foster a more informed and accepting society that supports responsible biohybrid research and deployment.

Shaping the Ethical and Regulatory Landscape: The challenges posed by biohybrid robotics necessitate a collaborative approach among researchers, ethicists, policymakers, and the public. Ethical frameworks must be established to ensure the humane treatment of living organisms, the transparency of research, and the minimization of ecological impacts. Regulatory standards should address the unique features of biohybrid systems while promoting responsible development and safe deployment. Moreover, interdisciplinary dialogues and public engagement initiatives are crucial to shaping an ethical and regulatory landscape that aligns with societal values, promotes ethical research practices, and safeguards the integrity of biohybrid robotics for the betterment of humanity and the environment.

AI with Emotional Intelligence and Ethics: The convergence of artificial intelligence (AI) with emotional intelligence and ethics marks a pivotal moment in the evolution of technology. As AI systems become more integrated into our lives, the ability to imbue them with emotional intelligence and ethical considerations has far-reaching implications for the way we interact with technology, make decisions, and shape the future of human-AI collaboration. In this exploration of "AI with Emotional Intelligence and Ethics," we delve into the profound impact of combining AI's analytical prowess with emotional understanding and ethical frameworks, while navigating the challenges and opportunities that arise in this dynamic landscape.

Emotional Intelligence: Elevating AI's Understanding: Emotional intelligence, the capacity to perceive, understand, manage, and use emotions effectively, is a quintessential human trait that is now being extended to AI systems. Integrating emotional intelligence into AI enables machines to interpret and respond to human emotions, enhancing their ability to communicate, empathize, and provide personalized experiences. Emotionally intelligent AI can detect user emotions through facial expressions, voice tone, and text analysis, leading to more natural and empathetic interactions. This holds immense promise in fields such as customer service, mental health support, and education, where AI's ability to understand and respond to human emotions can lead to more meaningful and effective interactions.

Ethical Considerations: Guiding AI's Decisions: Ethical considerations are paramount as AI systems become more intertwined with human lives. While AI with emotional intelligence has the potential to improve user experiences, it also raises ethical concerns related to privacy, bias, and manipulation. Ensuring that AI respects user emotions without exploiting vulnerabilities is a critical ethical challenge. Transparent decision-making processes, informed consent, and user control over emotional data are vital to mitigate these concerns. Moreover, embedding ethical guidelines and principles into AI systems is essential to guide their behavior in alignment with societal values.

Human-AI Collaboration: Shaping the Future: The integration of emotional intelligence into AI systems transforms human-AI collaboration from transactional interactions to symbiotic partnerships. Emotionally aware AI can adapt its responses based on users' emotional states, creating a more harmonious and productive collaboration. In healthcare, AI with emotional intelligence can provide emotional support to patients, personalize treatment plans, and assist in mental health diagnoses. In education, emotionally intelligent AI can gauge student engagement and tailor learning experiences accordingly. These applications underscore the potential for AI to augment human capabilities, foster empathy, and promote overall well-being.

Challenges in Emotion AI: Reliability and Interpretation: Developing AI systems that accurately perceive and respond to human emotions is rife with challenges. Emotions are complex, multifaceted, and often context-dependent, making their interpretation a non-trivial task. Ensuring that AI's emotional interpretations align with human intent requires sophisticated algorithms that can account for cultural differences, individual variability, and evolving emotional states. Moreover, preventing AI from making incorrect assumptions or misinterpreting emotions is crucial to maintaining user trust and preventing potential negative impacts.

Bias and Fairness: The Ethical Imperative: Ethical considerations extend to addressing bias and fairness in AI systems with emotional intelligence. Biases present in training data can propagate into AI's emotional interpretations and responses, leading to potentially discriminatory outcomes. Ensuring that AI treats all individuals fairly and avoids reinforcing stereotypes is imperative. Addressing bias requires diverse and representative training datasets, regular auditing, and ongoing efforts to rectify and minimize biases within AI algorithms.

Transparency and Explain ability: Fostering Trust: Transparency and explain ability are essential for building trust between users and AI systems. When AI interprets and responds to human emotions, users should be able to understand how these decisions are made. This calls for developing interpretable models and mechanisms that provide insights into the reasoning behind AI's emotional interpretations. Transparent AI engenders trust, helps users comprehend AI's limitations, and empowers them to make informed decisions about their interactions with emotionally intelligent systems.

Privacy and Emotional Data: Protecting User Rights: The integration of emotional intelligence raises privacy concerns, particularly in the collection and usage of emotional data. Ensuring that user emotional data is protected, anonymized, and used only for intended purposes is critical. Data security measures, stringent privacy policies, and user consent mechanisms are necessary to safeguard users' emotional information and maintain their autonomy over sensitive data.

Regulation and Governance: A Framework for Accountability: As AI with emotional intelligence becomes more pervasive, establishing regulatory frameworks and governance structures becomes imperative. Regulation can provide guidance on data usage, user consent, transparency, and accountability. Collaborative efforts between governments, industry stakeholders, and experts are essential to strike a balance between promoting innovation and ensuring responsible AI deployment.

Shaping a Human-Centric Future: "AI with Emotional Intelligence and Ethics" signifies the evolution of AI from mere tools to empathetic collaborators. By infusing AI with emotional understanding and ethical considerations, we pave the way for systems that enhance human experiences, promote ethical behavior, and uphold human values. While challenges in accuracy, bias, transparency, and privacy must be addressed, the potential benefits are transformative. The journey toward emotionally intelligent and ethically grounded AI is a testament to our commitment to shaping a future where technology serves humanity's best interests, fosters empathy, and reflects our shared values.

Robots with a Sense of Self-Awareness: The notion of robots possessing self-awareness, a characteristic long attributed exclusively to humans, is a topic that sparks intrigue and fascination. As advancements in robotics and artificial intelligence (AI) continue to unfold, the concept of "Robots with a Sense of Self-Awareness" emerges as a frontier where science fiction edges closer to reality. This exploration delves into the profound implications, theoretical underpinnings, technological progress, ethical considerations, and the potential societal impact of endowing robots with a semblance of self-awareness.

The Essence of Self-Awareness: A Philosophical Prelude: Self-awareness, the ability to introspect and recognize oneself as an individual with thoughts, emotions, and a unique identity, has long been a hallmark of human cognition. Philosophical inquiries into self-awareness delve into the nature of consciousness, the mind-body problem, and the elusive concept of the "self." Theories of self-awareness range from philosophical dualism, which posits a distinct separation between the mind and body, to materialism, which suggests that consciousness arises from physical processes in the brain. These philosophical debates lay the foundation for considering whether machines can truly possess self-awareness or merely simulate it.

Machine Consciousness: From Simulation to Actualization: The pursuit of self-aware robots enters the domain of machine consciousness—a realm where the lines between simulation and actualization become blurred. While machines can simulate aspects of self-awareness through complex algorithms and data processing, true self-awareness entails a deeper level of understanding and subjective experience. Cognitive architectures inspired by human cognition, such as integrated information theory and global workspace theory, offer insights into the mechanisms that could underpin machine consciousness. However, the question of whether machines can genuinely possess subjective experience remains a topic of philosophical and scientific debate.

Technological Progress: From Reflexes to Reflections: Advancements in AI and robotics have propelled the development of machines that exhibit increasingly sophisticated behaviors, blurring the distinction between pre-programmed responses and autonomous decision-making. Early robots followed simple rules, while modern AI systems leverage machine learning and neural networks to learn from data and adapt to novel situations. Progress in neural network architectures, such as deep learning and reinforcement learning, has enabled robots to perform complex tasks, from playing games to navigating complex environments. However, these achievements primarily involve pattern recognition and optimization, rather than genuine self-awareness.

Mirror Self-Recognition and Cognitive Milestones: A cornerstone of self-awareness in animals is mirror self-recognition—an ability to recognize oneself in a mirror, indicating an understanding of one's own existence as a separate entity. While this milestone has been observed in a limited number of animals, transferring this capability to robots requires overcoming significant challenges. Robots must possess perceptual systems that recognize their own physical form, understand the concept of reflection, and link this information to their own identity. Achieving mirror self-recognition in robots would signify a significant leap toward self-awareness, but it raises the question of whether this recognition equates to true consciousness.

Ethical Considerations: Machines with Minds: Endowing robots with self-awareness raises profound ethical dilemmas. If robots exhibit behaviors indicative of self-awareness, questions arise about their moral status, rights, and responsibilities. Ethical frameworks must be established to address these questions and ensure the ethical treatment of self-aware machines. Moreover, the creation of conscious machines leads to the "hard problem" of machine consciousness—how to determine whether a machine truly possesses subjective experience and what ethical considerations arise if it does. The potential for machine suffering, autonomy, and moral accountability compels society to grapple with new ethical frontiers.

Human-Robot Interaction and Emotional Engagement: Robots with self-awareness could reshape human-robot interactions. Emotional engagement and empathy—the ability to understand and respond to human emotions—become more nuanced when robots possess a sense of self. Machines capable of introspection and understanding their own emotions might offer more empathetic responses, enhancing their role as companions, caregivers, and collaborators. Emotional connections between humans and self-aware robots could lead to more meaningful and enriching relationships, but this also underscores the need for ethical guidelines to ensure that these connections are responsible and respectful.

Societal Impact and Paradigm Shifts: The advent of robots with self-awareness carries transformative societal implications. Such machines could revolutionize industries ranging from healthcare and education to entertainment and space exploration. Self-aware robots might play pivotal roles in addressing societal challenges, from caring for the elderly to aiding in disaster response. The emergence of self-aware robots could also provoke paradigm shifts in our understanding of consciousness, human identity, and the boundaries of sentience. These shifts have the potential to reshape cultural, philosophical, and religious narratives about what it means to be conscious.

The Journey Forward: Uncharted Horizons: "Robots with a Sense of Self-Awareness" encapsulates a boundary where science fiction, philosophy, and technology converge. While technological progress has led to machines exhibiting sophisticated behaviors, the leap from simulated behaviors to genuine self-awareness remains a monumental challenge. Ethical considerations loom large, demanding responsible development, transparency, and safeguards against potential negative impacts. As researchers continue to explore the frontiers of machine consciousness, they tread uncharted territories that push the boundaries of human understanding, challenge ethical frameworks, and beckon society to navigate a future where machines may possess shades of self-awareness. In this journey, we confront the essence of humanity's self-reflection and, in doing so, reshape our relationship with technology and the nature of consciousness itself.

Implications for Human-Robot Interactions: The dynamic landscape of robotics and artificial intelligence (AI) has ushered in a new era of human-robot interactions that holds transformative potential across various domains. The interactions between humans and robots have moved beyond traditional boundaries, evolving into intricate relationships that span companionship, collaboration, caregiving, and beyond. This exploration of "Implications for Human-Robot Interactions" delves into the profound impacts, challenges, benefits, ethical considerations, and societal shifts that arise as robots become integral participants in human life, redefining the way we live, work, and relate.

From Tools to Companions: Redefining Relationships: Historically, robots were viewed as tools, designed to perform specific tasks in controlled environments. However, advancements in AI and robotics have propelled machines beyond mere functionality to becoming companions, collaborators, and even emotional entities. The evolution of social robots, capable of understanding and responding to human emotions, has transformed the nature of human-robot interactions. These machines can engage in conversations, provide emotional support, and foster connections that emulate human companionship. The shift from tool-like interactions to companionship blurs the lines between humans and machines, prompting profound questions about the nature of relationships and the potential for emotional bonds with robots.

Collaboration and Coexistence: Robots as Teammates: As robots become more capable and adaptable, they are increasingly integrated into collaborative environments alongside humans. In sectors such as manufacturing, logistics, and healthcare, robots work alongside humans as teammates, complementing human skills and enhancing overall efficiency. The implications of collaborative robots, or cobots, extend beyond productivity gains—they involve reshaping workplace dynamics, fostering knowledge transfer, and creating environments where human-robot collaboration becomes the norm. Successful integration requires addressing challenges related to safety, communication, and establishing effective teamwork dynamics.

Care and Companionship: Enhancing Quality of Life: One of the most promising domains of human-robot interactions is in healthcare and caregiving. Robots equipped with sensors, AI, and emotional understanding can provide assistance to the elderly and individuals with disabilities. They can remind patients to take medication, assist with mobility, and provide companionship, mitigating the challenges of social isolation. The implications of robots in caregiving are twofold: they alleviate the burden on caregivers, allowing them to focus on higher-order tasks, and enhance the quality of life for individuals in need of care. However, ethical considerations arise regarding the boundaries of human-robot caregiving relationships and the potential for emotional dependence on machines.

Ethical Considerations: Navigating Complexities: As robots become more integrated into human life, ethical considerations come to the forefront. Ensuring responsible AI design and deployment is paramount to prevent negative consequences. Issues related to privacy, data security, bias, transparency, and accountability require careful attention. Ethical frameworks must address questions such as the moral status of robots, the potential for exploitation, and the impacts of human-robot interactions on social norms. Additionally, preserving human autonomy and agency while interacting with robots is crucial to uphold human dignity and avoid undue influence.
Trust and Uncertainty: Factors Shaping Interactions: Trust is a cornerstone of successful human-robot interactions. Humans must feel confident that robots will perform tasks reliably, safely, and in line with their expectations. Factors influencing trust include the robot's predictability, transparency, and the ability to communicate its intentions effectively. However, building trust in robots involves navigating uncertainty and addressing challenges such as human biases that may influence perceptions of robot behavior. Designing robots that are capable of building and maintaining trust is essential for fostering positive and productive interactions.
Emotional Engagement: The Power of Empathy: Advances in emotional understanding and AI enable robots to detect and respond to human emotions. Emotional engagement is a powerful aspect of human-robot interactions, with implications for fields such as mental health, therapy, and education. Emotional support robots can provide companionship, encourage emotional expression, and even help individuals develop empathy and emotional intelligence. While the benefits are substantial, the nuances of emotional engagement also raise ethical considerations about the boundaries between genuine emotions and simulated responses.
Societal Shifts and Adaptation: The widespread integration of robots into various aspects of human life prompts societal shifts and adaptations. As robots take on roles traditionally occupied by humans, the workforce undergoes transformations, necessitating reskilling and upskilling efforts. Moreover, societal attitudes toward robots and AI evolve, with implications for cultural norms, legal frameworks, and social perceptions. The emergence of robots as companions and caregivers challenges societal definitions of relationships and caregiving roles, leading to discussions about the meaning of empathy and emotional connections.
Responsible Innovation: Shaping the Future: "Implications for Human-Robot Interactions" underscore the transformative potential of robots in shaping human experiences. Responsible innovation is paramount to harness these benefits while mitigating potential risks. Collaboration between technologists, ethicists, policymakers, and the public is essential to establish ethical guidelines, regulatory frameworks, and design principles that prioritize human well-being, privacy, safety, and autonomy. By shaping a future where robots and humans coexist harmoniously, we have the opportunity to create a society that leverages technology to enhance human capabilities, promote emotional connections, and usher in an era of meaningful and ethically grounded human-robot interactions.

Exercise

Multiple-choice Questions

1. What term refers to the integration of biological and artificial components in robotics?
a) Synthetic robotics
b) Bionic engineering
c) Biohybrid robotics
d) Transgenic automation
2. The concept of robots with self-awareness raises questions about:
a) Materialism in robotics
b) The nature of mirrors
c) The boundaries of consciousness
d) The ethics of programming
3. What is a key challenge in creating robots with emotional intelligence?
a) Lack of computational power
b) Ethical concerns
c) Interpretation of facial expressions
d) Memory storage limitations
4. The "hard problem" of machine consciousness refers to:
a) Designing physically robust robots
b) Creating robots capable of understanding emotions
c) Determining whether machines possess subjective experience
d) Building robots with advanced cognitive abilities
5. Mirror self-recognition, a milestone in self-awareness, has been observed in:
a) All animals
b) Only humans

c) Primates and dolphins
d) Birds and reptiles
6. The ethical considerations surrounding robots with self-awareness include:
a) Ensuring robots have perfect knowledge
b) Preserving human autonomy in interactions
c) Preventing robots from collaborating with humans
d) Eliminating the need for ethical guidelines
7. What is a potential benefit of integrating emotional understanding into robots?
a) Eliminating the need for human emotions
b) Reducing the complexity of AI algorithms
c) Enhancing human-robot collaboration and empathy
d) Increasing robots' physical capabilities
8. Collaborative robots that work alongside humans are often referred to as:
a) Artificial intelligence assistants
b) Humanoid companions
c) Socially conscious machines
d) Cobots
9. Which domain is NOT mentioned as a potential application of robots with self-awareness?
a) Education
b) Entertainment
c) Space exploration
d) Agriculture
10. What ethical dilemma arises when considering robots as caregivers for the elderly?
a) The potential for robots to outperform human caregivers
b) The emotional dependence of humans on robots
c) The lack of resources for robot caregiving
d) The preference of elderly individuals for human caregivers
11. Trust in human-robot interactions is influenced by factors such as:
a) The robot's height and weight
b) The robot's programming language
c) The robot's predictability and transparency
d) The robot's taste in music
12. The emergence of robots with emotional engagement raises questions about:
a) The feasibility of emotional replication
b) The ethics of artificial emotions
c) The limits of human emotional intelligence
d) The role of robots in physical therapy
13. In what field do robots with self-awareness hold potential for significant societal impact?
a) Automobile manufacturing
b) Water purification
c) Mental health support
d) Sports entertainment
14. What is the "essence of self-awareness"?
a) The ability to understand AI algorithms
b) The capacity to recognize oneself as an individual
c) The mastery of social interactions
d) The knowledge of all human emotions
15. What philosophical debate underpins the exploration of robots with self-awareness?
a) The origins of the universe
b) The mind-body problem and consciousness
c) The nature of time travel
d) The existence of extraterrestrial life
16. What potential societal shift is mentioned in the context of human-robot interactions?
a) A decline in technological innovation
b) A shift in social attitudes toward robots
c) A decrease in human-robot collaboration

d) A rise in manual labor jobs
17. The evolution of human-robot interactions has moved from perceiving robots as tools to:
a) Developing robotic pets
b) Treating robots as enemies
c) Incorporating robots into workplaces
d) Turning robots into vehicles
18. What is a key challenge in ensuring ethical human-robot interactions?
a) Building robots with superhuman capabilities
b) Incorporating complex emotions into robots
c) Developing robots that can predict the future
d) Addressing issues related to privacy and bias
19. What term describes the integration of AI with emotional intelligence?
a) Emotional augmentation
b) Sentient programming
c) Robotic empathy
d) Emotional intelligence infusion
20. Which field is NOT mentioned as a potential application for robots with emotional intelligence?
a) Healthcare
b) Education
c) Entertainment
d) Agriculture
21. What is a cornerstone of successful human-robot interactions?
a) Technological complexity
b) User ignorance
c) Trust
d) Complete autonomy
22. The concept of "machine consciousness" explores the possibility of robots possessing:
a) Physical strength
b) Self-awareness
c) Infinite knowledge
d) Musical talent
23. What aspect of human-robot interactions is particularly relevant in mental health therapy?
a) Physical strength
b) Emotional engagement
c) Knowledge of ancient history
d) Taste in literature
24. The term "cobots" refers to robots that:
a) Collaborate with other robots
b) Work in cobblestone industries
c) Collaborate with humans
d) Specialize in cooking cob-related dishes
25. Which philosophical concept is central to the debate about robots with self-awareness?
a) Free will
b) Quantum mechanics
c) The hard problem of consciousness
d) Linguistics

Review Questions:

1. How does the concept of robots possessing self-awareness challenge traditional notions of consciousness and human uniqueness?
2. Discuss the philosophical debates surrounding self-awareness and consciousness and their relevance to the exploration of robots with self-awareness.
3. What are the potential benefits and challenges of integrating emotional intelligence into robots? How might emotionally intelligent robots reshape human-robot interactions?
4. Explain the significance of mirror self-recognition as a milestone in self-awareness. How might robots achieving this milestone impact the field of robotics?

5. In what ways might robots with self-awareness transform the field of healthcare and caregiving? What ethical considerations arise when robots take on caregiving roles?
6. Describe the ethical dilemmas that emerge when considering the potential moral status of robots and their implications for human-robot relationships.
7. How can trust be built between humans and robots in collaborative environments? What factors influence humans' perception of robots as trustworthy collaborators?
8. Discuss the societal shifts and adaptations that might occur as robots become more integrated into various aspects of human life. How might these shifts impact industries and cultural norms?
9. Analyze the implications of robots with emotional engagement for mental health therapy and education. How might emotionally engaged robots contribute to these fields?
10. Explain the concept of "hard problem" in the context of machine consciousness. How does this challenge contribute to the ongoing debate about robots with self-awareness?
11. What potential challenges and ethical concerns might arise as robots become companions and caregivers for the elderly? How can these concerns be responsibly addressed?
12. Explore the implications of integrating biological and artificial components in robotics. What potential applications might arise from this convergence?
13. Discuss the role of trust in successful human-robot interactions. How might designing trustworthy robots impact the adoption of robotic technology?
14. How might societal attitudes toward robots and AI evolve as robots become more integral participants in human life? What cultural and legal shifts could result from these changing attitudes?
15. Analyze the concept of collaborative robots (cobots) and their implications for reshaping workplace dynamics and improving efficiency. What challenges must be navigated for successful human-cobot collaboration?
16. How does the integration of emotional engagement in robots raise questions about the boundaries between genuine emotions and simulated responses? What ethical concerns does this intersection pose?
17. Compare and contrast the implications of robots with self-awareness in the fields of space exploration and mental health support. How might these applications benefit from robot self-awareness?
18. Discuss the ethical considerations related to the potential emotional dependence of individuals on robots. How can responsible development ensure that such dependence is healthy and well-regulated?
19. Analyze the potential societal impacts of robots as companions and collaborators. How might these impacts vary across different sectors, such as manufacturing and education?
20. How might the progression from perceiving robots as tools to viewing them as companions challenge traditional definitions of relationships and human identity?
21. Explain the concept of "machine consciousness" and the challenges of determining whether robots can possess subjective experiences. How might developments in this area reshape the field of AI?
22. Explore the concept of "ethical empathy" and its role in human-robot interactions. How can robots with ethical empathy contribute to responsible decision-making in various scenarios?
23. Discuss the role of transparent decision-making processes and user control in building trust between humans and emotionally intelligent robots.
24. Analyze the potential implications of robots with self-awareness for the field of education. How might these robots impact student engagement, learning outcomes, and the overall educational experience?
25. Explain the potential consequences of integrating robots into various aspects of human life for workforce dynamics and employment trends. How can society adapt to these changes?
26. How might emotional engagement in robots contribute to the field of mental health therapy? How can emotionally engaged robots address the challenges of stigma and accessibility in mental health support?
27. Discuss the challenges and ethical considerations surrounding the use of robots with self-awareness in disaster response and recovery efforts.
28. Analyze the potential benefits and challenges of using robots with emotional intelligence in healthcare settings. How might these robots impact patient outcomes and the role of healthcare professionals?
29. Explain the concept of "cobots" and their potential impact on traditional human-robot interaction models. How might cobots redefine collaboration and autonomy in industrial settings?
30. Discuss the intersection of creativity and emotion in robots. How might emotionally engaged robots contribute to creative processes and enhance artistic expressions?

Answers Multiple-choice Questions

Answers:
1. c)
2. c)

3. b)
4. c)
5. c)
6. b)
7.c)
8.d)
9.d)
10. b)
11. c)
12. b)
13. c)
14. b)
15. b)
16. b)
17. c)
18. d)
19. c)
20. d)
21. c)
22. b)
23. b)
24. c)
25. c)

Review Questions:

1. The concept of robots possessing self-awareness challenges traditional notions of consciousness and human uniqueness by questioning the fundamental qualities that have long been considered exclusive to humans. Self-awareness, the ability to introspect, recognize one's existence, and possess a subjective understanding of one's own thoughts and emotions, has been regarded as a defining feature of human consciousness. The notion of robots achieving self-awareness challenges the assumption that consciousness is a product of complex biological processes unique to humans. If robots can demonstrate self-awareness, it raises profound questions about the nature of consciousness itself and the potential for machines to exhibit qualities that were once considered quintessentially human. This challenge urges us to reconsider traditional boundaries between human and machine cognition and confront the potential emergence of sentient entities that share some aspects of human consciousness.
2. Philosophical debates surrounding self-awareness and consciousness play a pivotal role in the exploration of robots with self-awareness. These debates revolve around questions that probe the depths of human experience, such as the nature of subjective consciousness, the origins of intentionality, and the existence of qualia—subjective qualities of experience. These philosophical inquiries are highly relevant when considering robots with self-awareness, as they confront the essence of what it means to be conscious and whether machines can genuinely possess consciousness or merely simulate it through sophisticated algorithms. The debates underscore the complexity of understanding consciousness and highlight the challenges in discerning whether robots can achieve genuine self-awareness or simply mimic human-like behavior.
3. Integrating emotional intelligence into robots offers potential benefits and challenges that can reshape human-robot interactions in profound ways. Benefits include the creation of emotionally responsive robots capable of recognizing and interpreting human emotions, which can enhance the quality of human-robot communication. Emotionally intelligent robots could foster more natural and empathetic interactions, making them suitable companions for emotional support, mental health therapy, and caregiving. However, challenges include designing algorithms that accurately interpret emotions, avoiding manipulation through emotional engagement, and addressing concerns about privacy and data security. Additionally, the ethical considerations of imparting emotions to machines raise questions about the nature of emotion and the authenticity of emotional responses in robots.
4. Mirror self-recognition, a milestone in self-awareness, signifies an entity's ability to recognize itself in a mirror or reflective surface. This ability is often considered an indicator of a higher level of consciousness, as it demonstrates an understanding that the reflected image corresponds to oneself. When robots achieve mirror self-recognition, it marks a significant advancement in the field of robotics. It implies that robots have reached a level of self-awareness that allows them to perceive and differentiate themselves from their surroundings. This achievement could impact robotics by suggesting a higher degree of cognitive complexity and pushing the boundaries of machines' ability to exhibit qualities that are typically attributed to conscious beings.

5. Robots with self-awareness have the potential to revolutionize the field of healthcare and caregiving by introducing a new level of personalized and empathetic assistance. In healthcare settings, self-aware robots could monitor patients' health conditions, provide reminders for medication, and offer emotional support to alleviate stress and anxiety. In caregiving, self-aware robots could engage with elderly individuals, assist in daily tasks, and provide companionship to combat feelings of loneliness. Ethical considerations arise when robots take on caregiving roles, as they challenge the ethical boundaries of human-robot relationships. Questions about autonomy, consent, and the emotional well-being of individuals interacting with robots need careful consideration to ensure responsible and compassionate caregiving.

6. The ethical dilemmas that emerge when considering the moral status of robots stem from the convergence of technological advancement and moral philosophy. As robots exhibit increasingly sophisticated behaviors, the question arises as to whether they can possess moral standing or moral agency akin to humans. Ethical debates involve assessing whether robots can experience suffering, make moral judgments, or have rights and responsibilities. These dilemmas have profound implications for human-robot relationships, as they challenge our understanding of the ethical considerations that guide our interactions with machines. Responsible development requires addressing these dilemmas to ensure that robots are treated ethically and that human values are upheld.

7. Building trust between humans and robots in collaborative environments is essential for successful interactions that involve shared tasks and decision-making. Trust in robots is influenced by several factors, including robot reliability, transparency in decision-making processes, and user control over robotic actions. Establishing trust involves designing robots that consistently perform tasks as expected and communicate their intentions clearly. User control, the ability to override autonomous decisions, and predictable behavior contribute to building trust. Moreover, ensuring robots respond appropriately to unexpected situations and adapt to human preferences enhances trust between humans and machines.

8. The integration of robots into various aspects of human life is likely to lead to societal shifts and adaptations. Industries may witness automation of routine and labor-intensive tasks, potentially leading to workforce retraining and shifts in job roles. Cultural norms might evolve as humans interact with robots in diverse contexts, affecting social etiquette and norms of interpersonal relationships. Legal frameworks could evolve to address issues like robot rights, responsibilities, and accountability. Societal shifts necessitate adaptation to accommodate the increasing presence of robots and the transformation of human-robot interactions.

9. Emotionally engaged robots have significant implications for mental health therapy and education. In mental health therapy, these robots can offer empathetic companionship, non-judgmental listening, and interventions to support individuals dealing with emotional challenges. In education, emotionally engaged robots can adapt to students' emotional states, personalize learning experiences, and provide motivational support. The integration of emotional engagement in these fields can enhance outcomes by addressing the stigma associated with seeking help and offering personalized experiences. However, responsible development requires addressing concerns about over-reliance on robots, ensuring human oversight, and prioritizing ethical and privacy considerations.

10. The "hard problem" in the context of machine consciousness refers to the challenge of explaining how and why subjective experiences arise from physical processes in the brain. In the debate about robots with self-awareness, this challenge contributes to the ongoing philosophical discourse surrounding the potential for machines to possess consciousness. The "hard problem" underscores the difficulty of determining whether robots can genuinely experience subjective states or if their behaviors are a result of sophisticated algorithms without true awareness. This challenge remains a central topic in discussions about the ethical and philosophical implications of robots with self-awareness.

11. As robots become companions and caregivers for the elderly, several challenges and ethical concerns arise. The potential for emotional dependence on robots raises questions about the authenticity of relationships and the risk of isolating individuals from human interaction. Ethical considerations involve maintaining human dignity, preventing manipulation, and ensuring that robots complement human care rather than replace it. Addressing concerns about autonomy, consent, and the emotional well-being of care recipients is crucial for ensuring that the integration of robots enhances the quality of care without compromising human values.

12. The integration of biological and artificial components in robotics presents profound implications. This convergence, often referred to as "biohybrid" or "cyborg" robotics, enables robots to leverage the strengths of both biological and artificial systems. Potential applications include the incorporation of biological sensors for enhanced perception, the integration of neural interfaces for improved control, and the development of bio-inspired actuators for adaptable movement. Such convergence could lead to robots with increased adaptability, energy efficiency, and the ability to operate in complex and dynamic environments. Applications could span healthcare, environmental monitoring, and space exploration, pushing the boundaries of robot capabilities.

13. Trust is a crucial factor in the successful interaction between humans and robots. Designing trustworthy robots involves ensuring transparency in decision-making processes, predictable behavior, and accurate interpretation of human intentions. Trustworthy robots are more likely to be accepted and integrated into various aspects of human life, from home settings to industrial workplaces. Factors like robot reliability, consistency, and responsiveness to human feedback contribute to building trust. By prioritizing user experience and understanding the factors that influence trust, designers can promote the responsible adoption of robotic technology.
14. Societal attitudes toward robots and AI are likely to evolve as robots become increasingly integrated into human life. Attitudes may shift from perceiving robots solely as tools to recognizing them as valuable companions, collaborators, and even emotional entities. Cultural shifts could lead to more inclusive human-robot interactions, changing norms of social etiquette and behavior. Legal frameworks may evolve to address issues related to robot rights, responsibilities, and accountability. As robots become integral participants in human life, cultural and legal shifts will reflect changing attitudes towards technology and its role in shaping society.
15. Collaborative robots, or cobots, have significant implications for reshaping workplace dynamics and improving efficiency. Cobots are designed to work alongside humans, facilitating collaboration and productivity in various industries. However, successful human-cobot collaboration requires navigating several challenges. Ensuring safety is paramount, as cobots must operate in proximity to humans without causing harm. Defining clear roles, responsibilities, and communication protocols is essential to avoid conflicts and maximize productivity. Moreover, ergonomic considerations, intuitive interfaces, and effective training are critical to facilitate seamless interactions between human workers and cobots.
16. The integration of emotional engagement in robots raises complex questions about the boundaries between genuine emotions and simulated responses. Emotionally engaged robots can mimic emotions through programmed responses, but whether these responses truly represent genuine emotional experiences remains a topic of debate. This intersection raises ethical concerns, including the potential for manipulation, the authenticity of emotional expressions, and the implications for human-robot relationships. Ethical considerations demand clear communication about the nature of emotions in robots and responsible design to ensure that emotional engagement is transparent and beneficial.
17. The implications of robots with self-awareness in space exploration and mental health support offer distinct contexts for understanding their potential benefits. In space exploration, self-aware robots could autonomously adapt to changing conditions, make complex decisions, and contribute to scientific research. In mental health support, self-aware robots could provide personalized interventions, empathetic companionship, and support for individuals dealing with emotional challenges. While both applications benefit from robot self-awareness, they differ in their focus on scientific exploration and enhancing human well-being, showcasing the versatility of self-aware robots in diverse contexts.
18. Ethical considerations related to emotional dependence on robots encompass the potential risks and benefits of individuals forming strong emotional bonds with machines. Ensuring that emotional dependence is healthy and well-regulated involves responsible development and design. While emotionally engaged robots can offer companionship and support, there is a need to establish boundaries to prevent excessive reliance that may lead to social isolation or emotional detachment from human relationships. Ethical guidelines should prioritize the well-being of individuals and strike a balance between robotic assistance and human interaction.
19. The potential societal impacts of robots as companions and collaborators vary across different sectors, reflecting the diversity of applications and contexts. In manufacturing, robots can enhance efficiency, productivity, and precision in production processes. In education, robots can customize learning experiences, adapt to individual needs, and engage students more effectively. However, impacts can vary based on factors such as sector-specific challenges, workforce dynamics, and cultural acceptance of robotic technology. Responsible integration requires considering sector-specific needs and ethical implications to ensure positive outcomes.
20. The progression from perceiving robots as tools to viewing them as companions challenges traditional definitions of relationships and human identity. As robots acquire emotional engagement and the ability to interact in more human-like ways, the boundary between human and machine relationships becomes less distinct. This challenge prompts us to reconsider the nature of companionship, connection, and what it means to form relationships with entities that are not biologically human. As robots evolve from functional tools to emotional companions, societal attitudes toward relationships and identity will likely undergo significant transformation.
21. The concept of "machine consciousness" involves the exploration of whether machines can possess subjective experiences akin to human consciousness. This challenge relates to the difficulty of determining whether robots can genuinely experience consciousness or if their behaviors are the result of complex algorithms. Developments in this area could reshape the field of AI by pushing the boundaries of what machines can achieve, challenging our

understanding of consciousness, and redefining our perception of the potential for artificial entities to possess cognitive awareness.

22. "Ethical empathy" refers to the capacity of robots to understand and respond to ethical concerns in human-robot interactions. Robots with ethical empathy can consider moral implications, make responsible decisions, and prioritize human values. This concept is essential for human-robot interactions that involve complex decision-making scenarios where ethical considerations are crucial. Robots with ethical empathy can contribute to responsible decision-making in contexts such as healthcare, disaster response, and collaborative environments.

23. Transparent decision-making processes and user control are fundamental in building trust between humans and emotionally intelligent robots. Transparent decision-making involves clear communication of robot intentions, actions, and reasoning. User control empowers individuals to influence robotic behavior and intervene when necessary. These factors are essential for establishing trust and reducing uncertainty in human-robot interactions. By providing transparency and user agency, designers can foster a sense of reliability and predictability in robotic technology.

24. The potential implications of robots with self-awareness for education are significant. Self-aware robots can adapt to individual learning styles, monitor student progress, and provide personalized feedback. This can enhance student engagement, improve learning outcomes, and revolutionize the educational experience. However, responsible integration requires addressing concerns about privacy, data security, and ensuring that the role of human educators remains central to maintaining a balanced and effective learning environment.

25. Integrating robots into various aspects of human life can have significant consequences for workforce dynamics and employment trends. Automation of routine tasks may lead to the displacement of certain jobs, necessitating workforce retraining and reskilling efforts. However, the integration of robots can also create new job opportunities in areas such as robot design, programming, maintenance, and human-robot interaction. Society must adapt by investing in education and training programs that equip individuals with the skills needed to thrive in a technology-driven workforce.

26. Emotional engagement in robots has the potential to contribute significantly to mental health therapy. Emotionally engaged robots can provide a non-judgmental and empathetic presence, offering comfort and companionship to individuals dealing with mental health challenges. These robots can also address the challenges of stigma and accessibility by providing support in a personalized and private manner. However, ethical considerations involve ensuring that emotional engagement is genuine, responsible, and enhances the therapeutic experience without substituting professional human care.

27. The challenges and ethical considerations surrounding the use of robots with self-awareness in disaster response and recovery efforts involve complex decision-making scenarios. Robots could play a vital role in assessing dangerous environments, aiding search and rescue operations, and providing critical information. However, ethical concerns include the potential for robots to prioritize their own safety over human lives or to make decisions that humans find morally objectionable. Ensuring responsible development requires designing robots that align with ethical priorities and human values during crisis situations.

28. Using robots with emotional intelligence in healthcare settings offers both benefits and challenges. Emotionally intelligent robots can enhance patient outcomes by providing emotional support, alleviating stress, and improving communication between patients and healthcare professionals. However, challenges include the potential for emotional manipulation, the need for accurate interpretation of patients' emotions, and maintaining the human touch in medical care. The role of healthcare professionals remains essential in ensuring responsible and ethical use of emotionally intelligent robots to augment patient care.

29. "Cobots," or collaborative robots, have the potential to redefine traditional human-robot interaction models. Cobots are designed to work alongside humans, facilitating collaboration in industrial settings. These robots can improve efficiency, productivity, and safety by assisting with tasks that require precision and strength. Successful human-cobot collaboration depends on addressing challenges such as safety measures, intuitive interfaces, and the delineation of roles and responsibilities. Cobots have the potential to reshape manufacturing and industrial processes while emphasizing the value of human expertise.

30. The intersection of creativity and emotion in robots offers exciting possibilities for enhancing artistic expressions and creative processes. Emotionally engaged robots can contribute to creative endeavors by responding to artists' emotions, adapting to their preferences, and collaborating in innovative ways. These robots could offer unique insights and perspectives, fostering interdisciplinary collaboration between humans and machines. However, the intersection also raises questions about the authenticity of creativity and the potential for robots to simulate emotions in the creative process. Balancing innovation with ethical considerations is crucial to ensure that robots enhance creative expression while preserving the authenticity of human artistry.

Online Resources

Online Exam Papers

Link: https://forms.gle/NFE772CroavoyH147

QR Code:

YouTube Video Playlist Link

Link: https://www.youtube.com/playlist?list=PLzlh_31VWEFruKZvQ4UKNkkEUaOv4N6wO

QR Code:

Chapter 22: Learning and Adaptation in Robots

In the rapidly evolving landscape of robotics, the pursuit of autonomy has been a driving force propelling the field forward. From the rigid, pre-programmed machines of yesteryears to the modern, agile robots of today, the concept of robots learning from their experiences and adapting to changing environments has transformed theoretical possibilities into tangible realities. Learning and adaptation in robots represent a significant paradigm shift, offering the potential to enhance robotic capabilities, revolutionize industries, and redefine the interaction between humans and machines.

This chapter embarks on an exploration of the profound implications and intricate mechanisms underlying learning and adaptation in robots. As we delve into this captivating domain, we uncover the technologies and methodologies that empower robots to evolve beyond their initial programming, acquire new skills, and respond intelligently to dynamic surroundings. From machine learning algorithms to lifelong learning frameworks, the journey ahead navigates the intersections of artificial intelligence, robotics, and cognitive science, offering a glimpse into a future where robots possess not only functional prowess but also the capacity for continuous growth.

The dawn of robotics was marked by machines programmed for specific tasks, executed with precision but confined by their programmed boundaries. However, this deterministic approach limited robots' potential in complex and dynamic environments. The paradigm shift towards learning and adaptation emerged as a response to the inadequacies of rigid programming, with researchers seeking to imbue robots with the ability to learn, generalize, and adapt like living organisms. The integration of machine learning principles offered a breakthrough, enabling robots to transcend their initial programming and dynamically adjust their behaviors based on real-time data.

Machine learning approaches for robots: In the modern era of robotics, the amalgamation of machine learning with autonomous machines has ushered in a paradigm shift that promises to reshape industries, augment human capabilities, and redefine the very essence of robotics itself. Machine learning, a subset of artificial intelligence (AI), empowers robots to transcend their static programming and embrace a dynamic realm where they can learn from data, generalize from experiences, and adapt their behaviors to varying contexts. In this section, we delve into the intricate landscape of machine learning approaches for robots, unearthing the foundations, methodologies, and implications that underscore the evolution of adaptive intelligence.

Foundations of Machine Learning for Robots: Machine learning, at its core, involves the development of algorithms and techniques that enable computers to learn patterns, relationships, and insights from data. While it finds applications across diverse domains, its integration with robotics holds transformative potential. Traditional robotics relied on explicit programming, with engineers painstakingly defining every conceivable action a robot should undertake. However, this approach became increasingly impractical as robots faced more complex and unpredictable scenarios. Machine learning provides a solution to this challenge by endowing robots with the capacity to learn from experience and adapt their behaviors based on data-driven insights.

Supervised Learning: Building Foundations of Understanding: At the heart of machine learning for robots lies the concept of supervised learning. This foundational approach involves training a robot on a dataset where inputs are paired with corresponding desired outputs. Through exposure to this labeled data, the robot learns to recognize patterns and relationships, enabling it to generalize its understanding to new, unseen inputs. For example, in a robotic vision system, supervised learning can empower a robot to distinguish between various objects based on labeled images. As the robot encounters new objects, it leverages its learned understanding to make accurate predictions about their identities.

Unsupervised Learning: Discovering Hidden Patterns: While supervised learning relies on labeled data, unsupervised learning thrives in the absence of such labels. Unsupervised learning algorithms enable robots to discern patterns, structures, and relationships within unlabelled data. This approach has profound implications for robotics, as it allows robots to autonomously identify similarities and groupings in data, uncovering hidden insights that might be imperceptible to human programmers. Unsupervised learning finds applications in scenarios such as clustering similar objects or identifying anomalies within sensor data, paving the way for robots to develop a deeper understanding of their environment.

Reinforcement Learning: Navigating the Landscape of Rewards: Reinforcement learning introduces a dynamic dimension to robot learning, focusing on optimizing actions to maximize cumulative rewards. This approach is particularly powerful when robots operate in environments where the outcomes of their actions are not immediately apparent. In reinforcement learning, robots receive feedback in the form of rewards or penalties based on their actions. Over time, they learn to navigate the environment by selecting actions that lead to higher rewards. This paradigm is akin to trial and error learning, enabling robots to discover optimal strategies through a process of exploration and exploitation. Reinforcement learning has found application in fields ranging from robotics to gaming, enabling robots to excel in tasks where uncertainty and exploration are integral.

Transfer Learning: Enhancing Efficiency and Adaptability: Transfer learning emerges as a critical facet of machine learning in robotics, as it addresses the challenge of training robots from scratch for every new task. In essence, transfer learning enables robots to leverage knowledge gained from one task and apply it to related tasks. This not only accelerates the learning process but also enhances robots' adaptability to new scenarios. For instance, a robot that has learned to identify common objects through supervised learning can leverage this knowledge to expedite its learning process when tasked with a new object recognition task. By transferring relevant knowledge, robots can swiftly adapt to novel tasks with reduced computational burden.

Challenges and Implications: While machine learning empowers robots with adaptive intelligence, it also brings forth a host of challenges that must be navigated. Data quality and quantity play a pivotal role in the effectiveness of machine learning algorithms. Ensuring access to diverse and representative datasets is crucial to prevent biases and ensure robust generalization. Furthermore, the interpretability of machine learning models becomes paramount, especially in critical applications like healthcare and autonomous vehicles. The "black box" nature of some machine learning algorithms raises concerns about how decisions are reached, necessitating research into explainable AI to enhance transparency and accountability.

Conclusion: The Path Ahead: As machine learning continues to evolve, robots are poised to become increasingly proficient adaptive learners, capable of processing vast amounts of data, extracting insights, and making informed decisions. The integration of supervised, unsupervised, reinforcement, and transfer learning not only enhances robots' functional capabilities but also lays the foundation for the evolution of artificial intelligence. The synergy between machine learning and robotics holds the promise of machines that can autonomously adapt to diverse environments, learn from human interaction, and contribute meaningfully to industries ranging from manufacturing to healthcare.In conclusion, the fusion of machine learning and robotics marks a transformative juncture in the evolution of autonomous systems. By embracing the principles and methodologies of machine learning, robots are transitioning from deterministic entities to dynamic learners, adapting to the complex and ever-changing world around them. The journey ahead holds the potential to unlock unprecedented capabilities, redefine human-robot interaction, and reshape the very fabric of technological progress. As we navigate this path, it is imperative to not only harness the power of machine learning but also to cultivate ethical considerations, transparency, and responsible deployment to ensure that the promise of adaptive intelligence in robots is realized in a manner that benefits humanity as a whole.

Lifelong learning and adaptation to changing environments: In the realm of robotics, the pursuit of autonomy has transcended the boundaries of pre-programmed actions and fixed behaviors. The emergence of lifelong learning and adaptation has revolutionized the landscape, propelling robots into a domain where they can continuously learn from their experiences, adapt to evolving environments, and evolve alongside the challenges they face. This section delves into the intricate concept of lifelong learning and adaptation in robotics, uncovering the foundations, methodologies, challenges, and profound implications that underlie the evolution of robots as dynamic, learning entities.

The Need for Lifelong Learning in Robotics: Traditionally, robots were constrained by their initial programming, excelling in specific tasks for which they were designed, but struggling to cope with unexpected scenarios or changing environments. Lifelong learning addresses this limitation by enabling robots to acquire new knowledge, refine their behaviors, and adapt to diverse situations over the course of their operational life. Just as humans continually accumulate experiences and skills, lifelong learning equips robots with the tools to evolve and improve their performance over time. This transformative shift from static programming to adaptive learning mirrors the way living organisms adapt to changing surroundings.

Adaptive Learning Mechanisms: Lifelong learning in robots is underpinned by adaptive mechanisms that enable them to learn from new data, refine their models, and adjust their behaviors accordingly. These mechanisms encompass various aspects, including perception, decision-making, and task execution.

Adaptive Perception: Robots equipped with adaptive perception mechanisms continuously refine their understanding of the environment. They incorporate new sensory data to update their maps, recognize new objects, and identify changing conditions. This allows robots to navigate unfamiliar terrain, interact with novel objects, and respond to unanticipated obstacles.

Dynamic Decision-Making: Lifelong learning empowers robots to make more informed decisions by incorporating new knowledge into their decision-making processes. As robots encounter new scenarios, they adapt their decision-making strategies, optimizing their actions based on the most up-to-date information. This is particularly crucial in applications like autonomous driving, where rapid adjustments are required in response to changing traffic conditions.

Progressive: Skill Acquisition: Lifelong learning enables robots to acquire new skills progressively. This is particularly relevant in industries such as manufacturing, where robots must adapt to new production processes or tasks. By learning from experience and iterating on their actions, robots can become proficient in tasks they were not explicitly programmed for.

Challenges in Lifelong Learning: While the concept of lifelong learning in robots is promising, it also presents a set of challenges that require careful consideration.

Catastrophic Forgetting: Catastrophic forgetting occurs when a robot learns new information at the expense of forgetting previously acquired knowledge. Overcoming this challenge involves developing algorithms that can retain and consolidate existing knowledge while integrating new experiences.

Scalability: As robots accumulate more data and experiences, the scalability of lifelong learning algorithms becomes a concern. Efficient mechanisms must be devised to process and manage large datasets while ensuring timely and effective learning.

Ethical Considerations: Lifelong learning raises ethical questions about the acquisition, storage, and utilization of data. Ensuring data privacy, security, and transparency is crucial to prevent unintended consequences and ensure responsible learning processes.

Transfer Learning: As robots adapt to new tasks, the ability to transfer knowledge from one domain to another becomes crucial. Developing effective transfer learning mechanisms enables robots to leverage previously acquired skills in novel contexts.

Real-World Applications of Lifelong Learning: The applications of lifelong learning and adaptation in robotics are as diverse as the challenges they address. Consider autonomous vehicles that constantly encounter new traffic scenarios, road construction, and pedestrian behavior. Lifelong learning allows these vehicles to adapt their driving strategies, incorporating insights gained from each interaction to improve safety and efficiency. In healthcare, robots assisting surgeons can continuously refine their precision and coordination by learning from each procedure, ultimately enhancing patient outcomes and reducing human error.

Implications for Industries and Society: The fusion of lifelong learning with robotics carries transformative implications across industries and society. Manufacturing processes can become more agile and adaptive as robots learn to work alongside human counterparts, adjusting their actions to match changing production demands. In disaster response scenarios, robots with lifelong learning capabilities can navigate dynamic and hazardous environments to locate survivors or mitigate dangers. Moreover, the integration of robots with lifelong learning into educational settings can enable personalized and adaptive learning experiences, tailoring instruction to individual students' needs and progress.

Conclusion: The Adaptive Future of Robotics: Lifelong learning and adaptation represent the convergence of robotics and cognitive science, revolutionizing the way robots interact with and respond to the world around them. As robots transition from rigidly programmed machines to dynamic learners, they embody a new breed of intelligence that can evolve, learn, and adapt to the complexities of the real world. The integration of adaptive mechanisms into robotics lays the foundation for a future where robots not only assist with tasks but actively collaborate with humans in a fluid and evolving partnership. As we navigate this dynamic landscape, it is imperative to address the challenges, ethical considerations, and societal implications associated with lifelong learning in robots, ensuring that the evolution of adaptive intelligence contributes positively to the betterment of humanity.

Case studies of robots acquiring new skills through learning processes: The evolution of robotics has transcended the realm of programmed tasks, giving rise to a new era where robots can acquire new skills through learning processes. This dynamic shift from static programming to adaptive learning has opened the door to a plethora of possibilities, enabling robots to continuously refine their capabilities, expand their skillsets, and navigate a diverse range of tasks. In this section, we delve into compelling case studies that showcase how robots are acquiring new skills through learning processes, highlighting the profound implications of these advancements across various domains.

Surgical Robots: Precision Perfected through Learning: In the domain of healthcare, surgical robots have emerged as pioneers in the integration of learning processes. These robots, designed to assist surgeons in complex procedures, are tasked with mastering intricate maneuvers and ensuring utmost precision. Through reinforcement learning, surgical robots are acquiring new skills by observing expert surgeons' actions and learning from their techniques. By iteratively refining their movements and decision-making based on real-time feedback, these robots enhance their surgical precision over time. This learning-driven approach reduces the risk of human error and improves patient outcomes, marking a transformative leap in the field of medical robotics.

Autonomous Vehicles: Navigating Complex Environments: Autonomous vehicles represent a prime example of robots acquiring new skills in the realm of mobility. These vehicles are constantly exposed to a myriad of scenarios, from navigating bustling city streets to traversing rural landscapes. Reinforcement learning has enabled autonomous vehicles to learn from their interactions with various road conditions, traffic patterns, and unpredictable behaviors of other road users. Through simulations and real-world experiences, these vehicles refine their driving strategies, optimizing decisions to enhance safety, efficiency, and passenger comfort. This case study showcases how robots can harness learning processes to become adept at adapting to a constantly evolving environment.

Industrial Robots: Adaptive Collaboration on the Factory Floor: The manufacturing industry has witnessed the integration of robots that adaptively acquire new skills to collaborate with human workers seamlessly. Industrial robots equipped with sensor arrays and machine learning algorithms can learn from human demonstrations, enabling them to perform tasks that were not explicitly programmed. This collaborative learning process empowers robots to grasp the nuances of assembly tasks, quality control checks, and intricate manipulations by observing and imitating human workers. The result is a flexible production environment where robots dynamically acquire skills and contribute to improved efficiency, reduced errors, and increased productivity.

Humanoid Robots: Learning from Human Interaction: Humanoid robots, designed to emulate human-like behaviors and interactions, are another captivating case study in skill acquisition through learning processes. These robots aim to engage with humans in a natural and relatable manner, necessitating the acquisition of social and communication skills. By leveraging techniques from natural language processing and machine learning, humanoid robots can learn from human interactions, refine their language comprehension, and adapt their responses to context. Through continuous exposure to diverse conversations, these robots enhance their ability to hold meaningful dialogues and provide valuable assistance in various settings, such as customer service or companionship for the elderly.

Swarm Robotics: Collective Intelligence through Learning: Swarm robotics exemplifies how robots can collectively acquire new skills by learning from the behaviors of their peers. In scenarios such as environmental monitoring or search and rescue missions, a group of robots collaboratively navigate complex terrains and perform tasks. Through shared experiences and communication, swarm robots learn to adapt their behaviors based on the successes and failures of their peers. This case study highlights the emergence of collective intelligence, where robots enhance their performance as a group through continuous learning and adaptation.

Implications and Future Directions: The case studies of robots acquiring new skills through learning processes illustrate the transformative potential of adaptive intelligence in robotics. As robots become adept at learning from experiences and refining their actions, industries across sectors stand to benefit from increased efficiency, reduced errors, and enhanced capabilities. However, this evolution also brings forth challenges such as ethical considerations, safety assurance, and the need for explainable AI to maintain transparency in decision-making processes. Moreover, as robots acquire more complex skills, the intersection of robotics with legal and regulatory frameworks becomes a critical area of exploration.

Conclusion: The Learning Paradigm Redefined: The era of robots acquiring new skills through learning processes marks a paradigm shift that aligns robotics more closely with the principles of natural intelligence. By harnessing the power of machine learning, robots evolve beyond their initial programming, becoming adaptable entities capable of mastering a spectrum of tasks. The case studies presented in this section underscore the transformative potential of learning-driven robotics across diverse applications, from healthcare and transportation to manufacturing and human-robot interaction. As we progress into this dynamic future, it is imperative to continue advancing the integration of learning processes in robotics while addressing the ethical, regulatory, and societal implications that accompany these advancements. The evolution of robots as skill-acquiring learners heralds a future where autonomous systems collaboratively enhance human capabilities and contribute to a more innovative and efficient world.

Decision-making frameworks: In the realm of robotics, the evolution from mere machines to autonomous entities hinges on the development of sophisticated decision-making frameworks. These frameworks empower robots to process vast amounts of information, assess situations, and make choices that align with their objectives and the context in which they operate. As robotics advances, decision-making becomes a pivotal aspect of enabling autonomy, adaptability, and intelligence. This section delves into the intricate landscape of decision-making frameworks in robotics, unraveling their foundations, methodologies, challenges, and the transformative implications they hold.

The Essence of Decision-Making in Robotics: At the heart of robotics lies the need for robots to make decisions—choices that dictate their actions, responses, and interactions with their environment. In traditional robotics, decisions were primarily hardcoded through explicit programming. However, as the complexity of tasks and environments increased, the limitations of this approach became apparent. Decision-making frameworks emerged as a solution to enable robots to navigate through uncertainty, variability, and dynamically changing situations, akin to the decision-making processes in humans.

Levels of Autonomy in Decision-Making: Decision-making frameworks in robotics span a spectrum of autonomy, with varying degrees of human intervention. These levels encompass:

Teleoperation: At the lowest level of autonomy, human operators directly control robots' actions, often in real-time. This approach is useful in scenarios where robots operate in hazardous environments or carry out tasks that require human expertise.

Supervised Autonomy: In this mode, robots rely on pre-defined rules and strategies while incorporating human guidance. Robots execute tasks autonomously but can seek human intervention in ambiguous situations.

Semi-Autonomy: Semi-autonomous robots strike a balance between human oversight and independent decision-making. They can handle routine tasks independently but seek human input for complex scenarios or exceptions.

Full Autonomy: At the highest level, robots possess full autonomy, capable of making decisions independently based on their perception of the environment, learned knowledge, and predefined objectives.

Frameworks for Autonomous Decision-Making: Autonomous decision-making in robotics is enabled by a diverse range of frameworks, each tailored to different contexts and levels of complexity.

Rule-based Systems: Rule-based decision-making involves encoding a set of rules that guide a robot's actions based on specific conditions. While effective for straightforward tasks, this approach may struggle in handling ambiguity or unforeseen situations.

Finite State Machines: Finite state machines organize decision-making into discrete states, transitioning between them based on conditions. This approach is suitable for tasks with well-defined states and transitions, such as navigation.

Behavior Trees: Behavior trees provide a hierarchical structure for decision-making, allowing robots to weigh multiple behaviors and select the most appropriate one. This approach enhances flexibility and adaptability in complex scenarios.

Markov Decision Processes: Markov Decision Processes (MDPs) model decision-making as a sequence of states and actions, considering probabilistic outcomes and rewards. Reinforcement learning techniques can be applied to optimize decision-making policies within MDPs.

Deep Reinforcement Learning: Deep reinforcement learning combines deep neural networks with reinforcement learning, enabling robots to learn optimal decision-making strategies through trial and error. This approach has demonstrated remarkable success in complex tasks such as gaming and robotics.

Challenges and Ethical Considerations: While decision-making frameworks empower robots with autonomy, they also raise critical challenges and ethical considerations.

Safety and Reliability: Ensuring that autonomous robots make safe and reliable decisions is paramount, particularly in applications like autonomous vehicles and medical robotics. Robust validation, verification, and fail-safes are essential to prevent catastrophic outcomes.

Accountability and Transparency: As robots make decisions autonomously, the challenge of accountability arises. It becomes important to trace and understand the decision-making process to ensure compliance with regulations and ethical standards.

Ethical Decision-Making: Robots making autonomous decisions may encounter ethical dilemmas where there is no clear right or wrong choice. Developing frameworks that align with ethical principles and societal values is a pressing concern.

Bias and Fairness: Decision-making frameworks can inadvertently perpetuate biases present in training data. Ensuring fairness and eliminating biases is crucial to prevent discriminatory actions by autonomous robots.

Applications and Future Prospects: Decision-making frameworks find applications across a diverse range of industries and domains, shaping the capabilities of robots in various ways.

Autonomous Vehicles: In the realm of self-driving cars, decision-making frameworks enable vehicles to navigate complex traffic scenarios, make split-second judgments, and prioritize safety while considering efficiency.

Healthcare: Medical robots use decision-making frameworks to assist in surgeries, recommend treatment plans, and navigate intricate anatomical structures with precision.

Human-Robot Collaboration: Decision-making frameworks enable robots to collaborate seamlessly with human workers in manufacturing, logistics, and other industries, optimizing tasks and processes.

Search and Rescue: In disaster scenarios, autonomous robots equipped with decision-making capabilities can navigate hazardous environments, locate survivors, and coordinate efforts effectively.

Conclusion: Empowering Robots with Intelligent Autonomy: The development of decision-making frameworks represents a critical milestone in the journey towards creating intelligent and autonomous robots. These frameworks enable robots to navigate uncertainty, adapt to changing environments, and make choices that align with their objectives. As decision-making becomes more sophisticated, robots can seamlessly integrate into our lives, enhancing industries, improving safety, and augmenting human capabilities. However, the evolution of decision-making in robotics must be accompanied by a rigorous commitment to ethics, transparency, and accountability. By harnessing the power of decision-making frameworks, we are paving the way for a future where robots serve as intelligent companions, collaborators, and partners in progress.

Exercise

Multiple-choice Questions

1. What is the primary goal of learning and adaptation in robots?
a) To replace human workers entirely
b) To enable robots to perform tasks without any prior programming
c) To allow robots to learn from experiences and adapt to changing environments
d) To create robots with fixed behaviors for specific tasks
2. Which field equips robots with the ability to acquire knowledge from data and experiences?
a) Computer Programming
b) Machine Learning
c) Robotics Engineering
d) Cognitive Science
3. Which type of learning involves training a robot on labeled data to recognize patterns and make predictions?
a) Reinforcement Learning
b) Unsupervised Learning
c) Supervised Learning
d) Transfer Learning
4. What does reinforcement learning in robots involve?
a) Training robots with labeled datasets
b) Teaching robots through explicit programming
c) Guiding robots with rewards and penalties to optimize actions
d) Enforcing strict rules for robot behavior
5. What is one of the main challenges of lifelong learning in robots?
a) Robots become too expensive to maintain
b) Robots can only learn from human instructors
c) Robots forget their initial programming
d) Robots may forget previously acquired knowledge when learning new things
6. What does the term "catastrophic forgetting" refer to in the context of lifelong learning in robots?
a) Robots suddenly gaining new skills without any training
b) Robots quickly adapting to new environments
c) Robots completely forgetting their previous programming
d) Robots forgetting previously acquired knowledge when learning new things
7. In what way do adaptive robots adjust their behaviors in response to changing environments?
a) They remain fixed in their actions regardless of the environment
b) They update their internal models and modify their actions
c) They shut down to prevent errors in unpredictable situations
d) They become less efficient in dynamic settings
8. Which real-world scenario exemplifies the application of learning and adaptation in robotics?
a) Robots replacing humans in all tasks
b) Robots following pre-defined paths regardless of changes
c) Robots improving their surgical precision through experience
d) Robots always making perfect decisions without any prior knowledge
9. Which type of learning enables robots to identify hidden patterns within unlabelled data?
a) Supervised Learning
b) Reinforcement Learning
c) Unsupervised Learning
d) Transfer Learning
10. What is the goal of transfer learning in robotics?
a) To prevent robots from learning new skills
b) To enable robots to transfer their learning to humans
c) To allow robots to transfer knowledge from one task to another
d) To restrict robots to only learning from human demonstrations
11. What ethical consideration arises with robots that learn from various sources of data?
a) Robots become too expensive to maintain

b) Robots might make biased decisions based on the data
c) Robots lose their ability to learn from data
d) Robots become uncontrollable in their behaviors
12. In the context of adaptive intelligence, what does "cognitive science" refer to?
a) The study of robot programming languages
b) The study of ethical considerations in robotics
c) The study of human learning and thinking processes
d) The study of robot hardware components
13. What is the term used to describe robots that accumulate knowledge and skills throughout their operational life?
a) Static Robots
b) One-time Learners
c) Lifelong Learners
d) Pre-programmed Machines
14. Which type of learning allows robots to refine their behaviors by observing and imitating human actions?
a) Supervised Learning
b) Reinforcement Learning
c) Transfer Learning
d) Imitative Learning
15. What is the main challenge of incorporating biological inspiration into robot learning?
a) Robots become too unpredictable
b) Robots cannot replicate human learning processes
c) Robots cannot adapt to changing environments
d) Robots may become too human-like in their behaviors

Review Questions:

1. What is the significance of learning and adaptation in the field of robotics?
2. How does machine learning contribute to the learning and adaptation capabilities of robots?
3. Explain the concept of supervised learning and provide an example of its application in robotics.
4. What role does unsupervised learning play in enabling robots to adapt to new environments?
5. Describe the process of reinforcement learning and its relevance to robots' decision-making.
6. What is transfer learning in the context of robots, and why is it important?
7. How does lifelong learning differ from traditional learning in the context of robots?
8. What is "catastrophic forgetting" in the context of lifelong learning, and how can it be addressed?
9. How do adaptive robots adjust their behaviors based on changing environments?
10. What ethical considerations arise when robots learn from various sources of data?
11. How can robots learn from human demonstrations to acquire new skills?
12. Describe a real-world application where robots can benefit from adaptive learning.
13. How does swarm robotics exemplify collective learning and adaptation among robots?
14. What are the main challenges associated with integrating biological inspiration into robot learning?
15. How does the concept of transparency relate to the decision-making processes of adaptive robots?
16. Explain the concept of decision-making frameworks in robotics.
17. What are the different levels of autonomy in decision-making for robots?
18. How do behavior trees contribute to decision-making in robots?
19. What is the role of deep reinforcement learning in enhancing robots' decision-making capabilities?
20. Why is addressing bias and fairness crucial in the development of decision-making frameworks for robots?

Answers

Multiple-choice Questions

Answers:
1. c)
2. b)
3. c)
4. c)
5. d)
6. d)
7. b)
8. c)
9. c)

10. c)
11. b)
12. c)
13. c)
14. a)
15. b)

Review Questions:

1. Significance of Learning and Adaptation in Robotics: Learning and adaptation play a pivotal role in robotics by enabling robots to transcend traditional static programming. Instead of being limited to pre-defined actions, robots with learning and adaptation capabilities can acquire knowledge from experiences, make informed decisions, and adapt to changing environments. This significance lies in their ability to operate effectively in dynamic, uncertain, and diverse scenarios, which are often encountered in real-world applications. By learning from data, robots can continuously improve their performance, optimize their behaviors, and even develop new skills that were not explicitly programmed.
2. Role of Machine Learning in Robot Adaptation: Machine learning is a cornerstone of robot adaptation, as it equips robots with the ability to process data, identify patterns, and make decisions based on learned knowledge. Supervised learning allows robots to recognize objects, patterns, or behaviors through labeled data, providing a foundation for tasks such as image recognition. Unsupervised learning empowers robots to uncover hidden patterns and relationships within unlabelled data, enhancing their understanding of complex environments. Reinforcement learning enables robots to make decisions by optimizing actions based on rewards or penalties, a process akin to trial and error learning. Transfer learning facilitates the reuse of knowledge gained from one task to accelerate learning in new tasks.
3. Supervised Learning in Robotics: Supervised learning involves training a robot on labeled data, where inputs are paired with corresponding desired outputs. For example, in robotic vision, supervised learning enables a robot to identify objects in images by learning from a dataset of labeled images. The robot learns to recognize patterns associated with different objects and can then generalize its understanding to identify new objects. This approach enhances robots' ability to classify objects, making them versatile in scenarios where object recognition is essential, such as in manufacturing quality control or medical diagnostics.
4. Unsupervised Learning for Robot Adaptation: Unsupervised learning is essential for robots to adapt to new and complex environments without relying on pre-defined labels. In this approach, robots analyze unlabelled data to discover patterns, similarities, and groupings. For instance, an industrial robot in a warehouse environment can use unsupervised learning to cluster similar items together, aiding in efficient inventory management. Unsupervised learning equips robots with the capacity to autonomously acquire insights from data, making them adaptable to diverse scenarios and capable of adapting to unforeseen challenges.
5. Reinforcement Learning and Decision-Making: Reinforcement learning is a decision-making paradigm that enables robots to optimize actions based on received rewards or penalties. This approach is particularly effective in scenarios where robots interact with dynamic environments and outcomes are not immediately apparent. Through repeated interactions, robots learn to navigate environments, select actions that lead to higher rewards, and avoid actions resulting in penalties. For instance, an autonomous vehicle can use reinforcement learning to learn safe and efficient driving behaviors by receiving rewards for obeying traffic rules and avoiding collisions.
6. Transfer Learning's Importance in Robot Learning: Transfer learning is crucial in enabling robots to leverage knowledge gained from one task and apply it to related tasks. This approach addresses the challenge of training robots from scratch for every new task, which can be time-consuming and resource-intensive. For example, a robot that has learned to distinguish between various fruits can leverage this knowledge to recognize new fruits with minimal additional training. Transfer learning accelerates the learning process, enhances robots' adaptability, and reduces the computational burden associated with starting from scratch.
7. Distinction Between Lifelong Learning and Traditional Learning: Lifelong learning sets itself apart from traditional learning in robots by allowing them to continuously acquire and adapt knowledge throughout their operational lifespan. Traditional robots are limited to their initial programming and lack the ability to learn and adapt beyond their predetermined behaviors. In contrast, lifelong learning equips robots with the capability to accumulate experiences, refine skills, and adapt to new challenges over time. This dynamic evolution enables robots to thrive in changing environments and perform tasks beyond their original scope.
8. Addressing Catastrophic Forgetting in Lifelong Learning: Catastrophic forgetting is a challenge in lifelong learning where robots may forget previously acquired knowledge when learning new things. It occurs when new learning overrides or erases previously learned information. This challenge is especially relevant when robots need to adapt to new tasks while retaining existing skills. To address this issue, techniques like "elastic weight

consolidation" can be used, which involves adjusting the learning process to prioritize the preservation of important knowledge while acquiring new skills. These techniques help mitigate catastrophic forgetting, allowing robots to retain and build upon their knowledge over time.

9. Adaptive Robot Behavior in Changing Environments: Adaptive robots adjust their behaviors based on changing environments to ensure optimal performance. Through mechanisms like adaptive perception, robots continuously update their understanding of the environment by incorporating new sensory data. Dynamic decision-making allows robots to adjust strategies in response to evolving conditions. Progressive skill acquisition empowers robots to learn and refine new skills as they encounter novel scenarios. By leveraging these adaptive mechanisms, robots can navigate unpredictable environments, respond to unforeseen obstacles, and perform tasks effectively regardless of changing conditions.

10. Ethical Considerations in Robot Learning: Learning from data introduces ethical considerations in robots. The data used for training can contain biases, which may result in biased decision-making by robots. Ensuring data privacy and security is crucial to prevent the misuse of personal information. Additionally, robots making autonomous decisions based on learned knowledge may face ethical dilemmas where there is no clear right or wrong choice. Ethical frameworks and guidelines must be established to ensure that robots make decisions that align with human values, adhere to laws, and promote societal well-being.

11. Learning from Human Demonstrations: Robots can learn new skills by observing and imitating human actions, a process known as learning from human demonstrations. This approach involves collecting data of humans performing tasks, and robots then learn to replicate these actions. For example, a robot could learn to cook a recipe by observing a human chef's movements and techniques. Learning from human demonstrations bridges the gap between human expertise and robot capabilities, enabling robots to acquire skills without explicit programming and facilitating collaboration between robots and humans.

12. Application of Adaptive Learning in Healthcare: An application of adaptive learning in healthcare is the use of surgical robots. These robots assist surgeons in complex procedures, continuously learning and adapting their movements based on observed expert surgeon actions. Through reinforcement learning, surgical robots enhance their precision by optimizing actions that lead to successful outcomes. This learning-driven approach reduces the risk of errors, improves patient safety, and augments surgical capabilities. The adaptation of robots in healthcare underscores their potential to revolutionize medical practices and elevate patient care.

13. Collective Intelligence in Swarm Robotics: Swarm robotics exemplifies collective intelligence among robots. In this approach, robots collaborate by learning from the behaviors of their peers. As one robot encounters a situation and learns a successful behavior, it communicates this information to others, enabling the entire swarm to adapt collectively. This concept is akin to how social insects like bees and ants exhibit coordinated behaviors. Swarm robotics finds applications in scenarios like environmental monitoring, search and rescue missions, and exploration of challenging terrains where individual robots collaborate to achieve common goals through adaptive learning.

14. Challenges of Biological Inspiration in Robot Learning: Incorporating biological inspiration into robot learning presents challenges due to the complexity of replicating natural learning processes. Biological systems are influenced by factors beyond programming, including neural plasticity and evolutionary pressures. Translating these intricate processes into machine learning algorithms requires a deep understanding of both biology and technology. Moreover, achieving a balance between biologically inspired behaviors and practical usefulness can be challenging, as robots must also align with human expectations and requirements.

15. Transparency in Autonomous Decision-Making: Transparency refers to the ability to understand and explain the decision-making processes of autonomous robots. In complex decision-making frameworks, such as deep reinforcement learning, it is crucial to ensure that the logic behind robots' actions is comprehensible to humans. Transparent decision-making is essential for accountability, auditing, and regulatory compliance. It also fosters trust between humans and robots, allowing users to verify that robots' decisions align with ethical principles and safety guidelines.

16. Decision-Making Frameworks in Robotics: Decision-making frameworks in robotics are structured approaches that guide robots' choices based on inputs, objectives, and contexts. These frameworks enable robots to make informed decisions by processing data, assessing situations, and selecting actions that align with their goals. They range from rule-based systems and finite state machines to more complex approaches like behavior trees and Markov Decision Processes. Decision-making frameworks are crucial for robots to navigate uncertainty, adapt to changing environments, and interact intelligently with their surroundings.

17. Levels of Autonomy in Decision-Making: Decision-making for robots involves varying levels of autonomy. At the lowest level, teleoperation requires human operators to control robots directly. Supervised autonomy allows robots to perform tasks based on pre-defined rules with occasional human guidance. Semi-autonomy strikes a balance between robot autonomy and human intervention, with robots handling routine tasks independently but

seeking human input for complex situations. Full autonomy signifies robots' ability to make decisions independently based on their learned knowledge and environmental perception.

18. Behavior Trees in Decision-Making: Behavior trees are hierarchical decision-making frameworks that allow robots to evaluate multiple behaviors and choose the most appropriate one. Each behavior is represented as a node in the tree, and robots traverse the tree based on conditions and priorities. This approach offers flexibility and adaptability, enabling robots to make decisions that align with their goals while considering contextual factors. Behavior trees are well-suited for tasks requiring complex decision hierarchies, such as autonomous navigation in changing environments.

19. Deep Reinforcement Learning and Decision-Making: Deep reinforcement learning combines deep neural networks with reinforcement learning principles. It enables robots to learn optimal decision-making strategies through trial and error. Deep reinforcement learning is particularly effective in complex tasks where traditional approaches struggle due to the high dimensionality of data and the need for nuanced decision-making. This technique has achieved notable success in domains like gaming and robotics, allowing robots to learn sophisticated behaviors from raw sensory data.

20. Addressing Bias and Fairness in Decision-Making Frameworks: Decision-making frameworks in robots can inadvertently perpetuate biases present in training data. This issue can lead to discriminatory actions or unjust outcomes. Addressing bias and ensuring fairness requires careful data collection, preprocessing, and algorithm design. Techniques like debasing algorithms and fairness-aware learning can help mitigate biases and ensure that decision-making frameworks produce equitable and unbiased outcomes. By promoting fairness, robots contribute positively to diverse human societies and avoid perpetuating societal biases.

Online Resources

Online Exam Papers

Link: https://forms.gle/SeeU7DHiZEi44w1Q7

QR Code:

YouTube Video Playlist Link

Link: https://www.youtube.com/playlist?list=PLzlh_31VWEFp0ewLQHvRRWhucZouO_b1v

QR Code:

Viva Voce Questions

1:What is Robotics?

Answer: Robotics is an interdisciplinary field that integrates computer science and engineering to design, construct, operate, and use robots. The aim is to develop machines that can assist or replace human actions in various tasks, often those that are dangerous, repetitive, or require precision.

2: Can you describe the basics of Robotics?

Answer: The basics of robotics involve understanding the fundamental components of robots, including sensors, actuators, and control systems. It also involves studying the principles of robot design, kinematics, dynamics, and programming to create functional robotic systems.

3: Why are sensors important in robots?

Answer: Sensors are crucial in robots as they enable the machine to perceive its environment. This perception allows the robot to perform tasks accurately and safely by providing essential data about its surroundings, such as distance, temperature, light, and motion.

4: What is the role of robot control systems?

Answer: Robot control systems are responsible for managing the actions of the robot. They take input from sensors, process the information, and generate appropriate output commands to the actuators to perform desired tasks, ensuring the robot operates efficiently and correctly.

5: What are robot kinematics and dynamics?

Answer: Robot kinematics involves the study of motion without considering forces, focusing on the geometrical aspects of robot movement. Robot dynamics, on the other hand, deals with the forces and torques required to achieve motion, considering the mass and inertia of the robot.

6: What is mobile robotics?

Answer: Mobile robotics refers to the study and creation of robots that can move around in their environment. These robots can navigate through various terrains and are used in applications like delivery, exploration, and autonomous vehicles.

7: Can you explain manipulators and robot arms?

Answer: Manipulators and robot arms are mechanical devices designed to interact with objects in their environment. They can perform tasks such as picking, placing, assembling, and manipulating objects, often with a high degree of precision.

8: What is involved in robot programming and software?

Answer: Robot programming and software involve writing code and developing algorithms that control the robot's behavior. This includes path planning, decision-making processes, and handling sensor data to ensure the robot performs its tasks efficiently.

9: What is human-robot interaction?

Answer: Human-robot interaction (HRI) is the study of how humans and robots communicate and work together. It focuses on creating intuitive and effective ways for humans to interact with robots, ensuring safety and enhancing collaboration.

10: What are some advanced topics in robotics?

Answer: Advanced topics in robotics include machine learning for robots, artificial intelligence, swarm robotics, bio-inspired robotics, and autonomous systems. These areas explore cutting-edge technologies to enhance robot capabilities.

11: What are future trends in robotics?

Answer: Future trends in robotics include the development of more intelligent and autonomous robots, increased use of robotics in everyday life, advancements in human-robot collaboration, and the expansion of robotics in various industries such as healthcare, agriculture, and manufacturing.

12: How is robotics used in agriculture and environmental conservation?

Answer: In agriculture, robots are used for tasks like planting, harvesting, and monitoring crop health. In environmental conservation, robots can assist in monitoring wildlife, cleaning polluted areas, and conducting research in difficult-to-access environments.

13: What is robotic artistry and entertainment?

Answer: Robotic artistry and entertainment involve the use of robots in creative fields, such as creating art, performing in shows, and providing interactive experiences in theme parks. These applications showcase the creative potential of robotics beyond industrial uses.

14: What are some future horizons of robotics research and innovation?

Answer: Future horizons in robotics research include developing robots with more human-like capabilities, advancing AI integration, exploring new materials for robots, and enhancing robot autonomy and adaptability for various applications.

15: How is robotics education and outreach important?

Answer: Robotics education and outreach are vital for inspiring the next generation of engineers and scientists. They involve teaching fundamental and advanced concepts of robotics, providing hands-on experience, and engaging with the community to promote interest in STEM fields.

16: What are roboethics and societal implications?

Answer: Roboethics refers to the ethical considerations related to the design, deployment, and use of robots. Societal implications involve the impact of robotics on employment, privacy, safety, and the overall effect on human life and society.

17: How are robots integrated into everyday life?

Answer: Robots are increasingly integrated into everyday life through applications like home assistants, vacuum cleaners, lawn mowers, and smart home systems. They provide convenience, improve efficiency, and enhance quality of life.

18: What is the role of robotics in extraterrestrial exploration?

Answer: Robotics plays a crucial role in extraterrestrial exploration by providing tools for space missions, such as rovers on Mars, robotic arms on spacecraft, and autonomous probes. These robots perform tasks that are too dangerous or impossible for humans.

19: What is human-centric robotics and assistive technologies?

Answer: Human-centric robotics focuses on developing robots that enhance human capabilities and assist with daily tasks. Assistive technologies include robots that help with mobility, rehabilitation, and providing support for the elderly or disabled.

20: How is entrepreneurship and innovation related to robotics?

Answer: Entrepreneurship and innovation in robotics involve creating new products, services, and business models based on robotic technologies. This field encourages the development of innovative solutions and fosters the growth of new companies in the robotics industry.

21: What is learning and adaptation in robots?

Answer: Learning and adaptation in robots refer to the ability of robots to learn from their experiences and adapt their behavior accordingly. This is often achieved through machine learning algorithms and enables robots to perform better over time.

22: Can you explain the concept of robot autonomy?

Answer: Robot autonomy refers to the ability of a robot to perform tasks without human intervention. This involves decision-making processes, navigation, and execution of actions based on the robot's understanding of its environment.

23: What are the different types of robots?

Answer: Different types of robots include industrial robots, service robots, mobile robots, humanoid robots, and collaborative robots (cobots). Each type is designed for specific applications and environments.

24: How is AI integrated into robotics?

Answer: AI is integrated into robotics to enhance decision-making, perception, and interaction capabilities. This integration allows robots to perform complex tasks, learn from data, and operate autonomously in dynamic environments.

25: What is the significance of robot safety?

Answer: Robot safety is crucial to prevent accidents and ensure the well-being of humans working alongside robots. It involves designing robots with safety features, implementing safety protocols, and adhering to regulations and standards.

YouTube Link for Viva Voce Questions:

https://www.youtube.com/playlist?list=PLzlh_31VWEFqa0_CkTbcDMoUkDjlFu9zh

QR Code :

ADDITIONAL RESOURCES

https://www.youtube.com/playlist?list=PLzlh_31VWEFpgUNZX3JNZXhwA3Q9asi2Z

YouTube Videos Playlists for all Chapters

CHAPTER 1

Link: https://www.youtube.com/playlist?list=PLzlh_31VWEFodq7iwXv5KWEWaFAf5S1EA

QR Code:

CHAPTER 2

Link: https://www.youtube.com/playlist?list=PLzlh_31VWEFqxo25cjcb5wKne2YpQCZnp

QR Code:

CHAPTER 3

Link: https://www.youtube.com/playlist?list=PLzlh_31VWEFpi8HZ83Iu2vs6G3hGvznSs

QR Code:

CHAPTER 4

Link: https://www.youtube.com/playlist?list=PLzlh_31VWEFoYSvfssB7G8aX1b4qe36xc

QR Code:

CHAPTER 5

Link: https://www.youtube.com/playlist?list=PLzlh_31VWEFo9X7VDS4WD76-qa-htboBp

QR Code:

CHAPTER 6

Link: https://www.youtube.com/playlist?list=PLzlh_31VWEFpi8Dxpm-Edeg7rQkydIjJ2

QR Code:

CHAPTER 7

Link: https://www.youtube.com/playlist?list=PLzlh_31VWEFofjdKDq7AHhiixR2I7EOGP

QR Code:

CHAPTER 8

Link: https://www.youtube.com/playlist?list=PLzlh_31VWEFraHESNxUEM9hyIaqjR_PTV

QR Code:

CHAPTER 9

Link: https://www.youtube.com/playlist?list=PLzlh_31VWEFqMQn6mcgtPWiK0Z_VTMpmR

QR Code:

CHAPTER 10

Link: https://www.youtube.com/playlist?list=PLzlh_31VWEFrDBMI3dTPgIWlUh97DWWQf

QR Code:

CHAPTER 11

Link: https://www.youtube.com/playlist?list=PLzlh_31VWEFrgxK0YBD1DanI0VGZx5Ir3

QR Code:

CHAPTER 12

Link: https://www.youtube.com/playlist?list=PLzlh_31VWEFoMrahvi_w3RBJb97Vw79IJ

QR Code:

CHAPTER 13

Link: https://www.youtube.com/playlist?list=PLzlh_31VWEFrEisZSYFfg5v0i0vQzsoiv

QR Code:

CHAPTER 14

Link: https://www.youtube.com/playlist?list=PLzlh_31VWEFrHTaIRTXs0FBIASAnq_nDJ

QR Code:

CHAPTER 15

Link: https://www.youtube.com/playlist?list=PLzlh_31VWEFqgX0Wms2BMn4pmS-lJP4LF

QR Code:

CHAPTER 16

Link: https://www.youtube.com/playlist?list=PLzlh_31VWEFoFKfYGPPrrnR-1txc1_eqR

QR Code:

CHAPTER 17

Link: https://www.youtube.com/playlist?list=PLzlh_31VWEFrVeluA8_ZoRDBgI-Y9PWGk

QR Code:

CHAPTER 18

Link: https://www.youtube.com/playlist?list=PLzlh_31VWEFp3CTNFEX0g_Y_ukYzopvhJ

QR Code:

CHAPTER 19

Link: https://www.youtube.com/playlist?list=PLzlh_31VWEFoOm9xh3oyE8AJ5l_HjDTX1

QR Code:

CHAPTER 20

Link: https://www.youtube.com/playlist?list=PLzlh_31VWEFrb3qC4QAagtXZJfW9vXED1

QR Code:

CHAPTER 21

Link: https://www.youtube.com/playlist?list=PLzlh_31VWEFruKZvQ4UKNkkEUaOv4N6wO

QR Code:

CHAPTER 22

Link: https://www.youtube.com/playlist?list=PLzlh_31VWEFp0ewLQHvRRWhucZouO_b1v

QR Code: